T0350968

Sustainable Surface Water Management

Sustainable Surface Water Management
A Handbook for SuDS

Edited by

Susanne M. Charlesworth

Professor, Urban Physical Geography,
Centre for Agroecology, Water and Resilience,
Coventry University, UK

Colin A. Booth

Associate Head of Research and Scholarship
for the School of Architecture and the Built Environment,
and Deputy Director of the Centre for Floods,
Communities and Resilience,
University of the West of England, Bristol, UK

WILEY Blackwell

This edition first published 2017
© 2017 by John Wiley & Sons, Ltd

Registered Office
John Wiley & Sons, Ltd, The Atrium, Southern Gate, Chichester, West Sussex, PO19 8SQ, United Kingdom

Editorial Offices
9600 Garsington Road, Oxford, OX4 2DQ, United Kingdom
The Atrium, Southern Gate, Chichester, West Sussex, PO19 8SQ, United Kingdom

For details of our global editorial offices, for customer services and for information about how to apply for permission to reuse the copyright material in this book please see our website at www.wiley.com/wiley-blackwell.

Library of Congress Cataloging-in-Publication Data

Names: Charlesworth, Susanne. | Booth, Colin (Colin A.)
Title: Sustainable surface water management : a handbook for SUDS / edited by Susanne M. Charlesworth, reader in urban physical geography, and director of sustainable drainage applied research, Department of Geography, Environment and Disaster Management Coventry University, Colin A. Booth, associate professor of sustainability, and associate head of research and scholarship, Construction and Property Research Centre, University of the West of England
Description: Chichester, West Sussex, United Kingdom : John Wiley & Sons, Inc., 2016. | Includes bibliographical references and index.
Identifiers: LCCN 2016023163 (print) | LCCN 2016029925 (ebook) |
  ISBN 9781118897706 (cloth) | ISBN 9781118897676 (pdf) | ISBN 9781118897683 (epub)
Subjects: LCSH: Urban runoff. | Watershed management. | Water quality management.
Classification: LCC TD657 .S868 2016 (print) | LCC TD657 (ebook) | DDC 628.1/6–dc23
LC record available at https://lccn.loc.gov/2016023163

A catalogue record for this book is available from the British Library.

Wiley also publishes its books in a variety of electronic formats. Some content that appears in print may not be available in electronic books.

Cover image: Gettyimages/Mara Palmen/EyeEm

Set in 9/11pt Sabon by SPi Global, Pondicherry, India
Printed and bound in Malaysia by Vivar Printing Sdn Bhd

10 9 8 7 6 5 4 3 2 1

May the thrill of jumping up and down in muddy puddles never be replaced by the misery of flooding!

This book is dedicated to

Douglas Ella–Rose, Aidan and Rónán
Esmée, Edryd and Efren

Contents

List of Contributors

Valerio C. Andrés-Valeri
Construction Technology Applied Research Group, Department of Transports, Projects and Processes Technology, Universidad de Cantabria, ETSICCP, Avenida de los Castros 44, 39005 Santander, Cantabria, Spain

Ignacio Andrés-Doménech
Instituto Universitario de Investigación de Ingeniería del Agua y Medio Ambiente (IIAMA). Universitat Politècnica de València. Spain

Stella Apostolaki
Department of Science, Technology and Mathematics, The American College of Greece – DEREE, 6 Gravias street, GR-153 42 Aghia, Paraskevi, Greece

Neil Berwick
Urban Water Technology Centre, University of Abertay Dundee, DD1 1HG, UK

Elena Blanco-Fernández
Construction Technology Applied Research Group, Department of Transports, Projects and Processes Technology, Universidad de Cantabria, ETSICCP, Avenida de los Castros 44, 39005 Santander, Cantabria, Spain

Colin A. Booth
School of Architecture and the Built Environment, and the Centre for Floods, Communities and Resilience, University of the West of England, Bristol, UK

David Butler
Centre for Water Systems, University of Exeter, North Park Road, Exeter, EX6 7HS, UK

Jaime Carpio-Garcia.
Construction Technology Applied Research Group, Department of Transports, Projects and Processes Technology, Universidad de Cantabria, ETSICCP, Avenida de los Castros 44, 39005 Santander, Cantabria, Spain

Daniel Castro-Fresno
Construction Technology Applied Research Group, Department of Transports, Projects and Processes Technology, Universidad de Cantabria, ETSICCP, Avenida de los Castros 44, 39005 Santander, Cantabria, Spain

Susanne M. Charlesworth
Centre for Agroecology, Water and Resilience, Coventry University, Priory Street, Coventry, CV1 5FB, UK

Stephen J. Coupe
Centre for Agroecology, Water and Resilience, Coventry University, Priory Street, Coventry, CV1 5FB, UK

Alison Duffy
Urban Water Technology Centre, School of Science Engineering and Technology, Abertay University, Dundee, DD1 1HG, UK

Ignacio Escuder-Bueno
Instituto Universitario de Investigación de Ingeniería del Agua y Medio Ambiente (IIAMA). Universitat Politècnica de València. Spain

Mark Everard
Geography and Environmental Management, Faculty of Environment and Technology, University of the West of England, Coldharbour Lane, Bristol, BS16 1QY, UK

Glyn Everett
Centre for Floods, Communities and Resilience, Faculty of Environment and Technology, University of the West of England, Coldharbour Lane, Bristol, BS16 1QY, UK

Amal Faraj-Lloyd
Coventry University, Priory Street, Coventry, CV1 5FB, UK

Bruce K. Ferguson
College of Environment and Design, University of Georgia, 285 Jackson Street, Athens GA 30602, USA

Andy Graham
Wildfowl & Wetlands Trust, Slimbridge, Gloucestershire, GL2 7BT, UK

Hazem Gouda
Geography and Environmental Management, Faculty of Environment and Technology, University of the West of England, Coldharbour Lane, Bristol, BS16 1QY, UK

Jessica E. Lamond
Architecture and the Built Environment, Faculty of Environment and Technology, University of the West of England, Coldharbour Lane, Bristol, BS16 1QY, UK

Craig Lashford
Faculty of Engineering and Computing, School of Energy, Construction and Environment, Coventry University, Priory Street, Coventry, CV1 5FB, UK

Tom Lavers
Centre for Agroecology, Water and Resilience, Coventry University, Priory Street, Coventry, CV1 5FB, UK

Lian Lundy
School of Science and Technology, Natural Sciences, Middlesex University, The Burroughs, London, NW4 4BT, UK

Larry W. Mays
Civil, Environmental, and Sustainable Engineering Group, School of Sustainable Engineering and the Built Environment, Arizona State University, Tempe, Arizona, USA

Robert J. McInnes
RM Wetlands and Environment Ltd., 6 Ladman Villas, Littleworth, Oxfordshire, SN7 8EQ, UK

Anne-Marie McLaughlin
Centre for Agroecology, Water and Resilience, Coventry University, Priory Street, Coventry, CV1 5FB, UK

Neil McLean
MWH, Eastfield House, Associate WSP-Parsons Brinckerhoff, Newbridge, Edinburgh, EH28 8LS, UK

Peter Melville-Shreeve
Centre for Water Systems, University of Exeter, North Park Road, Exeter, EX6 7HS, UK

Margaret Mezue
Centre for Agroecology, Water and Resilience, Coventry University, Priory Street, Coventry, CV1 5FB, UK

Marcelo Gomes Miguez
Polytechnic School and COPPE – Universidade Federal do Rio de Janeiro, Brazil

Alan P. Newman
Faculty of Health and Life Sciences, Coventry University, Priory Street, Coventry, CV1 5FB, UK

Sara Perales-Momparler
Green Blue Management, Avda. del Puerto, 180 pta. 1B, 46023 Valencia, Spain

David G. Proverbs
Faculty of Computing, Engineering and the Built Environment, Birmingham City University, Millennium Point, Curzon Street, Birmingham B4 7XG, UK

Jorge Rodriguez Hernandez
Construction Technology Applied Research Group, Department of Transports, Projects and Processes Technology, Universidad de Cantabria, ETSICCP, Avenida de los Castros 44, 39005 Santander, Cantabria, Spain

Brad Rowe
Michigan State University, Department of Horticulture, A212 Plant and Soil Sciences Building, East Lansing, MI 48824, USA

Luis Angel Sañudo Fontaneda
Centre for Agroecology, Water and Resilience, Coventry University, Priory Street, Coventry, CV1 5FB, UK

Robyn Simcock
Landcare Research, Auckland, Private Bag 92170, Auckland Mail Centre, Auckland 1142, New Zealand

Aline Pires Veról
Faculty of Architecture and Urbanism – Universidade Federal do Rio de Janeiro, Brazil

Sarah Ward
Centre for Water Systems, University of Exeter, North Park Road, Exeter, EX6 7HS, UK

Frank Warwick
Faculty of Engineering and Computing, School of Energy, Construction and Environment, Coventry University, Priory Street, Coventry, CV1 5FB, UK

Sara Wilkinson
Faculty of Design Architecture and Building, University of Technology Sydney, POB 123 Broadway, Ultimo, NSW 2007, Australia

Kevin Winter
University of Cape Town, Environmental and Geographical Science, Cape Town, South Africa.

About the Editors

Susanne M. Charlesworth is Professor of Urban Physical Geography at Coventry University in the Centre for Agroecology, Water and Resilience.

Colin A. Booth is Associate Head of Research and Scholarship for the School of Architecture and the Built Environment and is Deputy Director of the Centre for Floods, Communities and Resilience at the University of the West of England, Bristol.

Section 1   Introduction to the Book

An Overture of Sustainable Surface Water Management

Colin A. Booth and Susanne M. Charlesworth

Colin A. Booth and Susanne M. Charlesworth

1.1 Introduction

With more than 80% of the global population living on land that is prone to flooding, the devastation and disruption that flooding can cause will undoubtedly worsen with climate change (Lamond *et al.*, 2011). The built environment has become more susceptible to flooding because urbanisation has meant that landscapes, which were once porous and allowed surface water to infiltrate, have been stripped of vegetation and soil and have been covered with impermeable roads, pavements and buildings, as shown in Figure 1.1 (Booth and Charlesworth, 2014).

Surface water policy, to address flooding-related issues, differs widely across various regions and countries. For instance, in the UK, which is made up of four individual countries (England, Scotland, Wales and Northern Ireland), Scotland has policies that have enabled sustainable drainage to be implemented as a surface water management strategy for about the past 20 years; whereas, England, Wales and Northern Ireland have yet to completely embrace sustainable drainage devices in their planning policies and guidance, and hence it is not yet widely implemented (Charlesworth, 2010).

1.2 Surface Water Management

The Victorians (1837–1901 in Britain) undoubtedly made remarkable strides towards innovative approaches to the water resource challenges of their day. Facing the dual contests of addressing rapid population expansion and industrial urbanisation, a need developed for high capacity systems to deal with societal water supply and treatment. Comparable approaches were exported or developed independently across the globe, as

Sustainable Surface Water Management: A Handbook for SuDS, First Edition.
Edited by Susanne M. Charlesworth and Colin A. Booth.
© 2017 John Wiley & Sons, Ltd. Published 2017 by John Wiley & Sons, Ltd.

Figure 1.1   An example of a flooded car park where the impermeable asphalt surface is retaining stormwater runoff.

other nations faced similar challenges. In the UK, by a combination of philanthropy, public subscription and corporate vision, the infrastructure that would provide the vastly increased urban areas with sufficient clean water and the ability to discharge the surplus was put in place; and with it came the notion of the management of water as a single problem with one overarching solution: the provision of drains. However, while the solutions created by the Victorian engineers were magnificent in their day, the legacy of putting water underground seems to have created a collective mental block for many (Watkins and Charlesworth, 2014).

Nowadays, as mentioned earlier, urbanisation has had a transforming effect on the water cycle, whereby hard infrastructure (e.g. buildings, paving, roads) has effectively sealed the urbanised area (Davies and Charlesworth, 2014). As a consequence, excessive surface water runoff now exacerbates river water levels and overloads the capacity of traditional underground 'piped' drainage systems; this in turn contributes to unnecessary pluvial flooding. To many people, the solution is simply to replace the existing pipes with higher capacity ones. However, as Water UK (2008) states, bigger pipes are not the solution for bigger storms. Therefore, society should be encouraged to look towards more sustainable solutions.

1.3   Sustainable Surface Water Management

'Sustainable drainage' means managing rainwater (including snow and other precipitation) with the aim of: (a) reducing damage from flooding; (b) improving water quality; (c) protecting and improving the environment; (d) protecting health and safety; and (e) ensuring the stability and durability of drainage systems (Flood and Water Management Act, 2010).

Based on an understanding of the movement of water in the natural environment, sustainable drainage systems (SuDS) can be designed to restore or mimic natural infiltration patterns, so that they can reduce the risk of urban flooding by decreasing runoff volumes and attenuating peak flows. The choice of phrase or term that is applied to describe the approaches used can vary between countries, contexts and time. In the UK, for instance, SuDS is the most widely used term; whereas elsewhere in the world other relevant terms include surface water management measures (SWMMs), green infrastructure, green building design, stormwater control measures (SCMs), best management practices (BMPs), low impact development (LID) and water sensitive urban design (WSUD) (Lamond *et al.*, 2015). However, whichever term is used, the benefits and challenges are similar (Tables 1.1 and 1.2).

The typical design of any SuD system follows a step-wise hierarchy of various measures, commonly known as the 'surface water management train' (Figure 1.2), which minimises stormwater runoff and pollution via a series of devices/processes that store and convey stormwater at different scales: (i) prevention (e.g. land use planning); (ii) source control (e.g. green roofs, rainwater harvesting, permeable paving); (iii) site control (e.g. vegetation or gravel filtration); and (iv) regional control (e.g. retention ponds, wetlands) (Woods Ballard *et al.*, 2007, 2015). The primary goals of the original SuDS train placed equal emphasis on water quality and water quantity, together with amenity and biodiversity, which enabled the creation of the SuDS triangle (Figure 1.3a) (CIRIA, 2001). Subsequent iterations of the goals has enabled the creation of the SuDS square (Figure 1.3b) and, with much wider recognition of the role that SuDS can play in adapting to climate change challenges, the creation of the SuDS rocket (Figure 1.3c). The flexibility and multi-functional nature of SuDS are the main drivers pursued in the chapters of this book.

1.4   Organisation of the Book

This book emphasises the SuDS philosophy and elaborates the sustainable surface water management agenda with a wealth of insights that are brought together through the experts who have contributed. By integrating physical and environmental sciences, and combining social, economic and political considerations, the book provides a unique resource of interest to a wide range of policy specialists, scientists, engineers and subject enthusiasts.

Table 1.1   Examples of the benefits offered by sustainable drainage systems.

Sustainability	SuDS can provide an important contribution to sustainable development.
	SuDS are more efficient than conventional drainage systems.
	SuDS help to control and identify flooding and pollution at source.
	SuDS help to promote subsidiarity.
	SuDS can help to minimise the environmental footprint of a development.
	SuDS are a clear demonstration of commitment to the environment.
Water quantity	SuDS can help to reduce flood risk by reducing and slowing runoff from a catchment.
	SuDS can help to maintain groundwater levels and help to prevent low river flows in summer.
	SuDS help to reduce erosion and pollution, as well as attenuating flow rates and temperature by increasing the amount of interflow.
	SuDS can reduce the need to upgrade sewer systems to meet the demands of new developments.
	SuDS can help to reduce the use of potable water by harvesting rainwater for some domestic uses.
Water quality	SuDS can reduce pollution in rivers and lakes by reducing the amount of contaminants carried by runoff.
	SuDS can help to reduce the amount of wastewater produced by urban areas.
	SuDS can reduce erosion and thus decrease the amount of suspended solids in river water.
	SuDS can help to improve water quality by reducing the incidence of misconnection to foul sewers.
	SuDS can help to reduce the need to use chemicals to maintain paved surfaces.
	SuDS can prevent pollution by reducing overflows from sewers.
Natural environment	SuDS can help to restore the natural complexity of a drainage system and as a result promote ecological diversity.
	SuDS help to maintain urban trees.
	SuDS help to conserve and promote biodiversity.
	SuDS can provide valuable habitats and amenity features.
	SuDS help to conserve river ecology.
	SuDS help to maintain natural river morphology.
	SuDS help to maintain natural resources.
Built environment	SuDS can greatly improve the visual appearance and amenity value of a development.
	SuDS help to maintain consistent soil moisture levels.
Cost reductions	SuDS can save money in drainage system construction.
	SuDS can save money in the longer term.
	SuDS can allow property owners to save money through differential charging.
	SuDS can help to save money by reducing the need to negotiate wayleaves and easements.
	SuDS can save money through the use of simpler building techniques.

Source: List of benefits derived from CIRIA (2001).

The book comprises seven sections, which are collated into 29 chapters. Section 1 provides an *introduction to the book* and offers an initial background into surface water management issues and challenges (Chapter 1). Section 2 places *sustainable surface water management in context*, through its historical context, contemporary surface water strategy, policy and legislation and operations and maintenance (Chapters 2–4). Section 3 utilises the facets of the *functions of sustainable drainage systems*, to explore quantity and quality issues,

Table 1.2 Examples of the challenges posed by sustainable drainage systems.

Operational issues	There is no consensus on who benefits from SuDS. There is a belief that SuDS may present maintenance challenges. There may be concerns that the colonisation of SuDS may be too successful. SuDS may present a target for vandals.
Design and standards	SuDS are not promoted by the Building Regulations. There are no standards for the construction of SuDS. SuDS require input from too many specialists. SuDS may be seen as untried technology. The guidance on how to build SuDS is limited or unclear. It is difficult to predict the runoff from a site. SuDS can be difficult to retrofit to an existing development.
Management/operational framework	SuDS require new approaches to enable full participation. Planning, design and construction of SuDs will require better coordination. SuDS can require multi-party agreements that may be difficult to set up. SuDS present challenges in setting up long-term management and ownership agreements. SuDS can be difficult to implement because of the variability of roles and responsibilities within local authorities and other bodies. Sewerage undertakers may be reluctant to adopt foul sewers when they are only sewers serving developments using SuDS.

Source: List of challenges derived from CIRIA (2001).

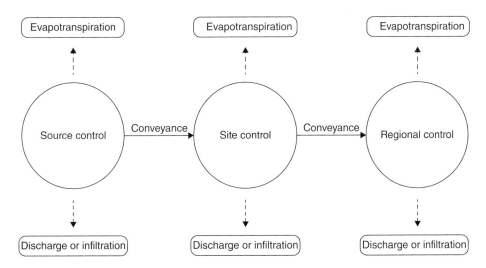

Figure 1.2 The SuDS surface water management train (adapted from CIRIA, 2001).

together with biodegradation, geosynthetics, biodiversity and amenity, (Chapters 5–11). Section 4 attempts to untangle the complex relationship of the *multiple benefits of surface water drainage systems*, through natural floodwater management, energy generation and reduction, carbon sequestration and storage, plus the use of rainwater harvesting as a water saving device and its use in ecosystem services (Chapters 12–16). Section 5 announces the

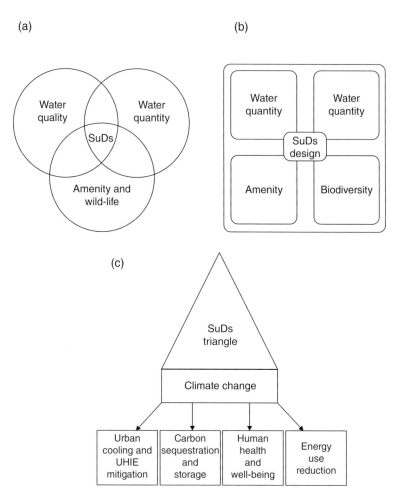

Figure 1.3 Goals of the SuDS management train (a) the SuDS triangle (CIRIA, 2001); (b) the SuDS square (Woods Ballard *et al.*, 2015); (c) the SuDS rocket (Charlesworth, 2010).

implementation of *integrating sustainable surface water management into the built environment*, through an interesting scrutiny of the cost benefits that can be derived, the possibility of sustainable drainage retrofit and conversion opportunities, and their use in the landscapes of motorway service areas, alongside human attitudes and behaviours towards sustainable drainage systems (Chapters 17–21). Section 6 contextualises *global sustainable surface water management*, through the use of examples from Brazil, New Zealand, South Africa and the USA, among others (Chapters 22–28). Section 7 congregates various aspects detailed in the earlier chapters by offering a *summary of the book* and propositioning many insights of the teachings that can be learnt for the future of sustainable surface water management (Chapter 29).

References

Booth, C.A. and Charlesworth, S.M. (2014) *Water Resources in the Built Environment: Management Issues and Solutions*, Wiley-Blackwell, Oxford.

Charlesworth, S.M. (2010) A review of the adaptation and mitigation of global climate change using sustainable drainage in cities. *Journal of Water and Climate Change*, 1, 165–180.

CIRIA (2001) *Sustainable Urban Drainage Systems: Best Practice Manual*. CIRIA Report C523, London.

Davies, J. and Charlesworth, S.M. (2014) Urbanisation and Stormwater. In: Booth, C.A. and Charlesworth, S.M. (eds) *Water Resources in the Built Environment: Management Issues and Solutions*, Wiley-Blackwell, Oxford, 211–222.

Lamond, J.E., Booth, C.A., Hammond, F.N. and Proverbs, D.G. (2011) *Flood Hazards: Impacts and Responses for the Built Environment*. CRC Press – Taylor and Francis Group, London.

Lamond, J.E., Rose, C.B. and Booth, C.A. (2015) Evidence for improved urban flood resilience by sustainable drainage retrofit. *Proceedings of the Institution of Civil Engineers: Urban Design and Planning*, 168, 101–111.

Watkins, S. and Charlesworth, S.M. (2014) Sustainable Drainage Systems – Features and Design. In: Booth, C.A. and Charlesworth, S.M. (eds) *Water Resources in the Built Environment: Management Issues and Solutions*, Wiley-Blackwell, Oxford, 283–301.

Woods Ballard, B., Kellagher, R., Martin, P., Jefferies, C., Bray, R. and Shaffer, P. (2007) *The SuDS Manual*. CIRIA Report C69, London.

Woods Ballard, B., Wilson, S., Udale-Clarke, H., Illman, S., Ashley, R. and Kellagher, R. (2015) *The SuDS Manual*. CIRIA, London.

Section 2 Sustainable Surface Water Management in Context

2

Back to the Future? History and Contemporary Application of Sustainable Drainage Techniques

Susanne M. Charlesworth, Luis Angel Sañudo Fontaneda and Larry W. Mays

History *"provides lessons from the past from which we can learn"*

Lucero *et al.* (2011)

2.1 Introduction

The early Babylonians and Mesopotamians in Iraq (4000–2500 BC) had surface water drainage systems, and regarded urban runoff as a nuisance, but also realised that it carried waste off with it and, for some, it was a resource (De Feo *et al.*, 2014). As these drainage systems developed, they relied mostly on hard infrastructure, for example, the Minoans (3200–1100 BC) used terracotta pipes to convey stormwater out of their settlements. However, these ancient civilisations also used water management techniques, which are included in the sustainable drainage suite of interventions and thus, as acknowledged in Chapter 1, SuDS as a technique is not new; it may not have been called 'sustainable drainage' in the past but, for example, water harvesting, storage and conveyance were all well-known and efficiently carried out by ancient cultures as long ago as the Early Bronze Age (*ca*. 3500–2150 BC, Myers *et al.*, 1992) in Crete. In the Mediterranean and Near East region, infrastructure for the collection and storage of rainwater was developed in the third millennium BC (Mays *et al.*, 2013). Water resource management dates back to the beginnings of early agriculture, whereby water was controlled in order to provide irrigation to enable crops to be grown in arid and semi-arid regions whose rainfall amount would not normally have supported it. As is stated by Lucero *et al.* (2011), rainfall extremes, too much or too little, result in failed crops and famine – water management was therefore a case of life or death in many instances, leading to the rise or fall of civilisations. While the majority of this chapter focuses on ancient rainwater harvesting techniques, since this was used extensively in antiquity, other 'sustainable urban water practices' were utilised

Sustainable Surface Water Management: A Handbook for SuDS, First Edition.
Edited by Susanne M. Charlesworth and Colin A. Booth.
© 2017 John Wiley & Sons, Ltd. Published 2017 by John Wiley & Sons, Ltd.

(Koutsoyiannis *et al.*, 2008), such as constructed wetlands, infiltration and non-structural approaches. For example, Ancient Greece *had* to develop water resource management techniques due to the lack of water and high evaporation rates, particularly during summer. They therefore had to efficiently capture what rain fell, provide for its safe storage with minimal losses, have the means to convey it for long distances and also bring in government structures and institutions to ensure its effective management (Angelakis and Koutsoyiannis, 2003). In fact, Apt (2011) compared the Inca drainage of Machu Picchu (built *ca.* 1400 AD) to that of present-day low impact development as described in Chapter 25. This chapter begins by considering the 'sustainable' part of drainage systems and goes on to explore whether the SuDS represented in this book is simply a case of history repeating itself, and whether techniques used in the past have any relevance today.

2.2   'Sustainability'?

Mays (2007a) defines water resources sustainability as: '*the ability to use water in sufficient quantities and quality from the local to the global scale to meet the needs of humans and ecosystems for the present and the future to sustain life and to protect humans from the damages brought about by natural and human caused disasters that affect sustaining life*'. Sustainable drainage uses the term to reflect its ability to mimic nature by managing surface water, such that the urban environment has minimal to no impact on the path of water through it, thus avoiding the 'human disasters', i.e. flooding caused by construction and impermeability. This section therefore considers the longevity of ancient drainage and whether it could be considered to be 'sustainable' and what lessons contemporary society could learn.

Street drainage was first used in the Mesopotamian Empire, Iraq (4000–2500 BC), but it was in Crete with the Minoan and Harappan civilisations that sewer and drainage systems were first developed which were well designed, organised and operated (De Feo *et al.*, 2014). Basic hydraulics was well understood, and great importance was given to the provision of sanitation in cities. While the Romans and Hellenes further refined these techniques, there was minimal further progress made during the 'Dark Ages' post 300 AD.

The next real advance in storm and sanitary sewerage systems was in London, in response to the 'Great Stink' of 1858 (Lofrano and Brown, 2010) and following cholera outbreaks (in 1831, 1848–49 and 1853–54) in which a total of 31,411 people died. Sir Joseph William Bazalgette was engaged to develop a piped stormwater sewer system during the late 19th century, much of which is still in use today. These systems in most cities are now not fit for purpose – this is mainly due to rapid urbanisation overwhelming the capacity needed for them to cope, leading to flooding and pollution (De Feo *et al.*, 2014). Combined sewers are of particular concern, due to their carrying a mixture of both foul and stormwater, the overflowing of which has significant health implications. The design period for modern water-related infrastructure is generally about 50 years (Koutsoyiannis *et al.*, 2008) and while it is perhaps a stretch to compare such infrastructure in the past with that in use today, nonetheless, ancient drainage operated for extensive periods, centuries even, and certainly for longer than 50 or 60 years. As an example, in Athens, water was supplied via the Hadrian Aquaduct (completed 140 AD) and was in use until the 1920s and partially up until the 1950s, and also the Peisistratean Aquaduct (built *ca.* 510 BC) is still used today to irrigate the National Garden in the centre of Athens (De Feo *et al.*, 2013). Thus, a modern city with a history of settlement of over 3000 years still has examples of ancient water infrastructure still in use. Ancient approaches can, therefore, have lessons that can be learnt for the present day, giving real meaning to the word 'sustainable' when used with respect to SuDS and giving credence to the idea that, if it is designed, installed and maintained correctly, there is no reason to suppose for SuDS to ever have an end-of-life (Bob Bray, SuDS Designer and Landscape Architect, UK, pers. comm.).

2.3   Rainwater Harvesting in Antiquity

Rainwater harvesting (RwH) has been defined as atmospheric precipitation collected and stored, usually in artificial reservoirs known as cisterns (Figures 2.1 and 2.2) (Angelakis, 2014). Methods of water harvesting are generally distinguished by the source of water they harvest, for example groundwater, surface water, rainwater or floodwater (Haut *et al.*, 2015); this section mainly focuses on rain as the source of water. RwH has played a decisive role in providing water resources for ancient civilisations across the world with its importance captured in manuscripts, hieroglyphs and religious texts through the millennia. Since the early Babylonians and Mesopotamians in Asia (4000–2500 BC) (De Feo *et al.*, 2014), and the Minoans in Europe (3000–1100 BC) (Angelakis and Durham, 2008), harvested rainwater was used extensively in urban areas as demonstrated by research carried out across the world. Mays (2008) lists RwH as the main source of urban water in antiquity which was associated with engineered infrastructure, such as canals and aqueducts, which were used to convey water from rivers to urban areas, to replenish wells and

Figure 2.1   Cistern complex near Chersonisos, Crete.

Figure 2.2   A large cistern from the Nabataean city of Little Petra.

to fill cisterns. In the city of Delos (the Cycladic culture in Greece) RwH was the main water supply dependent largely on the collection and storage of rainwater in cisterns (Koutsoyiannis *et al.*, 2008).

Historically, there is a correlation between heightened human efforts for the construction of water harvesting structures across regions and abrupt climate fluctuations, such as aridity, drought and floods (Pandey *et al.*, 2003). The objective was the safe and beneficial use of the harvested water, as well as the reduction of impacts on society associated with these climate fluctuations (Konig, 2001). Cultural factors and decisions around water management in antiquity included dealing with potable water supplies during annual droughts (Lucero *et al.*, 2011). Therefore awareness of rainfall seasonality was a defining factor in arid and semi-arid zones, as is clearly exemplified in Jawa and Petra (Jordan) (AbdelKhaleq and Ahmed, 2007), Palestine and the Mayan civilisation in Mesoamerica (Mays 2007b), as well as other cities. Seasonality was not the only factor that needed to be considered in the study of rainfall patterns, since the overall volume of water was important; this is related to the intensity and duration of individual storms, as well as the incidence of surface run-off, with other factors such as average temperature, solar radiation and wind strength and direction also being of importance (Imhoff *et al.*, 2007). The Mayan culture provides an excellent example of the importance of water as a cultural cornerstone. All aspects of Mayan life were rainfall dependent, helping the leaders to keep their power and control over the population through the provision of water for their daily activities; loss of power would ensue should water resource management be handled badly. Thus water had to be allocated for human and animal consumption, and the timely repair of any damage to water systems due to flooding had to be a prime consideration for community leaders (Lucero *et al.*, 2011).

In Ancient Greece, urban water management was carried out through the combination of large-scale public works such as reservoirs, but also the use of small-scale semi-public or private constructions such as cisterns and wells (Koutsoyiannis *et al.*, 2008) serviced by RwH. Cisterns were often spread throughout the whole city but they were also found in the backyards of private houses, so that each individual house was able to have its own facility for the storage of stormwater (Koutsoyiannis *et al.*, 2008). Rainwater stored in cisterns was used mainly for household purposes, such as bathing or washing (Mays *et al.*, 2013; Mays, 2014), washing dishes and laundering clothes (Lang, 1968), irrigation for agricultural purposes (Gikas and Angelakis, 2009; Beckers *et al.*, 2013), flushing lavatories (Antoniou 2007), for animal and human consumption (Beckers *et al.*, 2013) and aquifer recharge (Gikas and Angelakis, 2009). It was also stored for use in times of war and for other socio-political purposes (Cadogan, 2007), as was understood by Aristotle (385–322 BC) in which he stated that the water '*supply may never fail the citizens when they are debarred from their territory by war*' (quote taken from Politics, III, in Koutsoyiannis *et al.*, 2008).

2.3.1 RwH Infrastructure

The classic RwH system found in common in several ancient civilisations, such as the Minoans, Greeks and Romans, consisted of a combination of several techniques: rain was collected from rooftop catchments, which was then conveyed sometimes by means of terracotta pipes (Figure 2.3) to underground cisterns where the water was stored (Mays, 2007b, 2008, 2013). A good example of a complete RwH system can be found in the Amman Citadel, Jordan, where rainwater was collected from roofs and directed to storage areas through channels.

Other good examples of the use of cisterns in antiquity can be found in Jordan, one of the tenth poorest countries in the world in terms of water resources (AbdelKhaleq and Ahmed 2007), and also in Palestine. Both used cisterns to store rainwater in order to have enough water for the dry season. During the Early Iron Age (1200–1000 BC) the Palestinians modified cistern design by introducing a watertight plaster layer on the sides of the cistern to increase storage capacity and reduce losses through leakage (AbdelKhaleq and Ahmed, 2007) (Figure 2.4).

RwH was also carried out by constructing above-ground dams, very often complemented by the use of a network of cisterns, pools and conveyance canals similar to those described before for the Ancient Greeks. A good example of a combined system was found by archaeologists in the Jawa ruins (the oldest urban development in Jordan from the Bronze Age, 5000 years ago). Helms (1981) defined the water management area for the city, differentiating between macro-catchments controlled by deflection dams in the surface and underground reservoirs, and micro-catchments of cisterns, pools and deflection walls. This system was very advanced, and is considered a masterpiece of landscaping and water management for an arid zone. It provided multiple benefits and functionality, such as the provision of potable water for human consumption, water for crop irrigation and animal consumption, and the ability to collect additional rainfall runoff from other surrounding catchments.

Ponds were also often used, such as those found in Umm el-Jimal (Jordan, Early Roman Period) for irrigation and to water animals. Any excess water from the dams and cisterns was distributed via canals (Alkhaddar, 2005), the water having been previously treated by sedimentation to remove suspended solids before the water entered the main reservoirs.

Figure 2.3 Terracotta pipe for rainwater harvesting from roofs at Pompeii, Italy. The pipes would have directed water into cisterns to service individual buildings.

Stairs leading
down into cistern

Plaster on the
cistern walls

Inside the cistern

Figure 2.4 Cistern from Tylissos, Crete.

2.4   Water Quality Improvement

Modern SuDS incorporate infiltration, settling, sedimentation, biodegradation and trapping, which remove the pollutants associated with urbanisation, traffic and industry. Ancient societies also had to cope with the contamination of their water supplies, mainly due to suspended sediment, but also human and animal wastes, organic matter and excess nutrients. As the following sections show, there are many examples, both geographically and temporally distributed, in which these processes were also used to great effect, exemplified by the survival of some of these civilisations for hundreds and thousands of years, even in regions of the world where water was in very short supply.

2.4.1   Physical Treatment: Infiltration and Settling

Sand filters were used extensively where the supply of water was dependent on precipitation. There are many examples of it being designed in to the RwH infrastructure; for example, at Phaistos, Crete, a coarse sandy filter was used to remove silt and other pollutants from the water before storage in the cisterns (Antoniou *et al.*, 2014). Particular care was taken to ensure that the collection surfaces for the rainwater were kept scrupulously clean (Angelakis and Koutsoyiannis, 2003), although the water was not used for drinking apart from during times of war, for instance; but it was mainly used for clothes washing and for other cleaning tasks (Angelakis and Spyridakis, 1996). The Ancient Egyptians (2000–500 BC) made use of their surrounding landscape, and disposed of their wastewater by allowing it to infiltrate straight into the desert sands (De Feo *et al.*, 2014).

Suspended sediment was also removed by means of settling tanks (Figure 2.5) to avoid silt entering the water supply system. In Minoan and Mycenaean cities, RwH from roofs and

Figure 2.5   Stilling basin at the outlet of Wadi Jilf (Petra, Jorban). Bioretention is provided by the plants growing in the basin at the present day. Two outlets can be seen in the cross-drainage structure in the background.

Figure 2.6   Settling tank at Petra, Jordan, along the aqueduct in the Siq.

courtyards utilised cisterns to store the water. The palace at Tylissos, Crete (2000–1100 BC) had a stone tank before the main storage cistern to allow sedimentation of particulates and thus improve the water quality (Gorokhovich *et al.*, 2011). A drainage hole was used to empty the tank to allow it to be cleaned as necessary (Mays, 2008). Silting tanks and sediment settling (Figures 2.6 and 2.7) was also used at Tikal to remove any pollutants before the water entered the Temple Reservoir, and sand boxes were positioned at the inlets of several of the other reservoirs in order to remove suspended sediment (Scarborough *et al.*, 2012).

Rather than passively infiltrating the water through sand filters, at Tylissos, one of the important cities in Ancient Crete during the Minoan era (2000–1100 BC), infiltration devices of terracotta were found near the spring of Agios Mamas that were filled with charcoal as activated carbon (Mays, 2010; Gorokhovich *et al.*, 2011).

2.4.2   Biological Treatment

Reservoirs were also used to store captured surface and rainwater, and in areas where it was likely that evaporation would substantially reduce the volume, such as ancient Maya, floating aquatic plants were used. Their role was five-fold: (1) to reduce evaporation of water; (2) by covering the surface of the stored water they prevented disease vectors, such as mosquitoes, from breeding; (3) plants such as water hyacinth, water lilies (particularly *Nymphaea ampla*) (Figure 2.8) and ferns can clean the water, important during the dry season as water supplies became low, although the Classic Mayan civilisation was in

Figure 2.7   Sedimentation basin, Petra, Jordan, High Place to Triclinium.

Figure 2.8   Water hyacinth and water lilies in a moat at Angkor Thom.

existence for a millennium and, therefore, they must have managed a clean, potable source of water even during the drought; such floating plants remove nutrients such as nitrogen and phosphorus by metabolic processes and trap polluted particulates in their stems and roots – very much in the same way as wetlands function in SuDS; Ford (1996) even suggests that water treated in this way can be used for drinking; (4) these plants could provide an organic compost if they were harvested regularly – a necessary maintenance procedure to ensure the efficient functioning of what Lucero *et al.* (2011) call a 'constructed wetland'; (5) macrophytes, such as the water lily (Figure 2.9), would only thrive in clean water, and as such, they were an indicator of water quality; the underneath of the leaves are a blue colour that can reduce the development of algal blooms since this restricts the entry of light. Microscopic organisms, such as bacteria, can denitrify water but also feed on the spores of parasites. The fact that water lilies were found in these reservoirs also provides an indicator of the aquatic environment, because they do not grow in water deeper than 1–3 m, which must be still and cannot contain large amounts of algae. Water lilies are also intolerant of acidic conditions or water containing much calcium, but if the pond had been clay lined, this would avoid calcium ingress and assist in the stabilisation of pH conditions. Furthermore, it is unlikely that any sediment accumulating in the bottom of the ponds would have contained much in the way of decaying organic matter since its decomposition would have released compounds, such as methane and phenols, which would have killed the water lilies.

Figure 2.9   Water lilies, Angkor Wat.

2.5   Water Quantity Reduction: Sub-Surface Drainage

Infiltration was used at Machu Picchu, Peru, and, according to Wright and Valencia Zegarra (1999), this was a standard technique on the surrounding terraces, but it has also been found in plaza areas which infiltrated stormwater and therefore was able to store and dispose of it. Beneath some plazas, layers of loose rock and stone chips up to 1 m deep have been found, which provided sub-surface drainage (Wright *et al.*, n.d.; Mays and Gorokhovich, 2010). The rock chips were recycled from the stonecutters, as they constructed the buildings and walls at the site, but represent only a small portion of the thousands of cubic metres of rock waste. Water from major rainfall events that was able to percolate deep beneath the plaza was stored temporarily in the voids in the rock chip layer and then slowly released downstream, and thus avoided causing a high groundwater table and consequent instability of the plaza and its soils (Figure 2.10).

It is proposed by Apt (2011) that the terraces at Machu Picchu represent an early form of bioretention, since their structure is similar to modern versions of such devices, having gravel as the base layer, sand in the middle and a layer of topsoil as the surface course with

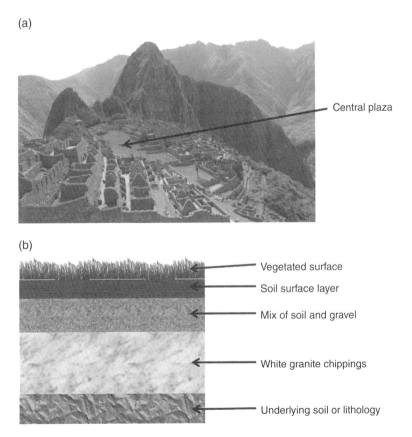

(a)

Central plaza

(b)

Vegetated surface

Soil surface layer

Mix of soil and gravel

White granite chippings

Underlying soil or lithology

Figure 2.10   (a) Machu Picchu, Peru, locating the central plaza; (b) cross-section through the drainage structure underneath the central plaza (adapted from Apt, 2011).

(a)   (b)

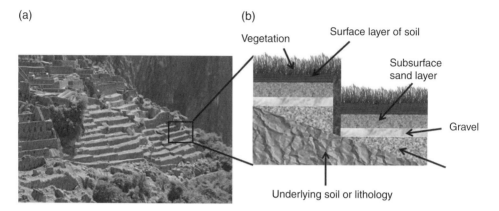

Figure 2.11   (a) Terraces, Machu Picchu, Peru; (b) cross-section through two terraces to show underlying drainage structure (adapted from Apt, 2011).

growing vegetation. This would provide physical trapping of pollutants, as well as biological and chemical treatment, thus improving water quality, and the plants would slow the water flow, hence attenuating the storm peak (Figure 2.11). It is thought that the water was directed to fountains on the lower terraces where it was used for drinking.

2.6   Water Storage

Water was stored in large reservoirs or barays at Angkor Wat with control structures at both the inlet and outlet. These enabled water to be stored in times of drought, but they could also have functioned as flood control measures during storms. Of the four barays at Angkor Wat, the one located in the west of the site held the most water, and still holds water today (Figure 2.12): it had a potential capacity of 48 million m³ followed by the east baray (37.2 million m³), Jayatataka baray (8.7 million m³) and Indratataka baray (7.5 million m³) (Coe, 2003). Figure 2.13 shows the remains of some of the barays at the Angkor complex. These barays were part of the intricate water infrastructure at the site, which included canals and moats, all of which would have required maintenance to keep them functioning.

2.7   Reduction in Water Demand: Greywater Recycling

It is acknowledged that greywater reuse is not a SuDS approach, but it has the potential to reduce water volume in storm sewers and hence attenuate the storm peak and also reduce potable water demand, so it is considered here.

According to Antoniou (2010), the semi-arid climate of the eastern Mediterranean and ancient Greece led to the reuse of water to flush toilets. This was carried out simply using a bucket, with, for example, greywater from kitchens or baths, and this was the case for the Minoan toilets on Amorgos. Leftover water, which had been used for ceremonial purposes in shrines, was also occasionally used, for example at the Askleipieion, Kos. Such uses of greywater continue today in the Aegean where, with the pressures of tourism seasonally increasing the demand for water and the expense of building desalination plants, flushing toilets with reused water is now common.

Figure 2.12   West baray, Angkor Wat, Cambodia, still containing water. (Wikimedia by Dario Severi: https://commons.wikimedia.org/wiki/File:WestBaray.jpg).

Figure 2.13   Prean Khan baray north-east of Angkor Thom and west of Jayatataka baray.

Crouch (1996), quoted in Angelakis *et al.* (2005), also suggests that water from cooking or bathing could have been used to water domestic animals, water indoor plants or wash floors in Minoan times. Stormwater was used for irrigation purposes as well as greywater, and in the villa of Hagia Triadha, surface water collected via the stormwater sewer system was directed into a cistern where it may have been used for washing, thus reducing use of water that could not be consumed, and could otherwise have been wasted.

2.8   Reducing Water Velocity

Cumbe Mayo is located near the City of Cajamarca, Peru, and features what remains of a Pre-Incan aqueduct, 9 km long, built *ca.* 1500 BC and excavated in volcanic rock. At times along its route, the channel meanders – so it is thought (e.g. De Feo *et al.*, 2013) – to reduce

Figure 2.14   Cumbe Mayo meandering channel. (Luis Padilla – own work, CC BY-SA 3.0, https://commons. wikimedia.org/w/index.php?curid=8069762).

flow velocity and hence prevent erosion (Figure 2.14). The technique of river restoration (Wohl *et al.*, 2005) in which meanders are reinstated into channelised urban rivers is used today to slow water down and attenuate the storm peak, and is arguably part of the sustainable drainage approach to the management of excess stormwater.

In some developments, the road can be used as a channel to direct excess stormwater. This is achieved via raised kerbs, as suggested by Becker *et al.* (2008), and is an approach found in Pompeii, where the kerbs were about 50–60 cm high. These could conceivably have been used to control stormwater flow, but they were actually open channels in which water from public fountains, stormwater and sewage mixed, and thus the flow would have contained some human waste (De Feo *et al.*, 2014). Stepping stones (or *pondera*) shown in Figure 2.15 were therefore installed at intervals along the street, so that the population did not have to walk through the foul water, and also so they did not have to step down into the road from the raised kerb.

2.9   Non-Structural Approaches to Sustainable Water Management

One of the present-day barriers to the implementation of SuDS is in arranging for its maintenance; it is not clear whose responsibility this is, and it is perceived as expensive. No one will therefore, take responsibility, or will take ownership. In the fifth century BC, Plutarch recorded the institutional, or non-structural, arrangements to ensure the efficient operation and maintenance of Athens' water system, which included the appointment of a 'superintendent of fountains', which according to Aristotle was filled by election rather than being a straight appointment, emphasising the importance of the role. The superintendent of fountains enforced regulation with regard to water resources and ensured the equitable distribution

Stepping stones

Cart wheel ruts

Raised kerbs

Figure 2.15   Pompeii raised kerbs with stepping stones.

of water in the city. In ancient Athens there must have been an obligation on citizens to maintain the city's stormwater cisterns, thus providing resilience in the case of excess stormwater, and water resource provision (Koutsoyiannis *et al.*, 2008). In fact, Ford (1996) suggests that at Mayan Tikal, Guatemala, particularly in the central area, the five-months dry season became a matter of public works, and such investment in water supply infrastructure actually became a key consideration in the control of the population.

Water supply and disposal in modern-day cities in the West has become the responsibility of others, whether the government, local authorities or stakeholder organisations, such as the Environment Agency or water companies. People have thus become distanced from water; drinking water comes from taps, water is flushed away through the toilet and wastewater is just that, a waste, to be disposed of elsewhere. Water is a now considered a right in most of the world, but in ancient times it was provided, or withheld, by the ruling elite, and this was one of the ways in which the populace was controlled. In the Negev, southern Israel, natural rainfall of 80 mm would not have been sufficient to support agriculture (Haiman and Fabian, 2009). However, during the Nabataean period (second century BC to second century AD) some of the people were encouraged to settle there initially by the Byzantine Empire. In order to survive, they constructed elaborate rainfall collection systems which concentrated water from an area five times the area to be irrigated. In this way, they were able to collect the equivalent of 400 mm annual rainfall, thus enabling agriculture to be carried out. The environment was a harsh one, but support was maintained by the Umayyad Empire with state subsidies in order to protect the border. Without subsidisation and support, settlement would have been impossible (Haiman and Fabian, 2009).

2.10   Conclusions

Angelakis *et al.* (2005) include a quote from Mosso (1907) in which he queries whether: *'our modern sewerage systems will still be functioning after even one thousand years'*. In today's modern cities much of their stormwater sewer systems are not fit for purpose after just 150 years; this is mainly due to population growth, urban expansion and the threat of climate change. To a certain extent, therefore, lessons can be learnt from the past, but in antiquity, populations were less dense and thus hard infrastructure drainage was far more efficient and was able to last for millennia. This chapter started with a quote from Lucero *et al.* (2011) that history *'provides lessons from the past from which we can learn'*; in fact, a lot of what history tells us, we have managed to forget over the millennia. Perhaps, therefore, we need reminding, rather than being taught these approaches since, while what has been discussed in this chapter is not, in the strictest sense, SuDS, nonetheless the tools, techniques and practices that were used in antiquity to manage water included infiltration, detention, storage and conveyance. These are all processes, as stated by Koutsoyiannis *et al.* (2008), of the Minoan period on Crete that can be classified as *'sustainable urban water management practices, which can be compared to modern-day practices'* and *'the entire regulatory and management system of water in Athens must have worked very well and approached what today we call sustainable water management'*.

It is true, however, that the ancients did not have the modern technologies and design methodologies for stormwater drainage (Mays, 2001), but they were able to effectively develop such systems to accommodate the needs of their societies. Water was viewed as a valuable commodity that was harvested, stored, treated and recycled, rather than hidden 'out of sight, out of mind' and essentially wasted as is much of current thinking. This mindset is exemplified by the importance of RwH in the past which, as stated by Gikas and Angelakis (2009): *'is an alternative freshwater source which to a large extent remains underexploited'*.

References

AbdelKhaleq, R.A. and Alhaj Ahmed, I. (2007) Rainwater harvesting in ancient civilizations in Jordan. *Water Science and Technology: Water Supply*, 7 (1), 85–93.

Alkhaddar, R., Papadopoulos, G. and Al-Ansari, N. (2005) Water Harvesting Schemes in Jordan, International conference on efficient use and management of urban water supply, paper 10087, Tenerife, Spain.

Angelakis, A.N. (2014) Evolution of Rainwater Harvesting and Use in Crete, Hellas, through the Millennia. *Water*, 6, 1246–56.

Angelakis, A.N. and Durham, B. (2008) Water recycling and reuse in EUREAU countries: Trends and challenges. *Desalination*, 218 (1-3), 3–12.

Angelakis, A.N. and Spyridakis, S.V. (1996) The status of water resources in Minoan times – A preliminary study. In A. Angelakis & A. Issar (eds) *Diachronic climatic impacts on water resources with emphasis on Mediterranean region* (pp. 161–191). Heidelberg: Springer-Verlag.

Angelakis, A.N., Koutsoyiannis, D. and Tchobanoglous, G. (2005) Urban wastewater and stormwater technologies in ancient Greece. *Water Res.*, 39 (1), 210–220.

Angelakis, A.N. and Koutsoyiannis, D. (2003) *Urban Water Engineering and Management in Ancient Greece*. Encyclopaedia of Water Science, Marcel Dekker Inc. p 999–1007.

Antoniou, G.P. (2007) Lavatories in ancient Greece. *Water Science and Technology: Water Supply*, 7 (1), 155–164.

Antoniou, G.P. (2010) Ancient Greek lavatories: Operation with reused water (book chapter). *Ancient Water Technologies* pp. 67–86.

Antoniou, G., Kathijotes, N., Spyridakis, D.S. and Angelakis, A.N. (2014) Historical development of technologies for water resources management and rainwater harvesting in the Hellenic civilizations, *International Journal of Water Resources Development*, DOI: 10.1080/07900627.2014.900401.

Apt, D. (2011) *Inca Water Quality, Conveyance and Erosion Control*. 12th International Conference on Urban Drainage, Porto Alegre, Brazil.

Beckers, B., Berking, J. and Schütt, B. (2013) Ancient water harvesting methods in the drylands of the Mediterranean and Western Asia. *Journal for Ancient Studies*, 2, 145–164.

Becker, M., Spengler, B. and Flores, C. (2008) Systematic disconnection and securing of areas as preventive measures providing protection against flooding. 11th International Conference on Urban Drainage, Edinburgh, Scotland, UK.

Cadogan, C. (2007) Water management in Minoan Crete, Greece: the two cisterns of one Middle Bronze Age settlement. *Water Science and Technology: Water Supply*, 7 (1), 103–111.

Coe, M.D. (2003) *Angkor and the Khmer Civilization*. Thames and Hudson, New York.

De Feo, G., Antoniou, G., Fardin, H.F., El-Gohary, F., Zheng, X.Y., Reklaityte, I., Butler, D., Yannopoulos, S. and Angelakis, A.N. (2014) The Historical Development of Sewers Worldwide. *Sustainability*, 6, 3936–3974.

De Feo, G., Angelakis, A.N., Antoniou, G., El-Gohary, F., Haut, B., Passchier, C.W. and Zheng, X.Y. (2013) Historical and Technical Notes on Aqueducts from Prehistoric to Medieval Times. *Water*, 5, 1996–2025.

Ford, A. (1996) Critical resource control and the rise of the classical period in Maya. In: S.L. Fedicl (Ed.) *The Managed Mosaic: Ancient Maya and Resource Use*. University of Utah Press, Salt Lake City. Ch. 18, pp. 297–303.

Gikas, P. and Angelakis, A.N. (2009) Water resources management in Crete and in the Aegean Islands, with emphasis on the utilization of non-conventional water sources. *Desalination*, 248 (1–3), 1049–1064.

Gorokhovich, Y., Mays, L.W. and Ullmann, L. (2011) A Survey of Ancient Minoan Water Technologies, *Water Science and Technology: Water Supply*, IWA, Vol. 114, 388–399.

Haiman, M. and Fabian, P. (2009) Desertification and Ancient Desert Farming Systems. Encyclopedia of Life Support Systems (EOLSS). *Land Use, Land Cover and Soil Science*, 5, 41–55.

Haut, B., Zheng, X.Y., Mays, L., Han, M., Passchier, C. and Angelakis, A.N. (2015) Evolution of Rainwater Harvesting in Urban Areas through the Millennia: A Sustainable Technology for Increasing Water Availability. In *Water & Heritage: Material, Conceptual and Spiritual Connections*, W.J.H. Willems and Henk P.J. van Schaik (eds), Leiden: Sandstone Press.

Helms, S.W. (1981) *Jawa, Lost City of the Black Desert*, Cornell University Press, NY.

Imhoff, J.C., Kittle, J.L., Jr, Gray, M.R. and Johnson, T.E. (2007) Using the climate assessment tool (CAT) in U.S. EPA basins integrated modelling system to assess watershed vulnerability to climate change. *Water Sci. Technol.* 56, 49–56.

Konig, K.W. (2001) *The Rainwater Technology Handbook*. WILO-Brain, Dortmund, Germany.

Koutsoyiannis, D., Zarkadoulas, N., Angelakis, A.N. and Tchobanoglous, G. (2008) Urban Water Management in Ancient Greece: Legacies and Lessons. *Journal of Water Resources Planning and Management*, ASCE, January/February, 2008, 45–54.

Lang, M. (1968) *Waterworks in the Athenian Agora*. Excavations of the Athenian Agora, Picture Book No 11, American School of Classical Studies at Athens, Princeton, NJ.

Lofrano, G. and Brown, J. (2010) Wastewater management through the ages: A history of mankind. *Science of the Total Environment*, 408, 5254–5264.

Lucero, L.J., Gunn, J.D. and Scarborough, V.L. (2011) Climate change and classic Maya water management. *Water*, 3, 479–494.

Mays, L.W. (2014) Use of cisterns during antiquity in the Mediterranean region for water resources sustainability. *Water Science and Technology: Water Supply*, 14(1), 38–47.

Mays, L., Antoniou, G.P. and Angelakis, A.N. (2013) History of water cisterns: Legacies and lessons. *Water* (Switzerland), 5 (4), 1916–1940.

Mays, L.W. (2001) *Stormwater Collection Systems Design Handbook*, L. W. Mays, Editor-in Chief, McGraw-Hill.

Mays, L.W. (2007a) *Water Resources Sustainability*. L.W. Mays, Editor-in-Chief. McGraw-Hill.

Mays, L.W. (2007b) Water sustainability of ancient civilizations in mesoamerica and the American Southwest. *Water Science and Technology: Water Supply* 7(1), 229–236.

Mays, L.W. (2008) A very brief history of hydraulic technology during antiquity. *Environ. Fluid Mech.*, 8, 471–484.

Mays, L.W. (2010) A brief history of water technology during antiquity. In: L.W. Mays (ed.) *Ancient Water Technologies*. Springer. Ch. 1, pp. 1–28.

Mays, L.W. and Gorokhovich, Y. (2010) Water technology in the ancient American societies.In: L.W. Mays (ed.) *Ancient Water Technologies*. Springer. Ch. 9, pp. 140–200.

Mays, L.W., Antoniou, G.P. and Angelakis, A.N. (2013) History of Water Cisterns: Legacies and Lessons. *Water*, 5, 1916–1940.

Myers, W.J., Myers, E.E. and Cadogan, G. (1992) The aerial atlas of ancient Crete. University of California Press, Berkeley and Los Angeles, CA.

Pandey, D.N., Gupta, A.K. and Anderson, D.M. (2003) Rainwater harvesting as an adaptation to climate change. *Current Science*, 85: 46–59.

Scarborough, N.P., Dunning, V.L., Tankersley, K.B., Carr, C., Weaver, E., Grazioso, L., Lane, B., Jones, J.G., Buttles, P., Valdez, F. and Lentz, D.L. (2012) *Water and sustainable land use at the ancient tropical city of Tikal, Guatemala*. PNAS, 109, 31, 12408–12413.

Wohl, E., Angermeier, P.L., Bledsoe, B., Kondolf, G.M., MacDonnell, L., Merritt, D.M., Palmer, M.A., Poff, N.L. and Tarboton, D. (2005) River restoration. *Water Resources Research*, 41, W10301, doi:10.1029/2005WR003985.

Wright, K.R., Valencia, A. and Lorah, W.L. (n.d.) *Ancient Machu Picchu Drainage Engineering*. Available at: http://www.waterhistory.org/histories/machupicchu/machupicchu.pdf

Wright, K.R. and Valencia Zegarra, A. (1999) Ancient Machu Picchu Drainage Engineering. *Journal of Irrigation and Drainage*. 125, 6, 360–369.

Surface Water Strategy, Policy and Legislation

Frank Warwick

This chapter provides an introduction to policy, legislation and strategy in relation to surface water management. It outlines how surface water legislation is implemented, using examples from a number of jurisdictions in the developed world, and highlights that a hierarchy of laws and regulations is employed in order to set a framework for managing water. The majority of the chapter consists of a case study that compares surface water legislation in England and Scotland, two constituent countries of the UK. The examples show that, from the same starting point, implementation of legislation varies across geographies and organisations, and is influenced by local circumstances and drivers, prior regulation and the political complexion at the time.

The title of the chapter is surface water strategy, policy and legislation. *Policy* identifies government intentions and principles, often in broad outline. *Legislation* is the legal operating framework for a national or federal state, consisting of laws, which are mandatory statements of procedures to follow, monitored and enforced by regulatory organisations. *Strategy* covers the actions and resources employed to achieve specific goals, and it defines how policies will be achieved. The term *surface water* encompasses a range of meanings and intentions. In broad international and national legislation, it covers water at the surface of the land. For example, the European Union (2000) defines surface water as all inland water except groundwater, plus transitional (estuarine) and coastal waters, but excluding open oceans. Similarly, the Agriculture and Resource Management Council of Australia and New Zealand (1998) include catchments and coastal waters. The USA's Environmental Protection Agency (2013) definition is broader: surface water is 'all water naturally open to the atmosphere' including estuaries and seas; US regulation utilises the narrower term stormwater, incorporating rainfall runoff, snow melt runoff, and surface runoff and drainage, for water bodies subject to environmental concerns (40 CFR 122.26(b)(13), US Government Publishing Office, 2015).

Sustainable Surface Water Management: A Handbook for SuDS, First Edition.
Edited by Susanne M. Charlesworth and Colin A. Booth.
© 2017 John Wiley & Sons, Ltd. Published 2017 by John Wiley & Sons, Ltd.

Increasing concerns about poor water quality and flooding in recent decades have resulted in the implementation of a range of water-related legislation, progressing from a focus on individual issues to a broader emphasis on the wider water environment and joined-up management of water resources. Greater awareness of the natural hydrological cycle has led to the promotion of a range of techniques to manage surface water especially in urban areas, variously referred to as water sensitive urban design (WSUD) in Australia, stormwater best management practices (BMPs) in the USA and sustainable drainage systems (SuDS) in the UK. The SuDS approach promotes solutions that balance water quantity, water quality and biodiversity/amenity issues (Charlesworth, 2010). Despite the progress of recent years, legislation often addresses these issues separately, and as a result the regulatory processes, and the bodies responsible for implementing and enforcing them, can also differ. This chapter illustrates these ideas with reference to a number of countries. First, examples of the legislative hierarchies found in many developed countries are given, followed by a summary of water management in Germany. The bulk of the chapter comprises a case study of surface water management in England and Scotland. The similarity of UK legislative complexity with that of other countries is briefly highlighted at the end of the chapter.

3.2   Legislative Hierarchies

Surface water legislation in developed economies is typically not encapsulated in a single national law implemented consistently across government, but by a range of regulatory measures concerning flooding and water quality at national, regional and local levels. In federal government systems such as the USA, Germany and Australia, high-level policy is enshrined in law at the national level, then interpreted and implemented at the state and further at the regional and/or local level. Table 3.1 gives examples of legislation hierarchies illustrating that national laws are applied by further legislatures in more detailed geographies, enabling local circumstances to be taken into account. Surface water legislation in Germany is then explained in greater detail to show relationships across these levels.

3.2.1   German Water Management Institutions

The Federal Water Resource Management Act 2009 (German Federal Ministry of Justice and Consumer Protection, 2014) combines the implementation of European directives on both water quality and flooding. The Water Framework Directive 2000/60/EC (WFD) (EU, 2000) combines and updates previous Directives on the topic of water quality, while the Floods Directive 2007/60/EC (EU, 2007) addresses issues of excess water quantity. In Germany, both states (Länder) and local authorities, while taking account of the federal legislation, have the power to alter and supplement it to suit particular circumstances, and this right has been used, for example, to address funding for particular measures (Jekel *et al.*, 2013), and to stipulate runoff treatment standards that are absent from national legislation (Dierkes *et al.*, 2015). In each of the 16 German states, water management is notionally undertaken by a three-tier hierarchy (Jekel *et al.*, 2013), comprising:

- a supreme authority, typically the state's environment ministry who implement regulations
- intermediate authorities, who apply state policies at regional level
- lower tier authorities, typically urban and rural local authorities responsible for detailed management and monitoring.

Table 3.1   Example surface water quality legislation hierarchies.

	National legislation	Federal/state example	Regional/local example
USA	Federal Water Pollution Control Act (1972), known as the Clean Water Act (US Congress 2002). Water Quality Standards Regulation (40 Code of Federal Regulation (CFR) 131) requires states to establish water quality standards (US Government Publishing Office 2015)	California Water Code (California State Legislature 2015)	Nine Regional Water Quality Control Boards (California State Water Resources Control Board 2013)
Australia	Water Act 2007 (Australian Government 2015). National Water Quality Management Strategy (Australian Dept. of the Environment 2015)	New South Wales Water Management Act 2000 (New South Wales Government 2000)	Sydney Water Act 1994 (New South Wales Government 1994)
Germany	Federal Water Resource Management Act 2009 [Wasserhaushaltsgesetz] (German Federal Ministry of Justice and Consumer Protection 2014)	North-Rhine Westphalia State Water Act 1995 [Landeswassergesetz] (North-Rhine Westphalia State Office for Nature, Environment and Consumer Protection 2013)	54 Lower Water Authorities [untere Wasserbehörden] in districts and municipalities (North-Rhine Westphalia Water Network 2008) interpret and implement national and federal laws

Table 3.2   Summary of the number of water authorities at the different hierarchy levels in the German states. Data source: Jekel *et al.* (2013).

Number of authorities	Supreme authority	Intermediate tier	Lower tier
None		5	2
One	16	4	
Several		7	1
Many			13

In practice, while each state appoints a supreme authority, the existence of intermediate and lower tier authorities is not consistent across the 16 states (Table 3.2). The city states of Berlin, Bremen and Hamburg cover a relatively small spatial area, and have no need for a multi-level organisational structure. Seven of the larger states have devolved intermediate powers to regional authorities, but six have not implemented the intermediate level or have a single body at this tier, indicating that there are differing views on the structures necessary to manage surface water, with most variability shown at the intermediate level.

3.3   Case Study – The United Kingdom

This section considers the regulation of surface water in the United Kingdom as an example. The UK system of government is a constitutional monarchy, with national laws as well as those relating to England enacted in London. Increasingly, powers are devolved to national legislatures in Scotland, Wales and Northern Ireland. There is no overarching national surface water strategy for the UK as a whole, or in any of the constituent countries.

The hierarchy in Table 3.1 simplifies the structure of water regulation at different levels of government; the reality is often more complex. As in Germany, policy and legislation in the UK is enacted at a number of levels, from international guidance, through national and regional interpretations of that guidance enhanced by strategies, to practical implementation and monitoring at the local level. A comparison of surface water regulation in two constituent countries of the United Kingdom highlights how the same initial legislative drivers can result in varying interpretations, depending on government persuasion and the underpinning regulatory starting point. The hierarchy of legislation relating to surface water management across government in England and Scotland is shown in Figure 3.1, and this structure is examined in the remainder of this section.

3.3.1   International Policy and Legislation

Much of the environmental legislation in the UK in recent decades has been driven by European Union (EU) policy rather than national policy, and EU water policy has been implemented through two key Directives in the 21st century: the Water Framework Directive (WFD) in relation to water quality, and the Floods Directive addressing issues of excess water quantity. The WFD is principally concerned with improving the quality of water in all member states, both surface and groundwater as well as coastal water within territorial limits, to be achieved by regular monitoring of water bodies, supported by a programme of measures to make improvements and reduce pollution. The goal of the Floods Directive is to reduce potential adverse consequences of flooding for human health and activity and for the environment. It requires member states to document evidence of historical floods, to create maps for areas of low, medium and high probability flood hazard, flood risk maps identifying potential impacts and finally to use this information to build flood risk management plans.

3.3.2   National Legislation

EU directives are transposed into national legislation in each member state. Table 3.3 lists the UK national legislation used to implement the two EU water directives, and identifies the differing emphasis placed in the separate constituent countries simply by listing the number of pages used to translate the directives into national law. For both directives, Scotland has included more detail in its implementation, and these differences are explored later in this chapter.

The transposition of the WFD into law in England, Wales and Northern Ireland in 2003 followed the specific technical requirements of the Directive, focusing on the organisational responsibilities and mechanisms for implementation, and outlining the prescribed timescales for activities to be carried out and documents to be written, with the goal of understanding the current state of water quality in each country. Scotland took a more proactive approach to addressing water pollution (Hendry and Reeves, 2012), and the Water Environment and Water Services (Scotland) Act 2003 (WEWS) set a more aspirational tone than legislation in the other constituent countries of the UK, reiterating the WFD's goal of aiming to deliver good quality water and prevent pollution under the umbrella of achieving sustainable development.

A similar pattern can be detected in the transposition of the Floods Directive into UK law in 2009. Prior to this, in England there was a lack of clarity and coordination between

Organisation	Role	England	Scotland
EU Council & Parliament	Define international policy	**International** Water framework directive (2000/60/EC) Floods directive (2007/60/EC)	**International** Water framework directive (2000/60/EC) Floods directive (2007/60/EC)
National government departments	Define national policy	**National** Water environment (WFD) regulations 2003 Flood risk regulations 2009 Flood and water management Act 2010 National planning policy framework 2012	**National** Water environment & water services (Scotland) act 2003 Flood risk management (Scotland) act 2009 Water environment (controlled activities) 2013 Scottish planning framework 2014 Scottish planning policy 2014
Regulatory agencies	Define national Strategy	National flood and coastal erosion risk management strategy	Regulatory method 08 SUDS
Regulatory agencies	Define regional Strategy	**Regional** River basin management plans Catchment flood management plans	**Regional** River basin management plans
Local authorities	Define local policy and strategy	**Local** Local development framework Local flood risk management strategy & plans Surface water management plans Strategic flood risk assessment	**Local** Local development framework Local flood risk management strategy & plans Strategic flood risk assessment Planning advice notes
Developers	Develop using policies	**Site** Site flood risk assessment	**Site** Site flood risk assessment

Figure 3.1 The strategy and policy context for surface water management in England and Scotland. A hierarchy of organisations is related to their role and level in the policy- and strategy-making institutional structures. The Organisation column defines the bodies responsible for creating the policies/strategies/plans at that level. Examples of key development-related and surface water-related policies and strategies are identified. The term 'local authorities' simplifies the mixture of one-tier unitary and two-tier county/district councils present in England.

Table 3.3   Regulations used to transpose EU water directives into UK national law.

Geography	Water quality	Flooding
European Union	A framework for Community action in the field of water policy (2000/60/EC) [Water Framework Directive]. 72 pages	Assessment and management of flood risks (2007/60/EC) [Floods Directive]. 8 pages
England and Wales	The Water Environment (Water Framework Directive) (England and Wales) Regulations (Act of Great Britain Parliament 2003a). 12 pages	The Flood Risk Regulations (Act of Great Britain Parliament 2009a). 14 pages
Scotland	Water Environment and Water Services (Scotland) Act (Act of the Scottish Parliament 2003). 47 pages	Flood Risk Management (Scotland) Act (Act of the Scottish Parliament 2009). 73 pages
Northern Ireland	Water Environment (WFD) Regulations (Northern Ireland) (Act of Great Britain Parliament 2003b). 12 pages	The Water Environment (Floods Directive) Regulations (Northern Ireland) (Act of Great Britain Parliament 2009b). 13 pages

organisations dealing with flood risk owing to the large number of agencies with differing roles and responsibilities (Douglas *et al.*, 2010). The England Flood Risk Regulations addressed this issue by creating the role of lead local flood authority (LLFA) to coordinate flood responses at the local government level, and by establishing a duty for organisations to cooperate. However, the focus of the transposition remained a direct implementation of the requirements for risk mapping, planning and assessment. The Scottish Flood Risk Management Act (FRMA) reiterated the desire for sustainability and an integrated approach previously stated in the WEWS Act. It dealt with the basic requirements of the EU directive, but also considered wider interactions, for instance calling for an assessment of the contribution of sewerage systems in areas identified as vulnerable to flooding, and a review of whether existing natural features such as wetlands and floodplains could be enhanced to reduce flood risk. It also required developers of larger sites to submit a flood risk assessment if the development might lead to an increase in the number of buildings at risk of flooding. In England this requirement was already in place in Planning Policy Statement 25: Development and Flood Risk (Department for Communities and Local Government (DCLG) 2010), and was retained in the replacement National Planning Policy Framework (DCLG, 2012), which may explain its absence from the English regulations. However, the Scottish FRMA took a more transparently inclusive and consultative approach, requiring the establishment of flood risk advisory groups including all those with an interest in flooding at the local level. For both Directives, the Scottish legislation appears to have engaged more with the spirit rather than just the letter of the legislation.

3.3.3   Event-Driven Legislation

Legislation is also passed as a reaction to significant events in a particular country, and the floods of 2007 in England, which caused an estimated £4 billion of damage in that one year (Environment Agency England, 2007; ABI, 2008), led to a review (Pitt, 2008) calling for a more responsive and coordinated approach to water management, and ultimately to the passing of the Flood and Water Management Act (FWMA) (2010) applicable to England and Wales. To address issues of a lack of coordination in flood risk management, the FWMA designated the Environment Agency (EA), the environmental regulator in

England, as the body with overall responsibility for English flood risk management strategy, in addition to their role in managing flooding from major rivers and along coasts. The Pitt Review identified the benefits of more widespread use of SuDS to address surface water flooding and water quality, and as a result the FWMA included provisions in schedule 3 to make the use of SuDS mandatory in new developments in England and Wales. The Act tackled the lack of clear definition of responsibilities for SuDS adoption and maintenance by assigning that role to new Approving Bodies embedded in local authorities, and concerns over a lack of knowledge and experience in local government were addressed by the promise of a full set of SuDS national standards that could be followed, and a capacity building programme. However, the specific requirements of a piece of legislation are subject to further government interpretation, and the SuDS provisions of the FWMA were finally implemented in 2015, a five-year delay that contributed to inertia in achieving more sustainable surface water management in England. Measures in the 2010 FWMA directly relating to SuDS are listed in Table 3.4, and their ultimate interpretation in 2016 illustrates that initial intentions are not always realised.

The picture portrayed in Figure 3.1 concentrates on surface water management, but regulators and local authorities must take into account a broader palette of regulations encompassing factors such as planning and building law, pollution control, emergency planning, water supply and protection of species and habitats. For instance, Appendix 2 of the Scottish Surface Water Management Planning Guidance (Scottish Advisory and Implementation Forum for Flooding, 2013) highlights the wide range of legislation related to surface water management. Interested readers could also consult Ellis *et al.* (2009) for a detailed case study of Birmingham, UK, and Bettini *et al.* (2015) for a more theoretical perspective on institutional behaviour and potential adaptions to complexities in Australian water governance.

Table 3.4   Implementation in 2016 of some SuDS provisions in the Flood and Water Management Act 2010.

Provision	FWMA (2010) location	Current (2016) status in England
Removal of the right to connect to a public sewer	section 42	Not implemented
Publication of National standards for the design, construction, maintenance and operation of sustainable drainage	schedule 3 point 5	Non-statutory standards (2 pages)
Constitution of new Approving Bodies by the unitary authority or county council	schedule 3 point 6	Approval through existing planning processes
A requirement to obtain approval from the Approving Body for construction of any building or structure that covers land affecting the ability of that land to absorb rainwater	schedule 3 point 7	Planning consultation with LLFA for development of 10 or more properties
Approval must be granted if the drainage system complies with the national standards	schedule 3 point 11	Decision rests with the planning authority
An approving body must adopt approved drainage systems unless they drain a single property or publicly maintained roads	schedule 3 point 17–19	Requirement to ensure that proposals are properly funded
Adopted drainage systems must be maintained by the approving body according to the national standards	schedule 3 point 22	Requirement to ensure that maintenance is properly funded

3.3.4   Local Implementation of National Laws

Throughout the UK, the development planning process is the mechanism for controlling and promoting appropriate development in line with planning law. Implementation of surface water management in new developments is accomplished by means of policies and plans at a number of levels, implemented by a variety of organisations (Figure 3.1). Thus, policy and legislation relating to surface water are strongly linked to new development of properties and sites. This section explains how this is achieved in England and Scotland, and is followed by a comparison of the two systems.

3.3.4.1 England

Regulatory planning guidance for new developments in England is contained in the National Planning Policy Framework (NPPF) (DCLG, 2012), which gives priority to the appropriate use of SuDS for surface water management (point 103), supported by standing advice for flood risk assessments encouraging the use of SuDS (Environment Agency England, 2015). The FWMA implementation expects SuDS to be used in all larger developments for surface water management 'unless demonstrated to be inappropriate' (DCLG, 2014). The decision on what is inappropriate has been delegated to individual local planning authorities (LPAs), who should take into account the non-statutory SuDS technical standards (Department for Environment, Food and Rural Affairs (Defra) 2015) and bear in mind that design and construction costs should not exceed those of conventional systems (DCLG, 2015). The two pages of technical standards represent the fulfilment of the promised national standards that were to address local authority concerns about the need for detailed guidance on SuDS approval. Furthermore, the standards are only concerned with water quantity, and the goal that SuDS address *all* aspects of surface water management (Charlesworth, 2010) is ignored. The impact of government interpretation of the FWMA on LPAs has yet to be tested in detail, but clearly leaves scope for a range of interpretations in different locations.

Complexity is added by the existence of two different local authority structures in England, largely for historical reasons. In some places, local government is executed in two tiers: the lower tier district councils provide local operational services, with a number of districts coordinated by an upper tier county council. In other areas, functions of both tiers are performed by a single unitary authority. In principle, the upper tier authorities define policy, which is implemented by the lower tier authorities, although the upper tier authorities can, and do, delegate some of their duties to the lower tier organisations. The lower tier also makes and applies planning policy to surface water management in new developments. This complexity is exacerbated by the existence of a range of policies that address different forms of flooding. The EA manages flooding from major rivers and on coasts, and has a national coordinating role. The EA (2015) manages the workload for Flood Risk Standing Advice in England by adopting a process that small developments in low flood risk areas are provided with on-line guidance, while larger developments and those in higher flood risk zones must submit detailed applications which are reviewed through the planning system.

The Flood Risk Regulations (Act of Great Britain Parliament, 2009a) assigned the role of lead local flood authority (LLFA) to unitary and upper tier councils. LLFAs were tasked with creating and applying a local flood risk management strategy (LFRMS) to define the objectives, means and costs of managing local flood risk from surface runoff and smaller watercourses outside the remit of the EA.

Two-thirds of the floods in England in summer 2007 were caused by surface water drainage rather than fluvial flooding (Environment Agency England, 2007). As a result, English local authorities were also charged with creating Surface Water Management Plans (SWMPs) intended to enable cooperation between organisations to manage surface water in a local area over the longer term (Defra, 2010). The term surface water has a different interpretation in policy documents at this local scale, with a more specific focus on 'flooding from sewers, drains, groundwater, and runoff from land, small water courses and ditches that occurs as a result of heavy rainfall' (Defra, 2010). SWMPs were intended to inform the LFRMS, and envisage a role for SuDS to support a more strategic approach to surface water planning across a wider area (Defra, 2010).

The Environment Agency has the principal responsibility in England for addressing water quality issues, and employs a system of licensing and environmental permitting to manage point source pollution risks (Environment Agency England, 2013). The NPPF directs that the planning system should contribute to preventing water pollution, but that its management and control should rest with the EA (DCLG, 2012). Consequently, there is limited accountability at the local planning level for issues of water quality in England, which may partially explain the difficulties coping with diffuse pollution (Balmforth, 2011).

So there are a number of plans, policies, strategies and organisations that address surface water management at the local authority level in England. Emphasis is currently placed on local authorities defining their own implementation of national legislation. While this approach caters for consideration of local conditions, it does not allow for economies of scale, and potentially duplicates effort. It seems unlikely that the 2015 FWMA SuDS implementation will improve this situation, as the focus on local determination is retained. Organisations such as the parliamentary All Party Group for Excellence in the Built Environment (2015) have expressed reservations about whether the implementation of the SuDS provisions in the FWMA in England in their current form will achieve sustainable water management in any reasonable timescale.

3.3.4.2 Scotland

In Scotland, regulatory planning guidance for water management in new developments is contained in two key documents: Scottish Planning Policy (SPP) (Scottish Government, 2014a) defines policy on land development and use, and the Third National Planning Framework (Scottish Government, 2014b) details long-term infrastructure strategy. SPP requires that planning adopt a precautionary approach to flood risk, and explicitly equates surface water flood risk to pluvial flooding (i.e. due to precipitation) and endorses the use of SuDS to manage flooding from this source. However, there is little explicit mention of surface water *quality* management in either document, nor in the more detailed Surface Water Management Planning Guidance document (Scottish Advisory and Implementation Forum for Flooding, 2013), which concentrates on flood risk.

It is the WEWS Act (section 20 and schedule 2) that allows for regulation of 'controlled activities' that risk polluting, abstracting from or impounding water bodies, by means of regularly updated general binding rules (GBRs) defined in Controlled Activities Regulations (CAR) (Acts of the Scottish Parliament, 2011, 2013). Emphasis is placed on the role of the Scottish Environment Protection Agency (SEPA), the environmental regulator, to protect the water environment. CAR (2011 p. 7) states for instance that 'SEPA must impose such conditions as it considers necessary or expedient for the purposes of protection of the water environment' indicating that this duty is paramount. Statutory guidance on using

SuDS to safeguard urban water quality is provided using GBRs 10 and 11 to target pollution of surface water by runoff (diffuse pollution) and direct disposal of pollutants (point source pollution). GBR 10 requires any development constructed after 1 April 2007 to employ SuDS to prevent polluted surface water reaching freshwater bodies. Required numbers and types of SuDS devices are specified in Regulatory Method 08 (Scottish Environment Protection Agency, 2014), which covers housing, commercial, highway and industrial development. SEPA has managed the potentially large workload of reviewing all potential sources of pollution by defining a hierarchy for approval of increasing levels of pollution risk. GBRs apply to specific low risk activities and are monitored initially through the planning system. Medium and higher risk activities require explicit registration and licensing, for which charges are made. Regulatory Method 08 prescribes the number of SuDS features required for all types and sizes of development, and offers suggestions about the specific types of devices suitable in particular developments. SEPA also requires long-term maintenance be put in place.

3.3.5   Comparison of Approaches in England and Scotland

Both English and Scottish regulation of surface water separate the guidance for addressing quality and quantity issues. Flooding is controlled by the development planning process, whereas water quality is managed by the environmental regulator. Both countries assign a narrower definition to the term surface water at the local authority level compared to international law, but the two countries have taken different paths for improving surface water management. The differences between English and Scottish surface water and SuDS legislation are likely to influence the way that SuDS are implemented in the two jurisdictions, and Table 3.5 gives some examples.

Scotland provides more detailed specific centralised guidance to interpret surface water management legislation than in England, where much of the detailed interpretation is left

Table 3.5   Some key differences between detailed SuDS regulation in England and Scotland.

Element	England	Scotland
Regulator	Each unitary/upper tier local authority	Scottish Environment Protection Agency (SEPA)
Requirement for SuDS to manage surface water	Where reasonably practicable	Mandatory for all developments of more than one property after March 2007
Design guidance	The SuDS National Standards offer outline guidance on the volume and peak rate of runoff for the development site	Sewers for Scotland and Regulatory Method 08 provide detailed guidance about the specific types of SuDS that will be adopted, and their required design features
Adoptable public SuDS	SuDS in developments of 10 or more properties, and major commercial development	SuDS, serving two or more premises, that are detention ponds, detention basins or underground storage located in public open space, and are designed to reduce runoff rates up to a 1 in 30 year event
Adoption, operation and maintenance organisation	An organisation agreed during the planning approval process	Scottish Water, the local authority or a public body

to local authorities. In Scotland, the WEWS Act took the opportunity of transposing the Water Framework Directive into Scottish Law to make the use of SuDS mandatory in new developments in order to address poor water quality due to diffuse pollution. In England, by contrast, the SuDS National Standards utilise the concept of 'reasonably practicable' (Defra, 2015), limiting the need for compliance to the extent that SuDS construction should not be more expensive than an equivalent conventional drainage design.

One key reason for the wider implementation of SuDS for surface water management in Scotland is the timing of legislation. As a result, Scotland has over 10 years' more experience of implementing and regulating SuDS than England. The longer experience in Scotland through the planning, implementation and monitoring processes has informed regulation, so Scotland has more detailed and precise guidance on SuDS planning approval in Regulatory Method 08 (SEPA, 2014), compared with the brief SuDS National Standards in England. England does not have plans for a monitoring programme to determine the effectiveness of the revisions to the SuDS provisions in the FWMA (Stephenson, 2015), which, given the experience of Scotland in utilising monitoring to inform changes to legislation, may not generate the expected longer-term benefits.

Scottish legislation has assigned a much more precise definition to the meaning of SuDS than the English FWMA. The Sewerage (Scotland) Act (Act of Great Britain Parliament, 1968) defining the duties and powers of Scottish Water, was amended by the WEWS Act to identify references to sewers as including SuDS, whereas in England and Wales SuDS are regarded as distinct from public sewers. The WEWS Act (section 33) clarifies that SuDS facilitate attenuation, settlement or treatment of surface water from two or more premises. It names specific devices that are considered to be SuDS: inlet structures, outlet structures, swales, constructed wetlands, ponds, filter trenches, attenuation tanks and detention basins, and clarifies that associated pipes and equipment are to be treated as part of the system. In contrast, the English FWMA does not clarify the meaning of SuDS, implying the need for precise construction standards to define what would be acceptable, but the delivered SuDS technical standards (Defra, 2015) only outline broad functional criteria that should be applied, in the expectation that local standards will define what is desired and acceptable. In contrast, detailed standards are clearly defined in Scottish Regulatory Method 08, and Sewers for Scotland (Scottish Water, 2015) details specific construction standards for detention ponds and detention basins that the water and sewerage company will adopt.

In England, the FWMA assigns responsibility for approval and future maintenance of SuDS to an organisation approved by the planning process. While this approach allows flexibility, it is also likely to lead to inconsistencies in approach and standards across different planning authorities. Scottish legislation is on the whole more specific, and responsibilities are allocated to different organisations to those proposed in England. Overall, the approach in England may result in inconsistency and duplication of effort, while arguably the Scottish legislation promotes a more coherent view of integrated surface water management.

3.4   Comparison of UK Approaches with Other Countries

While the variations in legislative approach to surface water management in different constituent parts of the UK may appear surprising, this result mirrors findings internationally (Brown and Farrelly, 2009). In Australia, for example, each state and major city has different governance models in terms of the responsibilities and leadership of state and local government (Rijke *et al.*, 2013). The creation of the cross-state Murray-Darling Basin Authority (Australian Government, 2015) echoes the WFD's catchment-scale approach in the EU, but this has cut across historical state-level responsibilities and must consequently deal with challenges in coordinating implementation and monitoring across a range of bodies in the affected states (Connell and Grafton, 2011). Different approaches to implementing the WFD have also been identified in EU countries as a result of different governance structures and changes in political complexion (e.g. Liefferink *et al.*, 2011; Thiel, 2015). The more centralised path chosen by Scotland reflects WFD implementation in Denmark, while the devolution of responsibilities in England may lead to the lack of consistent coordination observed in Sweden (Nielsen *et al.*, 2013). In the USA, the existence of national, state and local administration and funding of stormwater management had led to noticeable variations across the country (US Environmental Protection Agency, 2011), which have slowed progress in achieving effective surface water management (Department of Trade and Industry, 2006). The hierarchical nature of US legislation has led to varying levels of rigour when pursuing improvements in surface water management, and the organisational separation of water quantity management from water quality management that occurs in the USA is also present in both England and Scotland, resulting in challenges to a coordinated approach to surface water management (National Research Council, 2008).

3.5   Conclusions

Policy, legislation and strategy for surface water management in developed countries is typically implemented across a number of levels, from international laws to local government regulations. International and national and legislation can be interpreted in different ways at these levels, even in the same country, based on local circumstances and government interpretation. In addition, flood risk and water quality in many developed countries are often addressed by separate legislation. As a result, consistent and comprehensive surface water management is a goal yet to be achieved.

References

ABI (2008) *The Summer Floods 2007: one year on and beyond.* Association of British Insurers, London.

Act of Great Britain Parliament (1968) *Sewerage (Scotland) Act 1968.* Office of Public Sector Information, London.

Act of Great Britain Parliament (2003a) *Statutory Instrument 2003 No 3242: The Water Environment (Water Framework Directive) (England and Wales) Regulations 2003.* Office of Public Sector Information, London.

Act of Great Britain Parliament (2003b) *The Water Environment (Water Framework Directive) Regulations (Northern Ireland) 2003.* Her Majesty's Stationery Office, London.

Act of Great Britain Parliament (2009a) *Statutory Instrument 2009 No 3042: The Flood Risk Regulations 2009.* Office of Public Sector Information, London.

Act of Great Britain Parliament (2009b) *The Water Environment (Floods Directive) Regulations (Northern Ireland) 2009.* Her Majesty's Stationery Office, London.

Act of the Scottish Parliament (2003) *Water Environment and Water Services (Scotland) Act 2003.* The Queen's Printer for Scotland, Edinburgh.

Act of the Scottish Parliament (2009) *Flood Risk Management (Scotland) Act 2009.* The Queen's Printer for Scotland, Edinburgh.

Act of the Scottish Parliament (2011) *The Water Environment (Controlled Activities) (Scotland) Regulations 2011.* The Queen's Printer for Scotland, Edinburgh.

Act of the Scottish Parliament (2013) *The Water Environment (Controlled Activities) (Scotland) Amendment Regulations 2013*. The Queen's Printer for Scotland, Edinburgh.

Agriculture and Resource Management Council of Australia and New Zealand (1998) *National Water Quality Management Strategy Implementation Guidelines*. Available at: http://tinyurl.com/zr9muwr

All Party Group for Excellence in the Built Environment (2015) *Living with water. Report from the Commission of Inquiry into flood resilience of the future*. Available at: http://tinyurl.com/h2j553z

Australian Government (2015) *Water Act 2007 as amended, taking into account amendments up to National Water Commission (Abolition) Act 2015*. Available at: http://tinyurl.com/zuooufo

Australian Dept. of the Environment (2015) *National Water Quality Management Strategy*. Available at: http://tinyurl.com/hrtnhe6

Balmforth, D. (2011) *Comparing the arrangements for the management of surface water in England and Wales to arrangements in other countries*. Available at: http://tinyurl.com/zzdjcc9

Bettini, Y., Brown, R.R. and De Haan, F.J. (2015) Exploring institutional adaptive capacity in practice: examining water governance adaptation in Australia. *Ecology and Society* 20(1), 47, doi.org/10.5751/ES-qwqw07291-qwqw200147

Brown, R.R. and Farrelly, M.A. (2009) Delivering sustainable urban water management: a review of the hurdles we face. *Water Science & Technology* 59 (5), 839–846.

California State Legislature (2015) *California Water Code. Division 7. Water Quality*. Available at: http://tinyurl.com/jm9x847

California State Water Resources Control Board (2013) *The Nine Regional Water Quality Control Boards in California*. Available at: http://tinyurl.com/7aq9jzh

Charlesworth, S.M. (2010) A review of the adaptation and mitigation of global climate change using sustainable drainage in cities. *Journal of Water and Climate Change* 1 (3), 165–180.

Connell, D. and Grafton, R.Q. (2011) Water reform in the Murray-Darling Basin. *Water Resources Research* 47, W00G03, doi:10.1029/2010WR009820

DCLG (Department for Communities and Local Government) (2010) *Planning Policy Statement 25: Development and Flood Risk*. The Stationery Office, London.

DCLG (2012) *National Planning Policy Framework*. Available at: http://tinyurl.com/o5s4ydt

DCLG (2014) *Written statement to Parliament. Sustainable drainage systems. 18 December 2014*. Available at: http://tinyurl.com/mdn32lz

DCLG (2015) *Planning Practice Guidance – Flood Risk and Coastal Change*. Available at: http://tinyurl.com/nn2fscu

Department for Environment, Food and Rural Affairs (Defra) (2010) *Surface Water Management Plan Technical Guidance*. Available at: http://tinyurl.com/nksw88e

Defra (2015) *Sustainable Drainage Systems. Non-statutory technical standards for sustainable drainage systems*. Available at: http://tinyurl.com/q4gqo8t

Department of Trade and Industry (2006) *Sustainable drainage systems: a mission to the USA*. Available at: http://tinyurl.com/nf5l9sw

Dierkes, C., Lucke, T. and Helmreich, B. (2015) General Technical Approvals for Decentralised Sustainable Urban Drainage Systems (SUDS) – The Current Situation in Germany. *Sustainability* 7, 3031–3051.

Douglas, I., Garvin, S., Lawson, N., Richards, J., Tippett, J. and White, I. (2010) Urban pluvial flooding: a qualitative case study of cause, effect and nonstructural mitigation. *Journal of Flood Risk Management* 3, 112–125.

Ellis, B., Scholes, L., Shutes, B. and Revitt, D.M. (2009) *Guidelines for the preparation of an institutional map for cities identifying areas which currently lack power and/or funding with regard to stormwater management*. Available at: http://tinyurl.com/znoujmt

Environment Agency England (2007) *Review of 2007 summer floods*. Available at: http://tinyurl.com/jejs5ks

Environment Agency England (2013) *Managing water for people, business, agriculture and the environment*. Available at: http://tinyurl.com/jduonl3

Environment Agency England (2015) *Flood risk assessment: local planning authorities*, Available at: http://tinyurl.com/gmvldjb

EU (European Union) (2000) *Directive 2000/60/EC of the European Parliament and of the Council of 23 October 2000 establishing a framework for Community action in the field of water policy.* Available at: http://tinyurl.com/gpfazrm

EU (2007) *Directive 2007/60/EC of the European Parliament and of the Council of 23 October 2007 on the assessment and management of flood risks.* Available at: http://tinyurl.com/nhawk7o

German Federal Ministry of Justice and Consumer Protection (2014) *Water Resource Management Act 2009 [Wasserhaushaltsgesetz]*, Available: http://www.gesetze-im-internet.de/bundesrecht/whg_2009/gesamt.pdf [30 July 2015]. In German.

Hendry, S. and Reeves, A.D. (2012) The Regulation of Diffuse Pollution in the European Union: Science, Governance and Water Resources Management. *International Journal of Rural Law and Policy* Occasional Paper Series, 13 pp.

Jekel, H., Arle, J., Bartel, H., Baumgarten, C., Blondzik, K. *et al.* (2013) *Water Resource Management in Germany. Part 1 – Fundamentals.* Available at: http://tinyurl.com/q5fa76w

Liefferink, D., Wiering, M. and Uitenboogaart, Y. (2011) The EU Water Framework Directive: A multi-dimensional analysis of implementation and domestic impact. *Land Use Policy* 28, 712–722.

National Research Council (2008) Urban Stormwater Management in the United States. Available at: http://tinyurl.com/ooufo5b

New South Wales Government (1994) *Sydney Water Act 1994 No 88.* Available at: http://tinyurl.com/ppdneq7

New South Wales Government (2000) *New South Wales Water Management Act 2000 No 92.* Available at: http://tinyurl.com/jjqz766

Nielsen, H.Ø., Frederiksen, P., Saarikoski, H., Rytkönen, A.-M. and Pedersen, A.B. (2013) How different institutional arrangements promote integrated river basin management. Evidence from the Baltic Sea Region. *Land Use Policy* 30, 437– 445.

North-Rhine Westphalia State Office for Nature, Environment and Consumer Protection (2013) *State Water Act [Landeswassergesetz].* Available at: http://tinyurl.com/hhfssuu In German.

North-Rhine Westphalia Water Network (2008) *Management Plan Chapter 13: Responsible Authorities.* Available at: http://tinyurl.com/o5ayxgd In German.

Pitt, M. (2008) *Learning lessons from the 2007 floods.* Available at: http://tinyurl.com/n6jwak8

Rijke, J., Farrelly, M., Brown, R. and Zevenbergen, C. (2013) Configuring transformative governance to enhance resilient urban water systems. *Environmental Science & Policy* 25, 62–72.

Scottish Advisory and Implementation Forum for Flooding (2013) *Surface Water Management Planning Guidance.* Available at: http://tinyurl.com/zx2u5c2

Scottish Environment Protection Agency (2014) *Regulatory Method (WAT-RM-08) Sustainable Urban Drainage Systems v5.2.* Available at: http://tinyurl.com/otdpkce

Scottish Government (2014a) *Scottish Planning Policy.* The Scottish Government, Edinburgh.

Scottish Government (2014b) *Scottish Planning Framework.* The Scottish Government, Edinburgh.

Scottish Water (2015) *Sewers for Scotland – A technical specification for the design and construction of sewerage infrastructure.* Third edition. Available at: http://tinyurl.com/hdk3kry

Stephenson, A. (2015) *New SuDS Regulations – Where Now for SuDS?* Available at: http://tinyurl.com/zncjuuy

Thiel, A. (2015) Constitutional state structure and scalar re-organization of natural resource governance: The transformation of polycentric water governance in Spain, Portugal and Germany. *Land Use Policy* 45, 176–188.

US Congress (2002) *Federal Water Pollution Control Act (33 U.S.C. 1251 et seq.).* Available at: http://www.epw.senate.gov/water.pdf

US Environmental Protection Agency (2011) *Summary of State Stormwater Standards*, Available at: http://tinyurl.com/zdjstqw

US Environmental Protection Agency (2013) *Vocabulary Catalog – Surface Water.* Available at: http://tinyurl.com/hqnqn2g

US Government Publishing Office (2015) *Electronic Code of Federal Regulations Title 40 –Protection of Environment.* Available at: http://tinyurl.com/jtrzq95

Sustainable Drainage Systems: Operation and Maintenance

Neil Berwick

4.1   Introduction

This chapter will consider the issue of sustainable urban drainage systems (SuDS) operation and maintenance. SuDS have undergone a transition from new, to an accepted technology in many countries. The past decade has seen considerable advancements in design guidance and a greater understanding of the multiple benefits that SuDS provide; however, there remains a deficit of detailed operation and maintenance guidance.

There is now a greater focus than ever on SuDS and their condition. This is particularly relevant in the UK, where SuDS have been commonplace since the 1990s, and their use has been further promoted by recent legislative change stipulating the use of SuDS for all new developments. In Scotland, SuDS have been the legally required norm to drain surface runoff from all new developments constructed after 1 April 2006. Similar legislation came into effect in England on 1 April 2015. A transition in the perception of SuDS has occurred over this period; initial studies (SNIFFER, 2005) undertaken at the first large-scale master-planned site in the UK (Dunfermline East Expansion, DEX) discovered that the main area of concern for residents living in close proximity to SuDS (ponds) was water safety. Revisiting the study several years later, Bastien *et al*. (2012) noted that local residents were most concerned with the maintenance of SuDS, illustrating a shift in perspective. SuDS can lend to saleability of property; houses located in the immediate vicinity of SuDS can be marketed at a higher value (SNIFFER, 2005; CNT, 2011), and sustainable development (for example, where SuDS lower flood risk) is becoming a more recognised process, particularly in light of flooding being a more newsworthy issue (Bryant, 2006). However, these positive effects of SuDS can be reduced if the SuDS and surrounding public areas are poorly maintained. It should be noted that maintenance of SuDS is not solely concerned with aesthetics; it is essential to ensure continued operation, delivering water quantity (reduced flood risk) and water quality (surface water discharged to the environment) benefits.

Sustainable Surface Water Management: A Handbook for SuDS, First Edition.
Edited by Susanne M. Charlesworth and Colin A. Booth.
© 2017 John Wiley & Sons, Ltd. Published 2017 by John Wiley & Sons, Ltd.

4.2   What is Operation and Maintenance and Why is it Important?

Operation and maintenance is the process of activities carried out to ensure the continued operation and prevent failure. As with all drainage systems, SuDS components should be regularly inspected and maintained to ensure efficient operation (Bell *et al.*, 2015), visual aesthetic and benefit to the local community.

The Oxford Dictionary (2015) defines the terms as:

Operation: The action of functioning or the fact of being active or in effect.

Maintenance: The process of preserving a condition or situation or the state of being preserved.

A range of SuDS techniques are available (Woods Ballard *et al.*, 2015; Highways Agency, 2015) and these can be used to match the catchment constraints and variables including: location, visibility, owning and/or maintaining body and design type (e.g. permanent water, dry attenuation or underground structures). The type of SuDS used will play a pivotal role in the type and frequency of maintenance and management activity.

There must be sufficient maintenance to reduce risk of each of the following:

Pollution of the water environment: In most developed countries, legislation exists to ensure the continued quality of surface and groundwater. In Europe, the Water Framework Directive (2000) is the overarching legislation, which is then enacted into national law for each country. For example, the Water Environment (Controlled Activities) (Scotland) Regulations 2011 (as amended) (otherwise referred to as CAR) makes it an offence to allow discharge of foul, industrial effluent, paint, oil or any other pollutants into a surface water system (General Binding Rule 10) or from a surface water system to the water environment (General Binding Rule 11) (SEPA, 2016).

Flooding: SuDS temporarily attenuate runoff and discharge it at a controlled rate, consequently reducing the likelihood of surface water flooding, within the constraints of the hydraulic level of service.

Personal injury: SuDS can present a range of hazards to the public and operatives, including slip/trip/fall and drowning, but regular inspection and maintenance can reduce risk.

Complaints from land owners/residents: Complaints regarding SuDS are most common where SuDS are located in residential developments. This is predominantly due to the proximity of houses and roads/paths to SuDS.

SuDS operation and maintenance is not solely the concern of facility managers and landscape contractors; knowledge of operation and maintenance is pertinent at all stages of the SuDS development process, from outline design to aftercare. The designer must have a clear understanding of how design can enable or inhibit operation and maintenance. A well thought out design will deliver water quantity and quality benefits, provide ecosystem services and will be safely maintainable. There is a recognised link between design for operation and maintenance and the survivability of the treatment train. Jefferies *et al.* (2009) identify the relationship between asset type and position in the SuDS treatment train to promote simple and cost-effective maintenance, limit the requirement for corrective maintenance and extend operational life.

4.3   Inspection, Reporting and Maintenance

Operation and maintenance can be split into two categories: (1) inspection and reporting; (2) maintenance activity.

These categories are closely related. Facility inspectors should possess a good understanding of the maintenance regime for the site so that they can assess their effectiveness, suitability and the standard of the maintenance carried out. Up-skilling of maintenance teams can improve the standard of maintenance, and operatives are more likely to identify operational issues at an early stage when they can be easily and cost-effectively resolved. This is particularly advantageous when the main facility inspection occurs on an extended interval (for example annually) maintenance teams are on site regularly and their observations can be an invaluable.

4.3.1   SuDS Inspection and Reporting

Inspection and reporting activity should begin during the construction phase and continue throughout the interim phase, to the handover and post-handover aftercare (Figure 4.1). Continued inspection will ensure that the SuDS are implemented as designed, protected from construction runoff and maintained so that planted vegetation becomes established.

4.3.1.1 Inspection

Inspection during the construction and handover phase should be carried out on a regular basis (e.g. monthly or every two months) allowing the effectiveness of the structure to be assessed, and lending familiarity to inspectors and maintenance teams. This interim inspection activity will also highlight any amendments required to the initial maintenance schedule.

Post-handover inspection and reporting for SuDS should be carried out on a regular basis. Current guidance for SuDS (Woods Ballard *et al.*, 2015) recommends (following

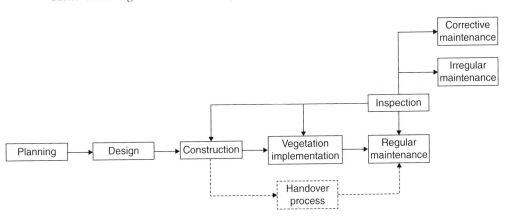

Figure 4.1   Integration of inspection and maintenance activity during the SuDS development process [author's image].

Table 4.1   Inspection skill level descriptions (adapted from the *Centre for Watershed Protection Stormwater Pond and Wetland Maintenance Guidebook*, 2004).

Skill level	Description
1	No special skills or prior experience
2	Inspector, maintenance contractor or citizen with prior experience of SuDS techniques
3	Inspector or maintenance contractor with extensive experience of SuDS maintenance issues
4	Professional engineering consultant/facility inspector

completion of the construction phase and establishment of vegetation) that assets are inspected frequently; this is general guidance, and applicable to the UK. Actual inspection frequencies can vary due to other factors such as geographic location, weather (local climate), site-specific conditions, the owning/maintaining organisation, planning agreements, among others.

Inspections should be conducted by persons with a skill level commensurate with the inspection type (Table 4.1). Maintenance teams can carry out basic inspections during visits, and ideally the operatives should possess a basic understanding of the engineered components of SuDS: what they are, how they operate and common operational issues. Specialist inspection will involve a more detailed assessment of the SuDS (including sub-surface appurtenances) and provide a report of condition and operation of the structure. The specialist inspection is carried out at a much longer interval (typically on an annual basis) and provides a snapshot of operational condition at a given time. It may not identify infrequent operational issues including greywater cross connections (typically, intermittent and evidence can be washed away during following rainfall events) and blocked or partially blocked outlets. A combination of specialist and maintenance inspections is the most effective means of reducing operational risk.

Input from non-technical observers should not be dismissed; reporting can be carried out by those with little or no knowledge of SuDS, homeowners commonly being a good example of this, where a complaint to the facility owner/maintainer is made, triggering an additional inspection or maintenance visit. SEPA (2006) recommends that new householders are provided with written information on the SuDS and how they work; this can increase the incidence of non-technical reporting.

4.3.1.2 Reporting

Reporting is the formal output of SuDS inspection (and maintenance) visits. Records of site visits provide three main benefits:

1. Due diligence: an audit trail confirming that the facility has been suitably maintained and inspected on a regular basis; this is of particular following an incident on site, for example localised flooding, pollution or an accident
2. Gathering maintenance data, which can be interrogated, and maintenance regimes amended accordingly
3. Evaluating operational risk of specific design details to inform future asset adoption strategy

Reporting should be carried out on every visit, usually in the form of a written record of the SuDS condition, complemented by photographs. The record can be in the form of a standard pro-forma (paper or electronic), written summary of observations, as-built drawings marked up with issues identified or a combination of all three.

The use of a standardised pro-forma provides the most consistent means of recording data, particularly where it is used in conjunction with a relational database management system. The pro-forma should be designed to capture a finite range of data, and will normally fall into one of three main categories:

1. Maintenance record: detailing maintenance activity carried out during the visit, usually in the form of a simple 'tick sheet' recording: litter removed, grass cutting detail, small-scale silt removal from structures
2. Operational record: a record of the condition and operation of the SuDS including outlet and flow control condition, water level variations
3. Maintenance and operational condition record: a combination of the previous two, typically used during the specialist inspection

The pro-forma should be a short non-complex document that can be quickly and easily completed by the user. Where maintenance pro-formas are lengthy, complex or use overly technical wording, the quality and consistency of the data gathered can be variable, and there is a risk that the completion of the record becomes a paper exercise. Pro-formas can be either specific to a given SuDS type or a generic form intended to solicit top line information from a range of SuDS techniques. For simplicity it is usually easier to use the latter.

The skill level necessary to provide valid and useable maintenance reporting is Level 2, and for the annual technical inspection Level 3 or 4 (Table 4.1), at the discretion of the owning/maintaining organisation or local legislation.

4.3.2 SuDS Maintenance

SuDS maintenance involves a range of activities which, when carried out to the required standard and at suitable intervals, will ensure the continued operation of the structure. SuDS are typically a combination of hard and soft engineered (or landscape) structures; these have different requirements for maintenance. Lampe (2005) and Woods Ballard *et al.* (2007) identify three categories of maintenance for SuDS; routine, infrequent and corrective.

Routine maintenance tasks are those of generally low specification, occurring on a frequent basis, as dictated by the maintenance schedule for the site (e.g. litter removal, grass cutting and inspections) and seasonal maintenance (e.g. pruning, leaf litter control and screen maintenance).

Infrequent maintenance tasks occur at frequencies greater than annually and are commonly larger in scale. Such activities are carried out on a predefined basis, but can also be triggered by the inspection regime. Examples of infrequent maintenance include management of emergent species around the perimeter of ponds and the emptying of engineered silt traps.

Corrective maintenance tasks are those which involve the repair, replacement or rehabilitation of existing structures/components or redesign of specifications to ensure the effectiveness of the SuDS. Corrective maintenance occurs on an ad hoc basis and is usually identified by the inspection process or following an operational incident.

Regular and (some) infrequent maintenance tasks are usually carried out by landscape contractors; the scope of the activity is predominantly limited to grounds maintenance work (horticultural and minor repair, such as fence repair) and reporting. Dependent upon the owning organisation, the landscape contractor may also carry out specialist inspection and reporting of the SuDS. More complex (irregular) works will normally require specialist contractors: for example replacement or reinstatement of items including pipework, flow controls, gabions and other ancillary structures.

4.3.2.1 Maintenance Levels

SuDS can be maintained to different levels, which vary dependent upon factors including the owning/maintaining organisation's maintenance policy, location and visibility of the SuDS and local/environmental considerations. Lampe (2005) identifies three levels of maintenance for SuDS: low, medium and high.

Low level: the basic level, to maintain function of the SuDS. Maintenance and inspection are carried out at extended intervals, with typical maintenance activities being litter removal and vegetation management to minimise risk of blockage to inlet/outlet structures. The inspection regime may also be at an extended interval and is used to identify any additional works required to ensure operation. This level of maintenance is most common for SuDS that are not in areas of high public footfall, or visible from main access routes, or for trunk road and motorway drainage. Maintenance tasks will be at a lower frequency than those for an accessible site.

Medium level: the level needed to ensure the function and appearance of the SuDS. This level of maintenance is commonly used for SuDS in areas of high footfall and visibility, such as housing developments, where there is increased focus on visual appearance and amenity space.

High level: enhanced maintenance activities for appearance and amenity. This level is common in areas with high footfall, such as commercial areas or public amenity space in dense urban areas. SuDS in these locations have been designed with additional emphasis on landscaping and planting specification to provide attractive surroundings.

It is essential that there is always enough maintenance to ensure that the SuDS operate as designed. All maintenance over and above this level will be as a requirement of operation or appearance or biodiversity or health, as illustrated in Figure 4.2. It should be noted that there can be a difference between perception of maintenance and actual maintenance carried out. Litter (or the absence of it) is often regarded as an indicator of effective maintenance (Bastien *et al.*, 2012), but this may not necessarily be the case: regular litter removal does not guarantee that the structural items are being maintained or inspected, although it will indicate that maintenance teams are on site on a regular basis. It should be noted that SuDS can also be over-maintained, reducing effectiveness, costing money and restricting the development of habitat diversity (Graham *et al.*, 2012).

4.4   Maintenance Schedules and Planned Maintenance

Maintenance schedules for SuDS should be prepared at the design stage and include both soft landscaped areas (e.g. basin banks, surrounding grass) and engineered structures (e.g. headwall, gabions, flow controls). Maintenance schedules will usually incorporate regular

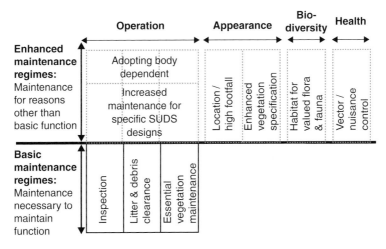

Figure 4.2   Maintenance requirements from SuDS (Adapted from Lampe, 2005).

and irregular maintenance only, listing the maintenance activity, specification and relevant supporting information. The schedule should be prepared by the design consultant (civil engineer) in consultation with the landscape architect or other person with detailed knowledge of plant species and good horticultural practice. Maintenance regimes for the hard engineered items should include all structures, above and below ground, including pipes, headwalls, screens, flow dissipation measures and flow controls. Maintenance regimes for the soft landscape will vary in accordance with the complexity of the landscape areas (both size and planting specification), from basic grassed structures and planted pods (e.g. for a detention basin) to ornamental planting within parkland or inner city areas. The schedule should also consider not only how the SuDS will be maintained, but by whom (Landscape Institute Technical Committee Water Working Group, 2014).

The planting specification and how the hard engineered structures are detailed should be considered because the relationship can impact on the maintenance regime. For example, where the SuDS area/surrounds include deciduous trees then additional maintenance visits for autumn and winter season to maintain screens, outlets and controls should be included. Where permanent water SuDS are used then the type of planted species (and management of self-seeding species post-implementation) used for marginal planting can influence maintenance regimes. For example, where ponds have monocultures of reeds (e.g. *Phragmites australis*) around the perimeter this can, with time, limit access for inspection, present a visual barrier of the water surface (impacting child safety) or encroach into other areas, so additional maintenance to crop and thin will be required.

When preparing the maintenance schedule the end user should be taken into account, which in most cases will be the maintenance contractor, and it should be clearly written using non-technical language.

Planned maintenance activity is normally carried out on either a frequency basis or on a performance basis (or a combination of both), whereas inspection activity should always be on a frequency basis (unless triggered by an operational incident). The type and extent of planned maintenance will be dependent upon the owning/maintaining organisation.

Frequency-based maintenance stipulates the number of visits that the contractor will make to the site per annum and the intervals by which specific maintenance activity will be carried out, for example grass cutting of amenity areas 18 times per annum. Frequency-based maintenance can be used for areas of low level maintenance (e.g. trunk roads or industrial areas) particularly where control of cost is paramount; this is typical of large areas with limited amenity value where contract periods are for a medium length of time (typically a five year contract).

Performance-based maintenance sets a standard (with upper and lower thresholds) that the contractor should work to. For amenity grass, this would mean a specification of 30–65 mm tolerance, where the contractor would cut initially to 30 mm and revisit before the sward exceeded 65 mm. Performance-based maintenance is commonly used where a consistently high standard of maintenance is required (e.g. premium residential developments or business districts). Experience of the type of work as well as knowledge of the location and climate (growing season) are necessary to submit accurate and realistic tenders for performance-based maintenance.

Whether maintenance is scheduled on a frequency or performance basis will normally be at the discretion of the owning organisation. Woods Ballard *et al.* (2007) states that SuDS maintenance '*tends towards a frequency requirement to ensure a predictable standard of care which can be recorded on site*' and that this '*provides a reasonable basis for pricing work*'. Where higher level maintenance (aesthetic drivers) is required then a combination of frequency and performance based maintenance is usually more appropriate. In such cases the most typical arrangement is for the hard engineered parts of the SuDS to use the frequency basis, for example inspection of the flow control on an annual basis, and the vegetated/landscaped areas of the SuDS (and surrounds) to use a performance basis. This combined method ensures continued operation of the SuDS (delivering hydraulic and pollutant removal benefits) with consistently well maintained and visually attractive landscaped areas maximising both amenity and biodiversity benefits.

4.5   Other Considerations that Will Impact on Maintenance

Maintenance type and specification for a SuDS scheme will be influenced by other factors, predominantly those that involve design, ownership and land type.

4.5.1   Asset Type and Design for Maintenance

The choice of SuDS technique and how it is designed and detailed will have an influence on the type, specification and cost of maintenance. The SuDS treatment train concept is where a series of SuDS are used in sequence to provide quality, quantity and amenity benefits. The number and type of SuDS used in the treatment train for a site will involve balancing the risks in the catchment (Woods Ballard *et al.*, 2015):

1. the type of land (i.e. use residential/commercial/industrial, in ascending pollution risk)
2. the extent of the land use (e.g. the number of houses or commercial parking)
3. the ecological sensitivity of the receiving water

Where more than one SuDS technique is used in sequence then the techniques used and their sequence in the train will influence maintenance requirements. Maintenance requirements for individual SuDS techniques are well documented (Woods Ballard *et al.*, 2015); in general, permanent water (wet) SuDS have a wider range of maintenance activities than non-permanent water (dry) SuDS. If ease and cost of maintenance is a driver then use of dry SuDS is preferable. However, when multiple SuDS are used in sequence then the interaction and maintenance requirement is less well understood. Treatment trains should be designed to promote *survivability* (i.e. maximising the operational life). Good design practice is to use dry SuDS upstream of wet SuDS to manage sediment (rather than proprietary silt traps or standalone ponds with sediment forebays), enabling silts to be easily monitored and cost-effectively managed, reducing the need for dredging equipment or drawing down the permanent water.

Consideration should also be made to the detailing of structures as this will impact on how maintenance is carried out. This includes access (for both operatives and plant), the operational suitability of the detail (e.g. vertical bar screens vs pitched screens, the former requiring a high frequency of maintenance in autumn and winter), whether key items are located above or below ground and material specification (aesthetic vs vandalism).

4.5.2 Balancing CAPEX and OPEX

Ownership solutions for SuDS can vary by region and can influence the cost of SUDS over their operational lifetime, a concept referred to as whole life cost (WLC). Whole life cost accounts for design, implementation and aftercare costs and can be split into two categories; capital expenditure (CAPEX), and operational expenditure (OPEX).

CAPEX includes all costs incurred before the asset is handed over. This includes costs such as scoping/feasibility studies, outline/detailed design, land acquisition, construction and landscape implementation. OPEX is the cost of maintaining the structure over its lifetime (i.e. regular, irregular and corrective maintenance) and may also include decommissioning costs. There may be an interim period (or defects liability period) before the SuDS is formally adopted, where the developer must maintain and make good any deficiencies in operation and condition.

Where SuDS are designed, constructed and maintained by a single party then both CAPEX and OPEX are borne by the owner. However, it is the norm for most SuDS to be built by one body (i.e. a house-builder) and ownership and maintenance passed to other bodies such as the local authority or a water utility or under private ownership. In such cases the relationship (balance) between CAPEX and OPEX is of interest from an operation and maintenance perspective. The design and detailing of the SuDS (i.e. which components are used, how they are constructed, material, access) will directly impact the complexity and cost to maintain. Where SuDS are highly detailed and include a range of hard engineered structures then the cost of construction is likely to be higher, but this can lead to easier and more cost-effective maintenance.

In certain circumstances it can be favourable for the final owner to stipulate design details so that future maintenance requirements, and their cost, are known and understood, thereby reducing operational and financial risk. This is typical of the situation in Scotland, where SuDS design must meet the technical standard (Scottish Water and WRc plc, 2015) of the national water authority (Scottish Water) in order to be adopted. The technical standard includes details to enable cost-effective maintenance. Examples include reinforced vehicular access around the margins of a pond, and concrete reinforced sediment forebays; these ensure that there is adequate space for corrective works to be carried out, that the type (and size) of machinery that can be used is known, and reinforcement of access points and forebay will minimise landscaped areas having to be reinstated and reduce the risk of liner breach. These examples will incur higher front-end costs – additional land acquisition for the access route and construction of the reinforced access – and they will be borne by the developer with the benefit passed to the final owner.

Conversely, the absence of design standards can present risk: if there is little control or focus of the SuDS detailing, the cost to construct can be minimised but this may be at the expense of operational cost to the final owner. For example, ponds that are detailed without a drawdown mechanism are costly to desilt, and limited or restricted access may require the use of larger plant to carry out corrective works.

4.5.3   Location of SuDS

Site and regional control SuDS should be located in areas of passive public open space. Where SuDS and the surrounding areas of public open space are owned and maintained by the same body, this lends economies of scale to maintenance regimes. Bray (2015) recognises the concept of passive SuDS, as those that are designed to integrate into available open space, and the principle of multi-functional space, similar to the concept of high performance landscapes (HPLs) 'landscapes that can perform many functions at once' (Design Trust for Public Space and the New York City Department of Parks and Recreation, 2010).

Integration of SuDS into public open space can enable cost-effective maintenance, allowing the use of larger equipment (e.g. ride-on mowers rather than walk-behind mowers) and reduced set-up time and cost. Where SuDS do not form part of the public open space, or where they are owned/maintained by a separate body (e.g. a water utility) then there is limited opportunity for cost efficiencies.

4.6   Conclusions

Maintenance of SuDS is an essential process and is not limited solely to the post-adoption phase. In order to maximise the operational life of SuDS, maintenance knowledge is necessary at the design stage and should inform the detailing of structures so that they can be safely and cost-effectively maintained. Maintenance and inspection during the construction phase is necessary to ensure that SuDS are constructed in accordance with the design and are fit for purpose. Post-adoption effective maintenance regimes can be encouraged by the up-skilling of maintenance operatives and facility inspectors; this also allows identification of operational concerns at an early stage when risk (financial or reputational) can be more easily mitigated.

Maintenance regimes are influenced by a range of factors including the adopting (owning) body, location, integration and visibility, and design and planting specification. Scheduled maintenance is necessary to maintain operation and there must be sufficient maintenance to ensure continued operation of the SuDS; maintenance in addition to this is for other reasons, predominantly for aesthetics, but it can include maintenance for biodiversity and habitat development.

Inspection processes will allow both the suitability and the (quality) standard of maintenance to be assessed, will provide an essential feedback loop to amend maintenance regimes, and will document an audit trail of inspection and maintenance.

References

Bastien, N.R.P., Arthur, S. and McLoughlin, M.J. (2012) Valuing amenity: public perceptions of sustainable drainage systems ponds. Water and Environment Journal 26.1 (2012), 19–29.

Bell, D., Ward, R., Kaye, G., Nowell, R. and Swales, P. (2015) South Yorkshire Interim Local Guidance for Sustainable Drainage Systems. Available from: http://tinyurl.com/h2woz7y

Bray, R. (2015) SuDS behaving passively – designing for nominal maintenance. Paper presented at: SUDSnet International Conference, Coventry, 3–4 September 2015.

Bryant, I. (2006) Sustainable urban drainage system (SUDS) – A new concept in total stormwater management solutions for new developments. New South West Australia: ROCLA Water Quality Publication.

Center for Neighborhood Technology (CNT) (2011) The Value of Green Infrastructure: A Guide to Recognizing its Economic, Environmental and Social Benefits. Center for Neighborhood Technology. Available from: http://www.cnt.org/repository/gi-values-guide.pdf

Center for Watershed Protection Stormwater Pond and Wetland Maintenance Guidebook (2004) Available from: http://tinyurl.com/najws2j

Design Trust for Public Space and the New York City Department of Parks and Recreation (2010) High Performance Landscape Guidelines: 21st Century Parks for NYC. New York: Design Trust for Public Space, Inc.

Graham, A., Day, J., Bray, B. and Mackenzie, S. (2012) Sustainable Drainage Systems: Maximising the Potential for People and Wildlife (A guide for local authorities and developers). Royal Society for the Protection of Birds and Wildfowl and Wetlands Trust.

Highways Agency (2015) Design Manual for Roads and Bridges: Volume 4 Geotechnics and Drainage. London: The Stationery Office.

Jefferies, C., Duffy, A., Berwick, N., McLean, N. and Hemmingway, A. (2009) Sustainable Urban Drainage Systems (SUDS) treatment train assessment tool. Water Science and Technology. 60 (5) (2009).

Landscape Institute Technical Committee Water Working Group (2014) Management and Maintenance of Sustainable Drainage Systems (SuDS) landscapes. Interim Technical Guidance Note. Available from: http://tinyurl.com/qg3jcpw

Lampe, L. (2005) Performance and Whole Life Costs of Best Management Practices and Sustainable Urban Drainage Systems: Final Report 2005, WERF Project 01-CTS-21T. Water Environment Research Foundation: Alexandria, Virginia.

Scottish Environment Protection Agency (SEPA) (2006) A Dos and Don'ts Guide for Planning and Designing Sustainable Urban Drainage Systems (SUDS).

Scottish Environment Protection Agency (SEPA) (2016) Regulatory Method (WAT-RM-08) Sustainable Urban Drainage Systems (SUDS or SUD Systems). Version: v6.

Scotland and Northern Ireland Forum for Environmental Research (SNIFFER) (2005) Social impacts of stormwater management techniques: including river management and SUDS.

Scottish Water and WRc plc. (2015) Sewers for Scotland 3rd Ed. Available from: http://tinyurl.com/q652arp

Woods Ballard, B., Kellagher, R., Martin, P., Jefferies, C., Bray, R. and Shaffer, P. (2007) The SuDS manual (C697). London: CIRIA.

Section 3   Functions of Sustainable Drainage Systems

Water Quantity: Attenuation of the Storm Peak

Craig Lashford, Susanne M. Charlesworth and Frank Warwick

5.1 Introduction

Since antiquity, the drainage of cities has been based on hard infrastructure whose purpose was to remove water as quickly as possible. As discussed in Chapter 2, in arid and semi-arid areas, rainwater harvesting was used extensively; however, with the development of pipes by the Minoans in the Bronze Age (Angelakis *et al.*, 2012) the scene was set for modern society's reluctance to allow open water to flow through their cities, preferring to hide it underground in pipes and conduits. This chapter compares the two contrasting water management methods whose purpose is to provide flood resilience in cities. With the potential changes to the rainfall regime caused by global climate change, cities need to adapt to the changes to come, and mitigate those changes that have already happened. Charlesworth (2010) argued that SuDS can provide extensive benefits in terms of mitigating and adapting to climate change, in particular reducing flood risk, but also providing a means to harvest water in areas subject to drought. The following sections compare the abilities of conventional drainage and SuDS to address these issues, their benefits and weaknesses.

5.2 Conventional Drainage, Water Flow and Volume

Conventional drainage is based on pipes and concrete channels designed in order to rapidly convey runoff from the impermeable area to the receiving water body (Kirby, 2005). Runoff in the sewer system typically flows underground via gullypots and pipes before reaching the watercourse (Stovin and Swan, 2007; Charlesworth, 2010), which poses an increased flood risk due to the reduced lag time and increased peak flow (Qin *et al.*, 2013). Additionally, the 'clogging' of conventional systems with debris inhibits their potential to

Sustainable Surface Water Management: A Handbook for SuDS, First Edition.
Edited by Susanne M. Charlesworth and Colin A. Booth.
© 2017 John Wiley & Sons, Ltd. Published 2017 by John Wiley & Sons, Ltd.

Table 5.1 Conventional drainage design storms for different land uses (adapted from Schmitt *et al.*, 2004).

Location	Design storm frequency
Rural areas	1 in 10 years
Residential areas	1 in 20 years
City centres	1 in 30 years

effectively drain the water, causing a backlog through the system, exacerbating flood risk as was evident in the 2007 UK summer floods (Oliver, 2009).

Table 5.1 outlines the design flood frequency for pipe-based systems, according to the European Standard EN 752 (CEN, 1996; CEN, 1997). All drainage systems in city centres should manage up to and including the 1 in 30 year storm. However, many cities in the industrialised, developed world are at risk of flooding due to insufficient capacity; this is exacerbated in less developed countries due to lower drainage standards (Mark *et al.*, 2004).

As well as the primary concern of increasing flood risk at both source and outfall, conventional drainage also results in water quality issues (Chapter 6 of this volume). Improvement of runoff quality prior to release into the watercourse is a neglected aspect of conventional drainage (Charlesworth, 2010), and consequently runoff transports a variety of urban pollutants without treatment (Zhang *et al.*, 2013) which has an impact on the biodiversity of urban streams (Charlesworth *et al.*, 2003).

The integration of pipe-based drainage at new build sites is still part of typical design culture in England and Wales, focusing on reducing the impacts of pluvial flooding. However, conventional drainage is often unable to manage the impact of fluvial driven flood events and therefore other strategies are often required.

5.3   Existing Flood Management

The need for flood management in the UK is driven by an increase in urban cover, resulting in more impermeable surfacing, coupled with development on floodplains. Flood management is controlled and managed by both the UK Environment Agency and Defra (Burton *et al.*, 2012), with annual damage due to flooding in the region of £1.1 billion (Bennett and Hartwell-Naguib, 2014). Hard infrastructure flood management focuses on engineered solutions that reduce flooding of the surrounding area and, as a result, many UK streams in towns and cities have been either culverted or brick lined, which has generated a dependence on these structures during periods of high rainfall (Werritty, 2006). As well as culverting, a number of hard abatement measures have been used both in the UK and internationally to manage high runoff; these are discussed in the following sections.

Dams are used globally as a method of controlling flow rates and reducing the potential for flooding (Higgins *et al.*, 2011). Although they can be an effective tool for reducing regional flood risk, they cause local environmental and social issues as a result of the disruption caused during the construction process (Yu, 2010). Dams are typically constructed to mitigate all events up to the 10,000 year return period (Sordo-Ward *et al.*, 2013), but with climate change impacting rainfall rates, the level of abatement is reduced (Veijalainen and Vehviläinen, 2008). Additionally, when dams fail, it results in large-scale flooding and sometimes loss of life, as they hold large volumes of water (Bosa and Petti, 2013).

Constructing flood walls is another hard engineering measure to reduce flood risk; however, they tend to be over-designed in order to mitigate the unknown impacts of climate change, sometimes requiring regular increases in height to ensure continued effectiveness (Pitt, 2008; Saito, 2014). Kenyon (2007) undertook a survey in Scotland to determine public perception of different flood management methods, finding flood walls to be the least popular option. Participants disliked the visual impact on the site, the need for redevelopment and the possibility of the barriers trapping water on the protected side if overtopped. Song *et al.* (2011) found that erosion of the flood walls during Hurricane Katrina in New Orleans compounded the damage caused.

Dredging accumulated silt from streams can increase their carrying capacity (Jeuken and Wang, 2010) and was widely used in the UK during the 1980s. However, it is expensive to carry out, has only short-term benefits since siltation continues, and has the potential to increase downstream flood risk, so the process was restricted to urban streams from the 1990s (Pitt, 2008). In fact, dredging has decreased internationally; for example, in Indonesia it is seen as inappropriate and unsustainable (Hurford *et al.*, 2010), while in Australia it is limited to estuarine environments (Wheeler *et al.*, 2010). It also has a significant negative impact on the local ecosystem by removing habitats and associated aquatic biota (Elliott *et al.*, 2007).

Hard engineering flood management solutions require large economic outlay and continued maintenance (Werritty, 2006); in addition, many of the strategies provide only short-term solutions. With the likely impacts of climate change altering rainfall patterns, a more sustainable approach is required to build for the future (Sayers *et al.*, 2014). In the UK, there has been a paradigm shift in thinking about drainage since the 1998 floods, with less reliance on unsustainable hard structures and more on sustainable techniques (Werritty, 2006; van den Hoek *et al.*, 2012). This is coupled with a realisation that absolute protection from flooding is not possible and that water should be utilised more efficiently in the built environment (Sayers *et al.*, 2014).

It is possible to flood proof at the individual building scale; this involves retrofitting the property to reduce the existing level of flood risk (Saito, 2014) and typically includes two foci:

1. dry flood proofing which stops water from entering the building
2. wet flood proofing which manages utilities if water does get into the building (Hayes and Asce, 2004)

In general, such flood proofing is underused, although it has featured in regional flood control plans since the 1980s in the USA. In the UK, the Royal Institute of Building Architects (RIBA) was challenged by Pitt (2008) to develop new, flood resilient housing. Their report (RIBA, n.d.), outlined structural measures, such as reinforced concrete walls, raising electrical sockets, using hard-wearing floor tiles and using either flood resistant furniture or materials that can be replaced reasonably cheaply. Also discussed is the potential use of amphibious buildings which are designed to float when the surrounding area floods. As well as flood proofing, Pitt (2008) also suggested the use of SuDS as a further solution for flexible adaptation to the impacts of flooding, particularly with the likelihood of more intense storm events in the future due to climate change.

5.4   Water Quantity

Water quantity forms a critical component of the SuDS triangle (Chapter 1), all three corners of which should be equally weighted in their importance (Charlesworth, 2010), but in practice this is rarely the case. However, water quantity reduction is generally acknowledged by stakeholders as the main factor for integrating SuDS into the drainage design of a site (Chahar *et al.*, 2012) followed by water quality improvements, with little regard paid to amenity or biodiversity benefits (Kirby, 2005; Zhou, 2014; Jose *et al.*, 2015). More emphasis is therefore placed on the ability to return the site to 'greenfield runoff rates' which should result in rates of infiltration and flow equivalent to that before development (Charlesworth and Warwick, 2012) up to and including the 1 in 100 year six-hour storm. The acknowledged flood resilience provided by SuDS is achieved by: promoting infiltration and ultimately groundwater recharge; recycling water; controlling peak flow; reducing the reliance on conventional pipe-based drainage; slowing down and retaining water in the drainage system (Charlesworth, 2010; Bastien *et al.*, 2010).

Table 5.1 showed that drainage systems in urban areas have been designed to manage events up to a 1 in 30 year storm with, according to Pitt (2008), some older systems only capable of dealing with even smaller events. On the other hand, the UK Environment Agency (2009) recommends designing SuDS for exceedance, ultimately up to the 1 in 100 year storm return period (with an additional 30% for climate change). The success of SuDS is also partially limited to site characteristics, most notably, the ability for infiltration (Woods Ballard *et al.*, 2007), but examples of SuDS best practice are limited, resulting in continued reliance on conventional pipe-based systems.

5.5   History of SuDS Implementation

Stahre (2008) suggested that the transition to more sustainable forms of drainage began in the 1970s, with the realisation that water *quality* needed to be addressed in conventional drainage, but it was not until the 1990s that urban stormwater was considered in a more positive light, as a resource rather than as a waste. Figure 5.1 shows this transition to the full SuDS triangle by the mid 1990s as envisioned by Stahre (2008).

Butler and Parkinson (1997) questioned the role that traditional drainage played in a developing urban environment, suggesting, not an overhaul of conventional methods, but the promotion of 'less unsustainable methods', focusing on long-term benefits. The earliest

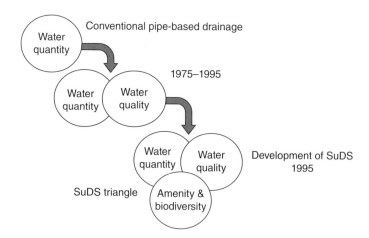

Figure 5.1   The evolution of SuDS from conventional drainage in the 1970s to its development in the mid 1990s (after Stahre, 2008).

implementations of SuDS focused on source control: the capturing and detaining of water early in the process at the building scale (Pompêo, 1999). Pratt *et al.* (1989) therefore investigated the potential of PPS to reduce both flow volumes and pollution, concluding that, in comparison with conventional drainage, PPS is more effective in reducing both factors. As a result of this change in drainage philosophy, and developments in knowledge of the discipline, Shaver and Hatton (1994) and CIRIA (1992) produced recommendations for the design of SuDS and the impacts of their implementation in the USA, and England and Wales, respectively.

In 1994, the United States Environmental Protection Agency (1994) produced a runoff control plan for Northern Virginia which represented one of the earliest uses of SuDS, installing open channels as opposed to pipe-based drainage and promoting infiltration across the area. These best management practices are still in place and continue to manage stormwater in Northern Virginia (British Water, 2006).

Field-scale implementation of SuDS in Europe also occurred in the mid 1990s. Household rainwater harvesting systems were retrofitted to buildings across Berlin in 1995, reducing overland flow, and in partnership with greywater recycling (GWR), decreased household water costs by reducing domestic potable water demand (Nolde, 1999). It was concluded that the total amount of water used through toilet flushing (approximately 15–55 l/person/day) could be replaced through GWR.

As a result of increased understanding of the benefits of SuDS and GWR, two Environment Agency demonstration sites were developed in the UK that incorporated a number of different devices: Wheatley Motorway Service Area (MSA), Junction 8A on the M40, Oxford (Figure 5.2) (Charlesworth 2010) and Hopwood MSA, Junction 2 M42, Bromsgrove, Worcestershire (Figure 5.3) (Heal *et al.*, 2009). Table 5.2 lists the SuDS devices used and some characteristics of the two sites.

The primary purpose of SuDS at both sites was to manage flood flows, while also enhancing water quality of the runoff leaving the site (SUSdrain n.d.). To the author's knowledge, neither site has flooded since they were constructed, even when surrounding areas were under water (B. Bray, Landscape Architect and SuDS management train

Figure 5.2   Wheatley MSA, Oxford.

Pond fed by runoff from
amenity building roof

Aeration fountain

Fringing vegetation

Figure 5.3   Hopwood MSA, Bromsgrove.

Table 5.2   Site characteristics and SuDS devices used in the management trains at Wheatley and Hopwood Motorway Service Areas.

Site	Hopwood MSA	Wheatley MSA
SuDS devices used in the management trains	SuDS devices used at the site Grass filter strips Constructed wetlands Balancing ponds Swales Infiltration trench	Permeable paving Filter drains Swales Filter strips Retention Pond Wetlands
Total area of site (ha)	34	16.7
Total area of SuDS	9	4.2
Return period (design storm)	1 in 25 years	Not known
Runoff design	5 l/sec/ha	3 l/sec/ha
Year built	1999	1997

designer, pers. comm.). Overall, the SuDS management train at Hopwood has been successful, both in terms of enhancing water quality and also financially. The average annual cost of the maintenance of the site was £2500, in comparison to £4000 for a similar sized conventionally drained site. The site also reduced between 70–90% of the total pollutants by the time runoff reached the outflow (Heal *et al.* 2009).

SuDS have been used extensively across Scandinavia since the late 1980s; in Malmö, Sweden, for example, SuDS have been employed in new developments where open channels were constructed, with water diverted into open sewers, engaging the public in the design process to ensure that water quantity, quality and aesthetics needs were met (Stahre, 2002). In addition, away from the city centre, pre-existing conventional drainage was directed into new open channels to reduce water quantity in the overloaded pipe-based systems. Wetlands, detention ponds and green roofs were also used (Forest Research n.d.). This use of SuDS has led to just 15% of the urban population in Sweden being served by sewers by the year 2000 (Mikkelsen *et al.*, 2002).

Not every use of SuDS is to reduce water, and in Australia, Water Sensitive Urban Design or WSUD, which utilises similar techniques to SuDS, has a greater focus on water due to drought issues and the efficient use of scarce water resources (Morison and Brown, 2011). Wong (2007) emphasised the requirement of WSUD to reuse and recycle water, thus promoting the use of rainwater harvesting systems (RwH), and such systems have been successfully retrofitted in both Sydney and Melbourne, to ensure that water is used effectively, primarily using RwH, swales, bioretention zones and detention ponds to store and capture runoff for use around the home, and in business premises (Landcom, 2009).

Whether devices are retrofitted or installed during new builds, SuDS can be incorporated into a system in two ways: as standalone devices or as part of a wider SuDS management train. While it is acknowledged that designing a SuDS management train is a viable strategy in comparison to conventional drainage (Stovin and Swan, 2007), there is little research regarding their ability to deal with high volumes of runoff.

5.6   The Management Train

A management train utilises a wide range of devices to reduce the overall level of pollution in runoff (Woods Ballard *et al.*, 2007) and also reduce the volume of water conveyed down the train and, therefore, entering the receiving watercourse (Figure 5.4). It conveys runoff down the train from one device to another where it is incrementally treated as it makes its way to the outflow (Bastien *et al.*, 2010). Apart from water quality improvements, extra flood resilience is provided as more devices are used, resulting in greater volumes of water

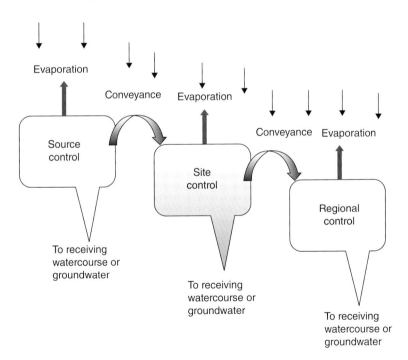

Figure 5.4   SuDS management train (adapted from Charlesworth, 2010).

Table 5.3 SuDS devices, their role in the management train, efficiency in reducing water quantity and potential to be retrofitted (Woods Ballard *et al.*, 2007).

Device	Source	Site	Regional	Conveyance	Effectiveness at reducing water quality	Potential for retrofit
Rainwater harvesting	X	X	—	—	Low*	Yes
Previous pavement systems	X	X	—	—	High	Yes
Filter strip	X	—	—	—	Low/Medium	Yes
Swale	X	X	—	X	Medium	Limited
Pond	—	X	X	—	Medium/High*	Unlikely
Wetland	—	X	X	O	Low/Medium	Unlikely
Detention pond	—	X	X	—	High*	Yes
Soakaway	X	—	—	—	Medium	Yes
Infiltration trench	X	X		O	Medium/High	Yes
Infiltration basin	—	X	X	—	Medium*	No
Bioretention device	X	X	—	—	High	Yes
Sand filter	—	X	O	—	Low	Yes
Green roof	X	—	—	—	Medium/High	Yes

*Depends on the size of the structure for water retention.
X = Most suitable O = Less suitable.
— = Not possible.

retained (Bastien *et al.*, 2010). Additionally, it is not always feasible to utilise one large device at a site, therefore a series of smaller linked devices in a management train can be more practical and better integrated into the urban landscape (Charlesworth, 2010).

An initial component of a successful management train is source control (Figure 5.4), SuDS that tackle water directly after precipitation, for example PPS (Zakaria *et al.*, 2003). Runoff is typically lost at this point into groundwater via infiltration, or into the overlying atmosphere via evaporation, particularly if green infrastructure is used. The remaining runoff is conveyed to a site control device, usually in a swale (Stovin and Swan, 2007). Such systems deal with greater amounts of runoff from multiple source control devices, and again runoff is lost via infiltration and evaporation (Kirby, 2005). What runoff remains can then be conveyed to another site for regional control, which deals with high volumes of water from a series of site control devices, representing the last element of the train. An example of a regional control device is a detention pond (Bastien *et al.*, 2010). By this point in the management train, much of the pollution should have been filtered out along with a substantial volume of the excess runoff, although moderate levels of pollutant removal does occur. After this step, water is either slowly released to a water body, infiltrated out of the system or evaporated away (Kirby, 2005).

The following sections assess the potential for each element of the management train for reducing water quantity, beginning with devices that provide source control. Table 5.3 gives examples of specific devices at each stage of the management train.

5.6.1 Source Control

Table 5.3 gives several examples of devices that can provide source control, including PPS, which both Kirby (2005) and Woods Ballard *et al.* (2007) suggest is 'highly effective' at dealing with runoff. PPS is most suited to either car parks or pedestrian areas due to its low

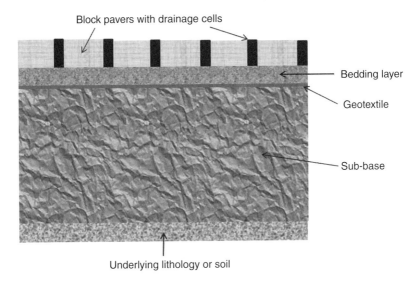

Figure 5.5 Cross-section through a typical PPS with block paver surface layer.

load capabilities (Scholz and Grabiowiecki, 2007), and it infiltrates water through the permeable surface layer, as shown in Figure 5.5 to sub-surface storage where the water is slowly transported to a water body (Woods Ballard *et al.*, 2007). Water moves through different layers of sub-base and geotextile, improving water quality (Scholz and Grabiowiecki, 2007; Charlesworth *et al.*, 2014).

Based on 150 different storm scenarios, Viavattene *et al.* (2010) calculated that PPS has the potential to reduce runoff flows by up to 75%, but this is dependent on the specific environment, that is to what extent infiltration is possible, the size of the PPS, its depth and the local topography.

Bioretention ponds (Figure 5.6) also fall under Woods Ballard *et al.* (2007)'s 'highly effective' category in terms of reducing water quantity using green infrastructure. These systems are also engineered to improve water quality at the outflow by utilising a combination of geotextiles and fine gravels to reduce pollutants (Woods Ballard *et al.*, 2007). Research by Debusk and Wynn (2011), on a system 4.6 m long, 7.6 m wide and 1.8 m deep, suggested that it was capable of dealing with all runoff, with no outflow, for events with an inflow rate up to 12.5 l/s. Also, the system only permitted infiltration into the topsoil; subsoil infiltration was restricted, showing the potential of installing bioretention devices into a management train.

Green roofs are categorised as having 'medium/high' effectiveness at reducing water quantity (Table 5.3). However, one of their primary benefits in terms of installation in the urban environment is that no additional land-take is required beyond the building footprint (Stovin, 2009). Based on the same 150 storm scenarios used to model the benefits of PPS (see Section 5.6.1), it was calculated that a green roof has the potential to reduce runoff by 45–60% (Viavattene *et al.*, 2010) depending on the storm. Green roofs slow down the time rainfall takes to reach the outflow through interception by the vegetation (Figure 5.7) (Fioretti *et al.*, 2010). Dependent on storm intensity, rainfall that has not been evaporated is then infiltrated into the substrate and either attenuated or conveyed out of the system (Stovin, 2009). However, if the storm intensity exceeds the infiltration rate,

Figure 5.6 Bioretention system, installed in a central reservation, Emeryville, California, USA (freely available).

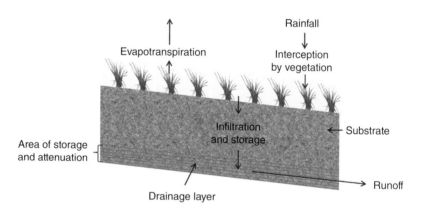

Figure 5.7 Schematic diagram of flow of water through a green roof (after Stovin, 2009).

overland flow will occur, reducing the impact of the green roof. Similarly, if the slope of the green roof is too steep, retention capacity is reduced, further promoting runoff (van Woert *et al.*, 2005). There is also a point at which the green roof becomes saturated, particularly in the case of low intensity, long duration rainfall, or repeated storms in fairly short order. However, while runoff from the roof does eventually occur, nonetheless, it slows its appearance and reduces the overall volume.

5.6.2 Site Control

Table 5.3 shows three SuDS devices capable being both source and site control devices: bioretention devices, infiltration trenches and swales, depending on their size (Woods Ballard *et al.*, 2007). Other highly effective devices for reducing runoff at the site scale are dry detention ponds and wet or retention ponds (van der Sterren, 2009), as they can store large amounts of water and encourage groundwater recharge through infiltration (Datry *et al.*, 2004). Strecker *et al.* (1999) estimated that retention and detention ponds were capable of reducing runoff by up to 30% based on 'significant storm events', but more detail of the nature of the modelled events is not provided. It should be noted, however, that their detention capabilities are relative to their size and the infiltration rate of the underlying soil type and lithology (Scholz, 2004).

5.6.3 Regional Control

As regional control devices are required to retain larger amounts of water, there are fewer devices at this scale, as is evident from Table 5.3. Detention ponds are most efficient for regional control, as van der Sterren (2009) suggests they reduce runoff to a 'high' standard.
    Retention ponds are also useful devices for storing water and ultimately reducing runoff levels but, as with detention ponds, their capability is dependent on size (Scholz, 2004).

5.6.4 Conveyance

Viavattene *et al.* (2010) suggest that swales are most suitable for conveyance, and this is shown in Table 5.3, which also shows their 'medium' ability at reducing flood flows. They mimic natural drainage by utilising vegetated channels for transporting water (Figure 5.8) (Kirby, 2005) and Table 5.4 details the design criteria for a swale, as given by Escarameia *et al.* (2006), in order for it to achieve optimum performance. Strecker *et al.* (1999) calculated that swales reduced peak flows by approximately 10%, on a storm-by-storm basis, but similar to detention ponds, the details of the modelled storm scenarios were not provided. It does, however, suggest that swales are not the most successful at reducing peak flows; their primary role is to transport runoff around the site. Other devices that could be considered in a conveyance role include infiltration trenches, wetlands and RwH, but this would not be their primary role, unlike swales (Woods Ballard *et al.*, 2007).

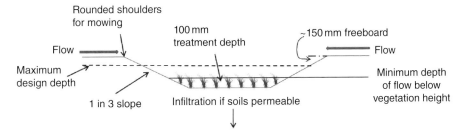

Figure 5.8   Schematic diagram of a swale (after Woods Ballard *et al.*, 2007).

Table 5.4   Swale design criteria (Escarameia *et al.*, 2006).

Long slope	Total depth	Height of grass	Design storm event	Velocity	Hydraulic residence time	Minimum length
<1:50 Vertical: horizontal	300–500 mm	100–200 mm	5 year/24 hour But check for: 10 year/24 hour	<0.25 m sec^{-1}	8–10 mins	60 m

Although some studies have been carried out to examine the storm attenuation capabilities of various individual SuDS devices (e.g. Strecker *et al.*, 1999; Ellis *et al.*, 2004; Kirby, 2005; Duchemin and Hogue, 2009; Viavattene *et al.*, 2010), there has been minimal attention given to management trains. However, MacDonald and Jefferies (2003) monitored the six ponds, wetland and associated upstream detention basins and swales of the Scottish SuDS train at DEX (Dunfermline Eastern Expansion) and found that the system provided significant lag times, and in a study of the EA demonstration SuDS trains at Hopwood MSA (see Section 5.5 for further details) Malcom *et al.* (2003) reported significant peak flow reductions in all but the largest events. However, as stated by Charlesworth and Warwick (2012), 80% of the dwellings required by 2040 in the UK are already in existence and if the flooding of these buildings is to be addressed, then thought needs to be given to retrofitting SuDS.

5.7   Retrofit

Retrofit into existing developments involves disconnecting the traditional drainage system and routing the stormwater into a SuDS device (Stovin and Swan, 2007). The process forms a tool for mitigating flooding in the built environment (Moore *et al.*, 2012). As pluvial flooding is increasingly becoming an issue in urban settings (Sharples and Young, 2008; Priest *et al.*, 2011), devices are required to reduce the risk (Environment Agency, 2007). Approximately 5.2 million houses are currently at risk of flooding in the UK (Committee on Climate Change, 2012) with new build contributing 1% of all buildings. Consequently, a combined strategy for dealing with both new and old buildings is essential to manage flooding (Environment Agency, 2007). Table 5.5 illustrates the potential for implementation of various retrofit devices across the UK.

Balmforth *et al.* (2006) have shown how integrating retrofit SuDS into the urban environment can prove troublesome, with the restraints of existing buildings, paths and roads limiting the space available for development. However, Stovin and Swan (2007) calculated that reductions of costs in terms of construction and over the whole life of the system would be experienced by retrofitting SuDS.

There are limited examples of SuDS retrofit across the UK, with Stovin and Swan (2007) suggesting that this is largely due to the complexity and disturbance associated with retrofitting SuDS. However, there is no reason why SuDS and conventional drainage should not be integrated as necessary, particularly in terms of retrofit. This perception of retrofit is not limited to the UK, with Shaver *et al.* (2007) suggesting that a lack of space in the urban environment as well as high land values make it expensive in the USA. Consequently, there are a limited number of examples where SuDS have been retrofitted to manage runoff from existing urban areas (Hyder Consulting, 2004; SNIFFER, 2006; White and Alarcon, 2009), although, where implemented, successful retrofit installations have solved issues of both

Table 5.5 Potential for a variety of retrofit SuDS devices (adapted from Environment Agency, 2007).

Technique	Coverage potential (conservative estimates)
PPS	50% of all off road hard surfaces
Rainwater harvesting	75% of commercial/industrial properties
	50% of public buildings
Water butts	90% semi-detached and detached houses
	40% terraced houses
Swales or infiltration ditches; filter drains	20% of roads in rural areas
	4% of roads in urban areas

water quality, such as at the Houston Industrial Estate in Livingston, West Lothian, Scotland (RCEP, 2007), and flooding at two schools in Worcestershire, UK (Atkins, 2004; SNIFFER, 2006). Successful retrofits have often been driven by a single organisation with the authority to implement solutions (Stovin *et al.*, 2007).

There are some international examples of retrofit projects including the design of a large-scale 15 km² SuDS retrofit project in Copenhagen, Denmark by Backhaus and Fryd (2012), the design methodology for which could be utilised elsewhere. They also highlight a series of challenges to retrofitting SuDS, which includes the complexity of designing a project at a range of scales to ensure that the solution is achievable and effective. Worldwide, cities such as Malmö (Sweden), Portland (Oregon), Seattle (Washington) and Tokyo (Japan) have demonstrated the viability and effectiveness of retrofitting SuDS (DTI, 2006; SNIFFER, 2006; Stahre, 2008).

In the UK, Stovin *et al.* (2013) provided an assessment of the potential for retrofitting a SuDS train in the Thames Tideway Catchment in order to reduce the cost of modifying the existing conventional drainage plan. A model is presented, but the research identified a number of challenges:

- the lack of pilot sites to determine challenges to implementation
- the large size of the study area resulting in significant disruption to residents
- a continued need to utilise conventional drainage alongside the SuDS system

From the point of view of those devices that are most suitable for retrofit installation, Table 6.5 shows that, apart from infiltration basins, all have the potential to be incorporated in a retrofit design; however, swales, ponds and wetlands are less so, being limited due to their size (Woods Ballard *et al.*, 2007).

5.8   New Build

Using SuDS in the design of new developments will ensure that a further increase of impermeable surfaces does not result in increased flood risk. This was acknowledged by the Government (Kellagher, 2013), in their response to Pitt (2008), in that flood mitigation needs to be in place for new developments that does not negatively impact greenfield runoff rates (Charlesworth, 2010; Charlesworth and Warwick, 2012; see Section 5.4).

When integrating SuDS into new build developments, source and site control devices should largely be sufficient to deal with storm return periods up to 1 in 30 year return rate (Bastien *et al.*, 2010). Once this is exceeded, larger attenuating devices such as ponds are required to ensure that the site deals with the extra runoff. However, ponds, wetlands etc reduce the space available for houses, which is typically an issue for developers, as houses provide the profit for the site; legislation and guidance are therefore needed to ensure that sufficient SuDS are included and this was covered in Chapter 3 .

Defra (2015) stipulates controlling runoff from both developed and 'previously developed' sites to greenfield runoff. If this is likely to be exceeded, then some means of throttling this back must be employed, therefore the use of flow control devices is encouraged.

5.9   Flow Control

Flow control devices are used to regulate runoff and outflow through the drainage system (Environment Agency, 2007). They aim to limit flow through a predetermined point, to a certain extent, causing water to back up through the system (Environment Agency, 2007), so provision must be made to deal with this (Woods Ballard *et al.*, 2007). There are a variety of devices that can be used to control flow rates throughout a SuDS management train and some of these are introduced in the following sections.

5.9.1   Hydro-brake®

Hydro-brake® controls flow by utilising an upstream hydraulic head through a vertical chamber to create a vortex which limits flow through the device (Hydro-International, 2006; Cataño-Lopera *et al.*, 2010). They are the most commonly used stormwater attenuation and flow control device (O'Sullivan *et al.*, 2012). In terms of site benefits, they have the ability to reduce the need for stormwater storage by up to 30% and due to the vertical vortex and size of the outlet, they reduce the chance of blockages (Hydro-International, 2011).

5.9.2   Weir

Weirs are overflow structures that are built perpendicular or parallel to a channel and are designed to limit flow through a certain point, reducing the risk of downstream flooding (Figure 5.9) (Zahiri *et al.*, 2013; Tullis and Neilson, 2008). They are widely used to regulate flood flow, but their role remains mainly associated with river channels, although adoption in a SuDS management train is also viable (Graham *et al.*, 2012). Semadeni-Davies *et al.* (2008) show how the potential implementation of SuDS in Helsingborg, Sweden could limit the impact of increased rainfall generated by climate change, suggesting a design that uses a network of weirs to regulate and control flow throughout the site.

5.10   Conclusions

In summary, the advantages of SuDS compared to conventional drainage methods include (National SuDS Working Group, 2004; Jones and Macdonald, 2007; MacMullan and Reich, 2007):

- decreased overall load on conventional drainage systems
- control of peak flows to prevent capacity overload and downstream flooding

Figure 5.9   Weir at the SuDS management train in Hamilton, Leicester, UK.

- removal of diffuse pollution
- increased groundwater recharge
- potential for reuse of water
- provision of aesthetic, ecological and educational benefits

SuDS can improve flood control, reduce the costs of upgrading conventional sewerage infrastructure to cope with greater demands and provide further hydrological benefits by preventing pollution reaching watercourses and retaining water in local groundwater stores, contributing to a reduction in water transportation requirements. There is a requirement for SuDS devices in new developments and retrofit in cities worldwide, but much of the research has been done on the abilities of individual devices, whereas a SuDS management train can provide a site with added resilience and coping capacity to deal with large storm events (Kirby, 2005).

References

Atkins (2004) Scottish Water SUDS Retrofit Research Project. Available at: http://www.gov.scot/resource/doc/921/0004694.pdf.

Angelakis, A.N., Dialynas, M.G. and Despotakis, V. (2012) Evolution of Water Supply Technologies in Crete, Greece through the Centuries. In Evolution of Water Supply throughout Millennia; IWA Publishing: London, UK, Chapter 9, pp. 227–258.

Backhaus, A. and Fryd, O. (2012) 'Analyzing the First Loop Design Process for Large-Scale Sustainable Urban Drainage System Retrofits in Copenhagen, Denmark'. Environment and Planning B: Planning and Design 39 (5), 820–837.

Balmforth, D., Digman, C., Kellagher, R. and Butler, D. (2006) Designing for Exceedance in Urban Drainage: Good Practice (C635) Available at: http://tinyurl.com/nk6vjka.

Bastien, N., Arthur, S., Wallis, S. and Scholz, M. (2010) 'The best management of SuDS treatment trains: a holistic approach' Water Science and technology, 61 (1) 263–272.

Bennett, O. and Hartwell-Naguib, S. (2014) Flood Defence Spending in England. Available at: http://tinyurl.com/o2uw6x3.

Bosa, S. and Petti, M. (2013) 'A numerical model of the wave that overtopped the Vajont Dam in 1963', Water Resource Management, 27 (6), 1763–1779.

British Water (2006) Sustainable drainage systems: a mission to the USA. Available at: http://tinyurl.com/npvmwn5.

Burton, A., Maplesden, C. and Page, G. (2012) 'Flood Defence in an Urban Environment: The Lewes Cliffe Scheme, UK'. Proceedings of the ICE – Urban Design and Planning 165 (4), 231–239.

Butler, D. and Parkinson, J. (1997) 'Towards sustainable urban drainage', Water, Science and Technology, 35 (9) 53–63.

Cataño-Lopera, Y., Waratuke, A. and Garcia, M. (2010) 'Experimental Investigation of a Vortex-Flow Restrictor: Rain-Blocker Performance Tests' Journal of Hydraulic Engineering, 136 (8) 528–533.

CEN (1997) Drain and sewer systems outside buildings – Part 4: Hydraulic design and environmental considerations, European Standard, European Committee for Standardization CEN, Brussels, Belgium 1997.

CEN (1996) Drain and sewer systems outside buildings – Part 2: Performance Requirements, European Standard, European Committee for Standardization CEN, Brussels, Belgium.

Chahar, B.R., Graillot, D. and Gaur, S. (2012) 'Stormwater Management through Infiltration Trenches'. Journal of Irrigation and Drainage Engineering 138 (3), 274–281.

Charlesworth, S. (2010) 'A review of the adaption and mitigation of global climate using sustainable drainage cities', Journal of Water and Climate Change, 1 (3) 165–180.

Charlesworth, S., Harker, E. and Rickard, S. (2003) 'A Review of Sustainable Drainage Systems (SuDS): A Soft Option for Hard Drainage Questions?'. Geography 88 (2), 99–107.

Charlesworth, S.M., Lashford, C. and Mbanaso, F. (2014) Hard SUDS Infrastructure. Review of Current Knowledge, Foundation for Water Research.

Charlesworth, S.M. and Warwick, F. (2012) Adapting and mitigating floods using Sustainable Drainage (SUDS). In: Flood Hazards, Impacts and Responses for the Built Environment'. Chapter 15, p 207–234. J. Lamond, C. Booth, F. Hammond and D. Proverbs (eds), Taylor and Francis CRC press.

CIRIA (1992) Scope for control of urban runoff, Report 124, Vol 1–4, CIRIA, London.

Committee on Climate Change (2012) Climate change – is the UK preparing for flooding and water scarcity? Available at: http://tinyurl.com/odpdf43.

Datry, T., Malard, F. and Gibert, J. (2004) 'Dynamics of solutes and dissolved oxygen in shallow urban groundwater below a stormwater infiltration basin' Science of the total environment, 329, 215, 229.

Debusk, K. and Wynn, T. (2011) 'Stormwater bioretention for runoff quality and quantity mitigation' Journal of Environmental Engineering, 137 (9) 800–808.

Defra (2015) Non-statutory technical standards for sustainable drainage. Available at: http://tinyurl.com/qem92y4.

DTI (2006) Sustainable drainage systems: a mission to the USA. Global Watch Mission Report. 148 pp. Available at: http://tinyurl.com/npvmwn5.

Duchemin, M. and Hogue, R. (2009) 'Reduction in agricultural non-point source pollution in the first year following establishment of an integrated grass/tree filter strip system in southern Quebec (Canada)' Agriculture, Ecosystems and Environment, 131, 85–97.

Elliott, M., Burdon, D., Hemingway, K.L. and Apitz, S.E. (2007) 'Estuarine, Coastal and Marine Ecosystem Restoration: Confusing Management and Science: A Revision of Concepts'. Estuarine, Coastal and Shelf Science 74 (3), 349–366.

Ellis, J.D., Scholes, L., Revitt, D.M. and Oldham, J. (2004) 'Sustainable urban development and drainage' Municipal Engineer, 157, 245–250.

Environment Agency (2009) Managing flood risk. Available at: http://tinyurl.com/plejuh3.

Environment Agency (2007) Cost–benefit of SUDS retrofit in urban areas. Available at: http://tinyurl.com/o3tj9ve.

Escarameia, M., Todd, A.J. and Watts, G.R.A. (2006) 'Pollutant removal ability of grassed surface water channels and swales: Literature review and identification of potential monitoring sites.' Published Project Report PPR169. TRL Limited. Available at: http://tinyurl.com/ns5l26c.

Fioretti, R., Palla, A., Lanza, L.G. and Principi, P. (2010) Green Roof Energy and Water Related Performance in the Mediterranean Climate. Building and Environment 45 (8), 1890–1904.

Forest Research (n.d.) Sustainable drainage systems. Available at: http://tinyurl.com/7fvm6wb.

Graham, A., Day, J., Bray, B. and Mackenzie, S. (2012) Sustainable drainage systems: maximising the potential for people and wildlife. Available at: http://tinyurl.com/l78q44z.

Hayes, B.D. and Asce, A.M. (2004) 'Interdisciplinary Planning of Nonstructural Flood Hazard Mitigation' 130 (1), 15–26.

Heal, K.V., Bray, R., Willingale, S.A.J., Briers, M., Napier, F., Jefferies, C. and Fogg, P. (2009) 'Medium-Term performance and maintenance of SUDS: a case-study of Hopwood Park Motorway Service Area', UK. Water, Science and Technology, 59 (12) 2485–2494.

Higgins, A.J., Bryan, B.A., Overton, I.C., Holland, K., Lester, R.E., King, D., Nolan, M. and Connor, J.D. (2011) 'Integrated modelling of cost-effective siting and operation of flow-control infrastructure for river ecosystem conservation', Water Resources Research, 47 (5).

Hurford, a.P., Maksimović, C. and Leitão, J.P. (2010) 'Urban Pluvial Flooding in Jakarta: Applying State-of-the-art Technology in a Data Scarce Environment.' Water Science and Technology: A Journal of the International Association on Water Pollution Research 62 (10), 2246–2255.

Hyder Consulting (UK) Ltd (2004) Retrofitting Sustainable Urban Drainage Systems: Case Study – Dunfermline. Report No NE02351/D1. Available at: http://www.gov.scot/resource/doc/1057/0004700.pdf.

Hydro-International (2011) Hydro-brake Flow Control: Superior vortex flow control. Available at: http://tinyurl.com/qg2l6us.

Hydro-International (2006) Hydro-brake Chamber User Manual. Available at: http://tinyurl.com/olggbub.

Jeuken, M.C.J.L. and Wang, Z.B. (2010) 'Impact of Dredging and Dumping on the Stability of Ebb-flood Channel Systems'. Coastal Engineering 57 (6), 553–566.

Jones, P. and Macdonald, N. (2007) 'Making Space for Unruly Water: Sustainable Drainage Systems and the Disciplining of Surface Runoff'. Geoforum 38 (3), 534–544.

Jose, R., Wade, R. and Jefferies, C. (2015) 'Smart SUDS: recognising the multiple-benefit potential of sustainable surface water management systems' Water, Science and Technology, 71 (2) 245–251.

Kellagher, R. (2013) Rainfall runoff management for developments. Report – SC030219 Available at: http://tinyurl.com/npvmwn5.

Kenyon, W. (2007) 'Evaluating flood risk management options in Scotland: a participant-led multi-criteria approach', Ecological Engineering, 64, 70–81.

Kirby, A. (2005) SuDS – innovation or a tried and tested practice? Municipal Engineer, 158 115–122.

Landcom (2009) Water Sensitive Urban Design Book 2: Planning and Development. Available at: http://tinyurl.com/z8wlfep.

MacDonald, K. and Jefferies, C. (2003) Performance and design details of SUDS. National Hydrology Seminar, 93–102. Available at: http://tinyurl.com/hqechpm.

MacMullan, E. and Reich, S. (2007) The Economics of Low Impact Development: A Literature Review. ECONorthwest, Eugene, OR. Available at: http://tinyurl.com/paz29mb.

Malcom, M., Woods Ballard, B., Weisgerber, A., Biggs, J. and Apostolaki, S. (2003) The hydraulic and water quality performance of sustainable drainage systems, and the application for new developments and urban river rehabilitation. National Hydrology Seminar. 83–92.

Mark, O., Weesakul, S., Apirumanekul, C., Aroonnet, S.B. and Djordjević, S. (2004) 'Potential and limitations of 1D modelling of urban flooding', Journal of Hydrology, 299, 284–299.

Mikkelsen, P.S., Viklander, M., Linde, J.J. and Malmqvist, P. (2002) 'BMPs in Urban Stormwater Management in Denmark and Sweden', Linking Stormwater BMP designs and Performance to Receiving Water Impact Mitigation. 354–368.

Moore, S., Stovin, V., Wall, M. and Ashley, R. (2012) 'A GIS-based methodology for selecting stormwater disconnection opportunities' Water Science and Technology, 66 (2) 275–283.

Morison, P.J. and Brown, R.R. (2011) 'Understanding the Nature of Public and Local Policy Commitment to Water Sensitive Urban Design'. Landscape and Urban Planning 99 (2), 83–92.

National SUDS Working Group (2004) Interim Code of Practice for Sustainable Drainage Systems. Available at: http://tinyurl.com/pda4l9f.

Nolde, E. (1999) 'Greywater Reuse Systems for Toilet flushing in Multi-Storey Buildings – Over Ten Years' Experience in Berlin'. Urban Water, 1, 275–284.

Oliver, R. (2009) The draft Flood and Water Management Bill. UKELA, 21 ELM. Available at: http://www.ukela.org/content/page/2273/06-ELM-qwqw21-qwqw3%20Oliver.pdf.

O'Sullivan, J., Bruen, M., Purcell, P. and Gebre, F. (2012) 'Urban drainage in Ireland – embracing sustainable systems' Water and Environment Journal, 26, 241–251.

Pitt (2008) Learning lessons from the 2007 floods. Available at: http://tinyurl.com/4nt35s.

Pompêo, C A (1999) 'Development of a State Policy for Sustainable Urban Drainage'. Urban Water 1 (2), 155–160.

Pratt, C.J., Mantle, J.D.G. and Schofield, P.A. (1989) 'Urban stormwater reduction and quality improvements through the use of permeable pavements', Water, Science and Technology, 21 (8) 769–768.

Priest, S.J., Parker, D.J., Hurford, A.P., Walker, J. and Evans, K. (2011) 'Assessing options for the development of surface water flood warning in England and Wales' Journal of Environmental Management, 92, 3038–3048.

Qin, H., Li, Z. and Fu, G. (2013) 'The Effects of Low Impact Development on Urban Flooding Under Different Rainfall Characteristics'. Journal of Environmental Management 129, 577–585.

RCEP (Royal Commission on Environmental Pollution) (2007) The Urban Environment. Available at: http://tinyurl.com/p4zslan.

RIBA (n.d.) 'Climate Change Toolkit: 07 Designing for Flood Risk' Available at: http://tinyurl.com/nfr2xn7.

Saito, N. (2014) 'Challenges for adapting Bangkok's flood management systems to climate change', Urban Climate, 9, 89–100.

Sayers, P., Galloway, G., Penning-Rowsell, E., Yuanyuan, L., Fuxin, S., Yiwei, C., Kang, W., Le Quesne, T. Wang, L. and Guan, Y. (2014) 'Strategic Flood Management: Ten Golden Rules to Guide a Sound Approach'. International Journal of River Basin Management (June), 1–15.

Scholz, M. and Grabiowiecki, P. (2007) 'Review of permeable pavement systems', Building and Environment 42, 3830–3836.

Scholz, M. (2004) 'Case study: design, operation, maintenance and water quality management of sustainable stormwater ponds for roof runoff' Bioresource Technology, 95, 269–279.

Schmitt, T., Thomas, M. and Ettrich, N. (2004) 'Analysis and modelling of flooding in urban drainage systems', Journal of Hydrology, 299, 300–311.

Semadeni-Davies, A., Hernebring, C., Svensson, G. and Gustafsson, L, (2008). The impacts of climate change and urbanisation on drainage in Helsingborg, Sweden: Suburban Stormwater. Journal of Hydrology 350 (1-2), 114–125.

Sharples, D. and Young, S. (2008) 'It never rains but it pours: pluvial flooding as a planning consideration' Journal of Planning and Environmental Law, 8, 1093–1097.

Shaver, E. and Hatton, C. (1994) 'Auckland Experience with BMPs Mitigating Adverse Impacts', in 'Linking Stormwater BMP Designs and Performance to Receiving Water Impact Mitigation', 387–402.

Shaver, E., Horner, R., Skupien, J., May, C. and Ridley, G. (2007) Fundamentals of Urban Runoff Management. Available at: http://tinyurl.com/q8uxd8c.

SNIFFER (Scotland and Northern Ireland Forum for Environmental Research) (2006) Retrofitting Sustainable Urban Water Solutions. Project UE3(05)UW5. Available at: http://retrofit-suds.group. shef.ac.uk/downloads/UE3(05)UW5%5B1%5D.pdf.

Song, C.R., Kim, J., Wang, G. and Cheng, A.H.D. (2011) 'Reducing erosion of earthen levees using engineered flood wall surface', Journal of Geotechnical Engineering, 874–881.

Sordo-Ward, A., Garrote, L., Bejarano, M.D. and Castillo, L.G. (2013) 'Extreme Flood Abatement in Large Dams with Gate-Controlled Spillways'. Journal of Hydrology 498, 113–123.

Stahre, P. (2002) Recent Experiences in the use of BMPs in Malmö, Sweden, in 'Linking Stormwater BMP Designs and Performance to receiving water impact mitigation, American Society of Civil Engineers Proceedings. 225–235, doi: 10.1061/40602(263)16.

Stahre, P. (2008) Blue green fingerprints in the city of Malmö, Sweden. Available at: http://tinyurl.com/nc4soy5.

Stovin, V., Dunnett, N. and Hallam, A. (2007). Green roofs – getting sustainable drainage off the ground. Proceedings of the 6th Novatech Conference, June 2007, Lyon, France, Conference Proceedings Vol 1, 11–18.

Stovin, V.R., Moore, S.L., Wall, M. and Ashley, R.M. (2013) 'The Potential to Retrofit Sustainable Drainage Systems to Address Combined Sewer Overflow Discharges in the Thames Tideway Catchment'. Water and Environment Journal 27 (2), 216–228.

Stovin, V.R. (2009) 'The potential of green roofs to manage urban stormwater' Water and Environment Journal, 24, 192–199.

Stovin, V.R. and Swan, A.D. (2007) 'Retrofit SuDS – cost estimates and decision – support tools' Water Management, 160, 207–214.

Strecker, E.W., Quigley, M.M. and Urbonas, B. (1999) 'A reassessment of the expanded EPA/ASCE national BMP database'. in Bisier, P. DeBarry, P. (ed.) Proceedings of the World Water and Environmental Congress 2003, held June 23–26 2003.

SUSdrain (n.d.) http://www.susdrain.org/.

Tullis, B. and Neilson, J. (2008) 'Performance of submerged ogee-crest weir head-discharge relationships' Journal of Hydraulic Engineering, 134 (4) 486–491.

United States Environmental Protection Agency (1994) Developing Successful Runoff Programs for Urbanized Areas. Available at: http://tinyurl.com/q2rozxx.

Van den Hoek, R.E., Brugnach, M. and Hoekstra, Y. (2012) 'Shifting to Ecological Engineering in Flood Management: Introducing New Uncertainties in the Development of a Building with Nature Pilot Project'. Environmental Science and Policy 22, 85–99.

Van der Sterren, M., Rahman, A., Shrestha, S., Barker, G. and Ryan, G. (2009) 'An overview of on-site retention and detention policies for urban stormwater management in the Greater Western Sydney Region in Australia' Water International, 34 (3) 362–373.

Van Woert, N.D., Rowe, D.B., Andresen, J.A., Rugh, C.L., Fernández, R.T. and Xiao, L. (2005) 'Green Roof Stormwater Retention: Effects of Roof Surface, Slope, and Media Depth.' Journal of Environmental Quality 34 (3), 1036–1044.

Veijalainen, N. and Vehviläinen, B. (2008) 'The effect of climate change on design floods of high hazard dams in Finland', Hydrology Research, 39 (5), 465–477.

Viavattene, C., Ellis, B., Revitt, M., Seiker, H. and Peters, C. (2010) 'The Application of a GIS-Based BMP Selection Tool for the Evaluation of Hydrologic Performance and Storm Flow Reduction'. NovaTech 2010, 7th International Conference on 'Sustainable Techniques and Strategies in Urban Water Management' held 27 June – 1 July 2010, Lyon, France.

Werritty, A. (2006) 'Sustainable Flood Management: Oxymoron or New Paradigm?' Area 38 (1), 16–23.

Wheeler, P.J., Peterson, J.A. and Gordon-Brown, L.N. (2010) 'Channel Dredging Trials at Lakes Entrance, Australia: A GIS-Based Approach for Monitoring and Assessing Bathymetric Change'. Journal of Coastal Research 26, 1085–1095.

White, I. and Alarcon, A. (2009) 'Planning Policy, Sustainable Drainage and Surface Water Management A Case Study of Greater Manchester' 35 (4).

Woods Ballard, B., Kellagher, R., Martin, P., Jefferies, C., Bray, R. and Shaffer, P. (2007) The SUDS Manual. Available at: http://tinyurl.com/od75lo3.

Wong, T.H.F. (2007) 'Water Sensitive Urban Design: the journey thus far', Australian Journal of Water Resources, 110 (3), 213–222.

Yu, C.F.W. (2010) Flooding and Eco-Build – strategies for the future: Dykes, Dams, SUDS and Floating Homes, Indoor and Built Environment, 19 (6) 595–598.

Zahiri, A., Azamathulla, H. and Bagheri, S. (2013) 'Discharge coefficient for compound sharp crested side weirs in subcritical flow conditions' Journal of Hydrology, 480, 162–166.

Zakaria, N., Ghani, A., Abdullah, R., Sidek, L. and Ainan, A. (2003) Bio-Ecological Drainage System (BIOECODS) for Water Quantity and Quality Control'. International Journal of River Basin Management 1 (3), 1–15.

Zhang, W., Zhang, X. and Liu, Y. (2013) 'Analysis and Simulation of Drainage Capacity of Urban Pipe Network'. Research Journal of Applied Sciences, Engineering and Technology, 6 (3), 387–392.

Zhou, Q. (2014) 'A Review of Sustainable Urban Drainage Systems Considering the Climate Change Urbanisation Impacts' Water, 6, 976–992.

Urban Water and Sediment Quality

Lian Lundy

6.1 Introduction

Urban growth and development, through the construction of buildings, roads and other impermeable surfaces, alters the natural hydrological cycle, changing peak flow characteristics and the volume and quality of runoff (Revitt *et al.*, 2014). As it travels over impermeable surfaces, the urban stormwater flows can mobilise and transport both dissolved and particulate pollutants deposited by a range of land-use activities during the antecedent dry period. On entering a watercourse, the quality and quantity of urban runoff can negatively impact the ecological, physico-chemical and hydro-geomorphological characteristics of the receiving water (Martínez-Santos *et al.*, 2015). It is within this context that this chapter outlines:

- sources, transport and behaviour of pollutants mobilised by urban runoff
- impacts of urban runoff on receiving water quality and sediment quality
- mitigation of urban runoff using SuDS
- the quality of sediments in SUDS

6.2 Sources of Pollutants Mobilised by Urban Runoff

Urban stormwater discharges are generated by runoff from impervious areas such as roads and roofs during rainfall and snowmelt events, as well as overland flows from compacted or saturated open spaces, parks, gardens, road verges and construction sites (Lundy *et al.*, 2011). Depending on land use characteristics (e.g. motorway, residential, industrial, parkland, etc.), surface runoff can mobilise and carry a range of pollutants including metals, hydrocarbons, particulate matter and litter. Figure 6.1 identifies the primary sources of pollutants within an urban catchment, together with an overview of the types of pollutants

Sustainable Surface Water Management: A Handbook for SuDS, First Edition.
Edited by Susanne M. Charlesworth and Colin A. Booth.
© 2017 John Wiley & Sons, Ltd. Published 2017 by John Wiley & Sons, Ltd.

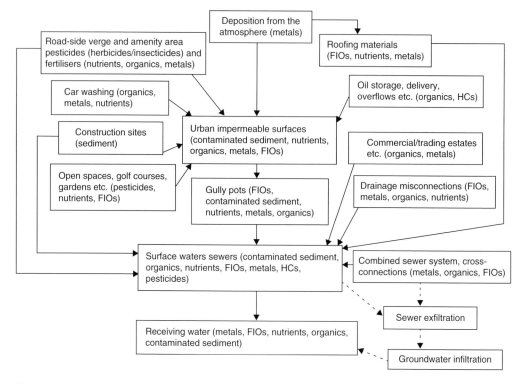

Figure 6.1   Principal stormwater pollutant sources and types (adapted from Revitt *et al.*, 2014).

these sources release. These include metals, organics and nutrients coming from a mix of vehicular wear and traffic emissions, roofing, highway activities, construction materials, commercial activities, litter and plant/leaf debris, spillages and animal/bird excreta in addition to atmospheric deposition. Building misconnections and in-sewer pollutant transformation can add to the 'cocktail' of pollutants discharging to the receiving waterbody via surface water drainage outfalls.

Figure 6.1 also illustrates some the key pathways along which pollutants can be transferred, indicating that initial receptors (e.g. road gullies) can subsequently also act as sources. The time taken for pollutants to move from such pollutant 'sources' to 'sinks' can vary on an event-by-event basis, with the potential for delays in transfer leading to opportunities for extensive pollutant transformation processes to occur. Under these conditions it can be difficult to track and identify sources of pollution, with the final composition of flow discharging to a receiving water body more closely representing the flow characteristics of a particular event and drainage system as opposed to those of the contributing sources (Lundy *et al.*, 2011).

6.3   Quality of Urban Runoff Originating from a Range of Land Use Types

Table 6.1 identifies the key sources of a range of organic and inorganic pollutants frequently reported to be present in urban runoff, event-mean concentrations reported on a land use by land use basis and loadings data (where available). While urban runoff pollutants may

Table 6.1 Pollutant concentrations and loadings determined in runoff from a variety of urban surfaces (taken from Revitt *et al.*, 2014).

Pollutant type	Source	Event mean concentrations	Loadings (kg/ha/yr)
Metals (µg/l)* Pb	Motorways and major roads	Pb: 3–2410; Zn: 53–3550; Ni:4–70; Cd: 0.3–13	Pb:1.1–13
Cd	Urban distributor roads	Pb:10–150; Zn: 410; Cd:0.2–0.5	Pb:0.17–1.9; Zn:1.15
Zn	Suburban roads	Pb:10–440; Zn: 300	Pb:0.01–1.91; Zn:1.15
Ni	Commercial estates	Ni: 2–493	
Cu	Residential	Cd: 0–5; Zn: 150; Pb: 0–140	Pb: 0.001–0.03
	Roofs	Pb:1–30	
	Gully Liquors	Pb:100–850	
Total suspended solids (mg/l)	Residential	55–1568	130–840
	High density Low density	10–290	50–183
	Motorways and major roads	110–5700	815–6289
	Urban roads	11–5400	
	Roadside gully chambers	15–840	409–1700
	Industrial	50–2582	620–2340
	Commercial	12–270	
	Roofs	12.3–216	
	Misconnections	300–511	
Hydrocarbons (µg/l)	Residential	Total HC:0.67–25.0	PAH: 0.002
	High density Low density	Total HC: 0.89–4.5	Total HC:1.8
	Motorways and major roads	Total HC:7.5–400; PAH:0.03–6	Total HC: 0.01–43.3; PAH:140
	Urban roads	Total HC: 2.8–31; PAH: 1–3.5	
	Commercial	Total HC:3.3–22; PAH:0.35–0.6	PAH:0.01–0.35
	Industrial	Total HC:1.7–20	PAH:0.07
Nutrients (mg/l)	Misconnections	Total P:39; NH$_4$:5	
	Residential	Total N:0–6; NH$_4$:0.4–3.8	
	Motorways and roads	Total N:0–4	NH$_4$: 7.2–25.1
	Commercial	NH$_4$:0.2–4.6	
	Industrial	NH$_4$: 0.2–1.1	
	Roofs	NH$_4$: 0.4–3.8	
	Gully Liquors	Total N:0.7–1.39	
E. coli (MPN/100 ml)	Misconnection	10^{3-}10^6	
	Roofs, roads and parks	40–10^6	1–4×10^8

*Key:= metalled roofs not included; HC = hydrocarbons; PAH = polyaromatic hydrocarbons; Total P = total phosphorous; Total N = total nitrogen.

be present in the dissolved or particulate phase, the majority of urban pollutants are reported to be associated with the particulate phase (Bjorkland, 2011; Selbig *et al.*, 2013).

Sources of particulate matter can include wear and tear of vehicles and road materials (e.g. abrasion of car bodies, engines, brakes and tyres), exhaust emissions, friction-assisted break-up of paving and road materials and degradation of street furniture, emissions from building materials (e.g. roof runoff), industrial emissions, street litter and soil erosion via both direct and indirect deposition routes (e.g. short- and long-distance aerial transport and re-suspension of previously settled particles) (Lundy and Wade, 2013). While the impacts of urban diffuse pollution on groundwater are not well understood in terms of processes, the fact that urban areas negatively impact on groundwater is well established

(Lerner and Harris, 2009). At locations heavily influenced by traffic, this source can be a more important source of key urban diffuse pollutants than industrial emissions in association with both direct deposition and subsequent re-suspension processes.

6.4   Quality and Behaviour of Sediment in Urban Receiving Water Bodies

The majority of urban rivers and watercourses routinely receive inputs from a wide variety of sources such as stormwater runoff from urban, highway and agricultural areas and combined surface water outfalls (CSOs), as well as receiving inputs originating from more episodic sources such as accidental spills and pollution incidents. Many of these inputs are known to carry significant loads of suspended solids, particularly during storm events (Martínez-Santos *et al.*, 2015). On entering a receiving watercourse, the transported particulate matter progressively settles out, as a function of both time and the hydrological conditions in the river system. This process can result in the build-up of substantial quantities of sediment, to such an extent that bed substrates may effectively disappear under an ever increasing sediment load.

This deposited sediment thus becomes a key environmental compartment of the urban river ecosystem, both in situ as a habitat for microorganisms, macro invertebrates and vegetation, and with regard to the future environmental quality of the aquatic ecosystem, because fine sediment can be a significant source of particulate matter and in-stream turbidity following a re-suspension event. However, although sediment-derived turbidity may be an issue with regard to the environmental status of a river, of much greater environmental concern is the association between particulate matter and a wide range of pollutants (Martínez-Santos *et al.*, 2015). Most of the pollutant load is reported to be transported by sediments, particularly metals (Coulthard and Macklin, 2003), and toxic effects have been reported in association with urban sediments (Selbig *et al.*, 2013) and runoff samples (Marsalek *et al.*, 1999). Table 6.2 gives an overview of metal concentrations reported in a range of urban water environments.

Table 6.2   Minimum and maximum total concentrations of metals measured in a selection of urban stream and river sediments (µg/g dry weight) (after Scholes *et al.*, 2008a).

	n	Cd	Cr	Cu	Ni	Pb	Zn
Scholes *et al.*, 1999	45	**3.0–10**	3–**169**	17–**178**	22–**187**	**33–332**	21–**1035**
Rhoads and Cahill, 1999	41		9–**328**	6–**55**	8–**244**	10–**225**	29–**528**
Wilson and Clarke, 2002	9			440.6	80.9		407.0
Filgueiras *et al.*, 2004[1]	33	0.37–0.41	**78–139**	30.5–**55.9**	**32.5–60.7**	**43.6–91.1**	
Tejeda *et al.*, 2006	32			9–**165**		12–64	38–**1467**
Thevenot *et al.*, 2007[2]		**1.70**	47	**31**		43	140
Samecka-Cymerman	21	0.20–0.58	4.9–**28.5**	2.1–10.6	7.5–**15.2**	15–57	6.8–**458**
and Kempers (2007)	24	0.24–**1.72**	17–**85.2**	9.5–**43.7**	14.5–**39.0**	17–97	22.9–**174**
Walling *et al.*, 2003[3]	51		8–17	**33–92**		**689–1471**	**775–1850**
	52		21–**181**	**118–198**		90–237	274–580
	17		65–**313**	141–235		199–343	397–907
Carpentier *et al.*, 2002	50[6]	**<0.8–6**	4–78	**<5–172**	**<5–30**	**<5–278**	39–**563**

Key: n = number of samples; values in bold exceed 1 or more of the guideline values presented in Table 3; [1] = range of values across 11 sites; [2] = estimated average metal contents over the time period 1995–2000 of dredged sediments; [3] = range of values reflect average concentrations at multiple sampling points on three different rivers sampled approximately bimonthly over 12 month period; rivers located close to metal.

Once in association with the sediment, the pollutant may be subject to a range of further processes as the physico-chemical conditions in the sediment can differ significantly from those in the water column. The pollutant may therefore not remain in its initial form but be transformed through a variety of physical, chemical or biological processes, and become associated with a different sediment phase. It is now appreciated that knowledge of the pollutant load of sediments, in terms of the total concentrations of the various pollutants present, is insufficient with regard to determining their full impact on receiving aquatic ecosystems (Stead-Dexter and Ward, 2004). Factors such as humic acid content, ionic strength, alkalinity, ammonia concentration and pH are only a few of the many factors reported to play a role in determining the mobility and bioavailability of metals (Shafie *et al.*, 2014).

Furthermore, urban watercourses are characteristically highly flashy in nature, so that deposited sediments are highly susceptible to further re-suspension processes (Old *et al.*, 2004). Sediments may be mobilised by dredging and/or by river processes (e.g. storm events, flooding, influx from groundwater and bioturbation) which can cause contaminants to be rereleased into the overlying water column and subsequently relocated many miles downstream from the original sources (Turner *et al.* 2008). Under such dynamic conditions, the change in physico-chemical conditions may result in the rerelease of previously bound pollutants into the water column. Thus, not only have elevated concentrations of a wide range of substances – including metals, hydrocarbons and faecal coliforms – been reported in urban streams and rivers (Chen *et al.*, 2004; Gasperi *et al.*, 2008; Martínez-Santos *et al.*, 2015), but the concentrations of these contaminants in deposited sedimentary structures also tend to be highly variable on both spatial and temporal scales (Faulkner *et al.*, 2000). Hence, sediments act as a sink and source for many substances and the interactions between water and sediment have major implications for compliance with the EU Water Framework Directive (EU WFD, 2000). Although under current EU WFD implementation plans sediment quality plays a relatively minor role, Article 3 of the EU Environmental Quality Standards Directive (2008) states that standards may be developed for sediment (in addition to water and biota). It calls for the long-term monitoring of priority substances that tend to accumulate in sediment, demonstrating awareness at a policy level that polluted sediments can be a problem for water quality across Europe. However, sediment Environmental Quality Standards (EQSs) have yet to be developed by any individual Members State.

6.5   Treatment of Urban Runoff Using SuDS

The continued and rapid growth of urban areas across Europe places increasing importance on the control of urban stormwater runoff. However, the criteria defining what constitutes effective stormwater management are themselves undergoing change. Comprehensive stormwater management plans in both new and existing urban areas should not only address stormwater quantity and quality but also need to consider sustainable development requirements (Revitt *et al.*, 2003). Furthermore, it is anticipated that the requirements for the control of stormwater, particularly with regard to the protection of receiving waters, are likely to become much more stringent through the implementation of the EU WFD. In order to meet these changing requirements a new approach to stormwater management is needed, which has led to increasing interest in the use of sustainable (urban) drainage systems (SuDS) (also known as stormwater best management practices or BMPs). SuDS encompass a wide range of solutions, which enable the

Table 6.3   Performance efficiency of a range of types of SuDS (after Revitt *et al.*, 2003). ND = no data; TSS = total suspended solids; TN = total nitrogen; HC = hydrocarbons; TM = total metals.

Individual SuDS device	% removal efficiency				
	TSS	TN	Bacteria	HC	TM
Filter drain	60–90	20–30	20–40	70–90	70–90
Infiltration basin	60–90	20–50	70–80	70–90	70–90
Swales	10–40	10–35	30–60	60–75	70–90
Sedimentation lagoon	50–85	10–20	45–80	60–90	60–90
Detention basin	60–80	20–40	20–40	ND	40–55
Detention basin (extended)	30–60	5–20	10–35	30–50	20–50
Retention basin	80–90	20–40	40–60	30–40	35–50
Wetland	70–95	30–50	75–95	50–85	40–75

planning, design and management of stormwater to be tackled equally from hydrological, environmental and public amenity perspectives (see Chapters 5, 7 and 8).

Table 6.3 provides an overview of the ranges of pollutant removal percentages that have been reported from various UK SuDS studies in urban areas. While total solids removal is generally good for most types of SuDS, there is considerable variation for other pollutant parameters with some showing rather poor removal capabilities. The efficiency performance data, being based on the average difference between inflow and outflow storm event concentrations, may be misleading, especially when inflow concentrations are low. For example, a retention basin experiencing 500 mg/l TSS in the inflow and 100 mg/l in the outflow would yield equivalent pollutant removal efficiency to a constructed wetland having 100 mg/l and 20 mg/l in the inflow and outflow respectively. Yet the final water quality for the latter SuDS device is clearly superior and provides more effective and efficient protection of the receiving waterbody (Revitt *et al.*, 2003). The use of a percentage removal term is probably only really appropriate for sites and SuDS facilities subject to high pollutant input concentrations. The US EPA National Stormwater Best Management Practices Database recommends the use of a normal probability plot of the inflow and outflow pollutant event mean concentrations (EMCs), with the EMC distribution matched against set (or target) receiving water quality standards (or against any discharge consent conditions). This would enable performance to be described in terms of exceedance probability of target standards for different flow conditions and/or return periods. This statistical methodology would also enable anomalous results (such as apparent negative efficiencies) to be identified, as well as determining whether a small number of large storms are biasing the resulting overall efficiency value.

It is clear that a comparative assessment of the performance of structural SuDS options is currently limited by a lack of data and the uncertainties associated with the simplified methodology used to calculate percentage performance efficiency. In addition, for most wetland/retention systems, given the dynamic nature of flow into and out of wet basins having a permanent pool, the recorded inflow and outflow concentrations are not normally contemporaneous, that is, they may not be generated by the same storm event. It is not yet feasible to provide definitive SuDS designs to meet specified and consistent performance requirements for given storm and catchment characteristics or to meet specific receiving water standards and storm return periods.

6.6   Pollutant Removal Processes that Occur in SuDS

The primary biological, chemical and physical pollutant removal mechanisms which result in the removal of pollutants from the water column and their transfer to the sediment in SuDS are identified as settling, adsorption to substrate, microbial degradation, filtration, volatilisation, photolysis and plant uptake (Scholes *et al.*, 2008b; Figure 6.2). The potential for these processes to occur in various types of SuDS varies on a system-by-system basis in relation to a broad range of factors from system design and age to operation and maintenance regime,

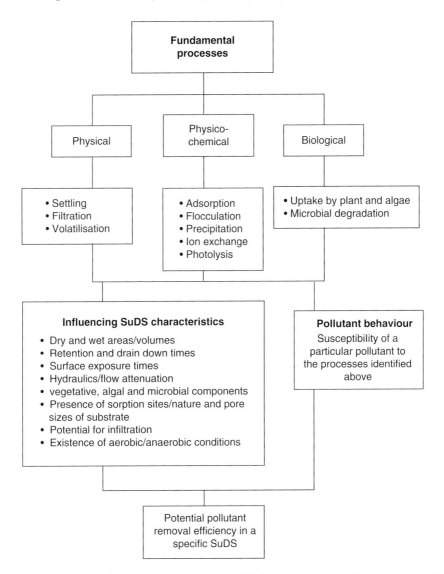

Figure 6.2   Fundamental unit processes in relation to SuDS characteristics and pollutant behaviour (after Scholes *et al.*, 2008b).

hydraulic retention and drain down times, type/level of vegetation and microbial consortia, surface exposure times, potential for infiltration and the existence of aerobic/anaerobic conditions.

Adsorption to substrate refers to the physico-chemical adherence of pollutants to an artificial substrate (e.g. the gravel matrix of a filter drain), a natural substrate (e.g. vegetation in a swale) or an introduced substrate (e.g. the deposited benthal sediment in a detention pond). It is an important removal process in SuDS such as filter drains, porous paving, sub-surface flow (SSF) constructed wetlands, infiltration basins, soakaways and infiltration trenches due to the close contact achieved between stormwater and substrate surface during the infiltration of an effluent through the relevant permeable material. The hydraulic pathways taken by stormwater in swales, filter strips, surface flow (SF) constructed wetlands, detention basins and extended detention basins will characteristically result in relatively lower direct contact times with the available substrate and therefore less potential for adsorption. Settling is the vertical movement of discrete or agglomerating suspended sediment particles to the base of a water column (Ellis et al., 2004) and is highly dependent on the retention of a quiescent water volume in a SuD system. Hence, it is an important mechanism in retention ponds, infiltration basins and extended detention basins. Although the presence of macrophytes in both types of constructed wetland contributes to the formation of quiescent conditions, the presence of dense stands of vegetation also effectively lowers the stationary water column volume, through which settling can occur. Filtration occurs by the same mechanisms as those present in conventional water treatment plant sand filtration units where physical sieving removes particulate pollutants as they pass through a porous substrate or hydraulic barrier (Ellis et al., 2004). Hence, the potential for filtration to occur is considered to be most effective in porous paving and porous asphalt due to surface filtration (Revitt, 2004). Infiltration trenches, infiltration basins, soakaways and SSF constructed wetlands involve the passage of stormwater through a sub-surface substrate, but filtration is relatively less efficient due to the greater void sizes in gravel, which is typically used as the substrate. Similar processes are possible in filter strips but shorter contact times between stormwater and the grassed surface result in a relatively lower potential for filtration.

Microbial degradation is facilitated by the availability of attachment sites and nutrients in a SuD system; both aerobic and anaerobic processes are enhanced by the occurrence of high contact ratios between stormwater and substrate material. Microbial degradation is therefore strongly encouraged in SuDS such as SSF constructed wetlands (Ellis et al., 2003) and infiltration basins. In contrast, the opportunity for prolonged contact of stormwater with an established microbial population is less feasible for detention basins (non-permanent water body), filter strips and swales (low retention times). The presence of terrestrial or aquatic vegetation provides the potential for plant uptake to occur, and is considered to be highest in SSF constructed wetlands due to the increased contact between stormwater and the elaborate root systems of aquatic macrophytes, followed by SSF wetlands, swales and filter strips (permanently vegetated structures).

Both volatilisation and photolysis processes are strongly dependent on surface exposure, but whereas photolysis requires direct exposure to sunlight, volatilisation can occur from the open spaces in a SuDS structure. Photolytic degradation is considered to be of negligible importance in SuDS, such as filter drains, porous paving, soakaways and infiltration trenches, due to the rapid incorporation of stormwater into the SuDS structure. The opportunity for photolysis to occur will be relatively highest in filter strips, swales, infiltration basins, retention ponds, detention basins and extended detention basins due to a combination of enhanced surface areas and increased exposure times. The degree of volatilisation is considered to be greatest in extended detention basins, detention basins, retention ponds, infiltration basins, SSF constructed wetlands and swales, where stormwater exposure times and surface area exposure to wind/ambient pressure differentials are optimised.

6.7 Quality and Behaviour of Sediment in SuDS

As noted earlier, the majority of the urban pollutant load is reported to be transported by sediments (Martínez-Santos *et al.*, 2015). Traffic and its associated infrastructure (road materials and street furniture) are frequently identified as major source of metal contaminants due to their presence in vehicle parts, tyres, brakes and road materials, and biocides are commonly used in roadside verges, parks as well as in a range of building materials (Donner *et al.*, 2010). The quiescent conditions associated with many types of SuDS facilitate the settlement of particles and associated pollutants, effectively leading to the accumulation or concentration of sediments with elevated levels of pollution at a single location. The quality of sediments in SuD systems can and does vary in relation to a range of factors including SuDS type, design, age and operation and maintenance regime. A further important influencing parameter is the type of land-use activities taking place in the surrounding catchment. For example, SuDS installed on industrial sites may contain relatively high levels of chemicals handled and stored on-site, while SuDS at airports or other transport facilities may have relatively high contents of hydrocarbons, detergents, de-icing chemicals, etc. in their sediments (Donner *et al.*, 2010).

Table 6.4 gives an overview of the maximum concentrations of a range of metals determined in 565 sediment samples collected from a range of SuDS located in the USA and the UK (WERF, 2005). Also included in Table 6.4 are the EU and US threshold values for the definition of hazardous waste. It is clear that, with respect to the metals identified, all samples (with the exception of one canal sediment sample from Scotland) fell well below both sets of guideline limits and thus did not need to be classified as hazardous waste. In the UK, this means that the dredged sediment can be disposed of on the banks surrounding the SuDS structure, which has the benefit of being a relatively low cost option.

Table 6.4 Maximum concentrations of metals in sediments from various SuDS sites (WERF, 2005).

Site structure	Maximum concentration (mg/kg)						
	As	Cd	Cu	Pb	Zn	Ni	Hg
Biofiltration strips (US)	2.9	1.2	60	144	337	13	0.05
Oil water separator (US)	5.0	1.7	106	189	702	27	0.07
Sand filter (US)	0.76	0.3	11	11	70	3.4	0.04
Sand filter (US)	1.2	0.3	8	25	61	3.1	0.04
Compost filter (US)	1.7	5.0	120	110	670	18	0.5
Pond (UK)	7.0	18	80	399	3718	175	2.0
Balancing pond (UK)	4.6	3.5	73		983	92	
Balancing pond (UK)	0.1	0.1	2.3		16	1.3	
Canal (Scotland)	98	21	451	8,275	6,671	114	2.7
Industrial stream	18	0.3	32	51	160	40	
Oil industry stream	4.6	1.3	441	81	407	81	
EU threshold values for hazardous waste	30,000	2500		2500		1000	2500
EU threshold values for hazardous waste	5000	1000	25,000	5000	250,000	20,000	

As metals are not biodegradable, their removal from the water column in SuDS does not involve any biodegradation processes (e.g. aerobic or anaerobic digestion) although there is potential for metals to be indirectly immobilised by microbial consortia as a result of processes such as biosorption and bioaccumulation (Valls and de Lorenzo, 2002). Metal removal in SuDS is understood to be primarily associated with adsorption, followed by sedimentation. Analysis of metal contents in pond sediments sampled by Marsalek *et al.* (1997), Färm (2002), Heal *et al.* (2006), and others has shown the deposition of metal-contaminated sediment over time and that sediment quality can be highly variable even within an individual SuDS. For example, Heal *et al.* (2006) showed that concentrations of metals tended to be higher near the pond inlet. While SuDS sediment quality data is much more limited than water quality, a review by Donner at al., (2010) reported that in most cases it was the rate of sediment infilling and subsequent reduction in the retention time for stormwater that determines the timing of sediment removal for SuDS maintenance, rather than the build-up of contaminants in SuDS sediments to unacceptable levels (Donner *et al.*, 2010).

References

Bjorkland, K. (2011) Sources and fluxes of organic contaminants in urban runoff. PhD thesis. Chalmers University, Gothenburg, Sweden.

Chen, B., Xuan, X., Zhu, L., Wang, J., Gao, Y., Yang, K., Shen, X. and Lou, B. (2004) Distributions of polycyclic aromatic hydrocarbons in surface waters sediments and soils of Hangzhou City China. Water Research 38 (16), 3558–3568.

Coulthard, T.J. and Macklin, M.G. (2003) Modelling long-term contamination in river systems from historical metal mining. Geology, 31 (5), 451–454

Donner, E., Seriki, K. and Revitt, D.M. (2010) Production, treatment and disposal of priority pollutant contaminated sludge. EU FP6 ScorePP Deliverable 5.5. Available from Middlesex University, The Burroughs, London, NW4 4BT.

Ellis, J.B., Chocat, B., Fujita, S., Rauch, W. and Marsalek, J. (2004) Urban Drainage: A Multilingual Glossary. IWA Publishing, London, UK, 512 pp (ISBN: 190022206X).

Ellis, J.B., Shutes, R.B.E. and Revitt, D.M. (2003) Constructed wetlands and links with sustainable drainage systems. Technical Report P2-159/TR1, Environment Agency, 178 pp (ISBN 1857059182).

EU Environmental Quality Standards Directive (2008). Directive 2008/105/EC of the European Parliament and of the Council of 16 December 2008 on environmental quality standards in the field of water policy. Available at: http://eur-lex.europa.eu/LexUriServ/LexUriServ.do?uri=OJ:L:2008:024:0008:0029:en:pdf

EU WFD (2000). Directive 2000/60/EC of the European Parliament and of the Council establishing a framework for Community action in the field of water policy. Available at: http://eur-lex.europa.eu/LexUriServ/LexUriServ.do?uri=CELEX:32000L0060:EN:HTML

Färm, C. (2002) Evaluation of the accumulation of sediment and heavy metals in a stormwater detention pond. Water Science and Technology 45 (7), 105–112.

Faulkner, H., Green, A. and Edmonds-Brown, V. (2000) Problems of quality designation in diffusely polluted urban streams – the case of Pymme's Brook, N. London. Environmental Pollution, 109, 91–107.

Filgueiras, A.V., Lavilla, I. and Bendicho, C. (2004) Evaluation of distribution, mobility and binding behavior of heavy metals in surfi cial sediments of Louro River (Galicia, Spain) using chemometric analysis: a case study. The Science of the Total Environment 330, 115–129.

Gasperi, J., Garnaud, S., Rocher, V. and Moilleron, R. (2008) Priority pollutants within heavily urbanized area: what about receiving waters and settleable sediments? Proceedings of the 11th International Conference on Urban Drainage, Edinburgh, Scotland, UK.

Heal, K.V., Hepburn, D.A. and Lunn, R.J. (2006) Sediment management in sustainable urban drainage (SUD) ponds. Water Science and Technology 53, 219–227.

Lerner, D. and Harris, B. (2009) The relationship between land use and groundwater resources and quality. Land Use Policy 26; 1, S265–S273.

Lundy, L., Ellis, J.B. and Revitt, D.M. (2011) Risk prioritisation of stormwater pollutant sources. Water Research, Water Research 46, 6589–6600.

Lundy, L. and Wade, R. (2013) A critical review of methodologies to identify the sources and pathways of urban diffuse pollutants. Stage 1 contribution to: Wade, R. *et al.* (2013) A Critical Review of Urban Diffuse Pollution Control: Methodologies to Identify Sources, Pathways and Mitigation Measures with Multiple Benefits. CREW, The James Hutton Institute. Available on-line at: crew.ac.uk/publications

Marsalek, J., Rochfort, Q., Brownlee, B., Mayer, T. and Servos, M. (1999) An exploratory study of urban runoff toxicity. Water Science and Technology, 39 (12), 33–39.

Marsalek, J., Watt, W.E., Anderson, B.C. and Jaskot, C. (1997) Physical and chemical characteristics of sediments from a stormwater management pond. Water Quality Research Journal of Canada 32, 89–100.

Martínez-Santos, M., Probst, A., García-García, J. and Ruiz-Romera, E. (2015) Influence of anthropogenic inputs and a high-magnitude flood event on metal contamination pattern in surface bottom sediments from the Deba River urban catchment. Science of the Total Environment 514, 10–25.

Old, G.H., Leeks, G.J.L., Packman, J.C., Stokes, N., Williams, N.D., Smith, B.P.G., Hewitt, E.J. and Lewis, S. (2004) Dynamics of sediment-associated metals in a highly urbanised catchment: Bradford, West Yorkshire. Water and Environment Journal, 18 (1), 11–16.

Revitt, D.M. (2004) Water pollution impacts of transport. In: Hester, R.E., Harrison, R.M. (eds), Transport and the Environment. Issues in Environmental Science and Technology, vol. 20, Royal Society of Chemistry, Cambridge, UK, pp. 81–109.

Revitt, D.M., Ellis, J.B. and Scholes, L. (2003) DayWater Deliverable 5.1. Review of the Use of Stormwater BMPs in Europe. Available from: Middlesex University, The Burroughs, Hendon, London, NW4 4BT.

Revitt, D.M., Lundy, L., Coulon, F. and Fairley, M. (2014) The sources, impact and management of car park runoff pollution. Journal of Environmental Management 146, 552–567.

Rhoads, B.L. and Cahill, R.A. (1999) Geomorphological assessment of sediment contamination in an urban stream system. Applied Geochemistry 14, 459–483.

Samecka-Cymerman, A.J. and Kempers, A. (2007) Heavy metals in aquatic macrophytes from two small rivers polluted by urban, agricultural and textile industry sewages SW Poland. Arch. Environ. Contam. Toxicol. 53 (2), 198–206.

Scholes, L., Faulkner, H.P., Tapsell, S. and Downward, S. (2008a) Urban rivers as pollutant sinks and sources: a public health concern to recreational river users? Water, Air and Soil Pollution Focus 8, 5-6, 543–553.

Scholes, L., Revitt, D.M. and Ellis, J.B. (2008b) A systematic approach for the comparative assessment of stormwater pollutant removal potentials. Journal of Environmental Management 88 (3), 467–478.

Scholes, L.N.L., Shutes, R.B.E., Revitt, D.M., Purchase, D. and Forshaw, M. (1999) The removal of urban pollutants by constructed wetlands during wet weather. Water Science and Technology. 40 (3), 333–340.

Selbig, W.R., Bannerman, R. and Corsi, S.R. (2013) From streets to streams: Assessing the toxicity potential of urban sediment by particle size Science of The Total Environment 444, 381–391.

Shafie, N.A., Aris, A.Z. and Haris, H. (2014) Geoaccumulation and distribution of heavy metals in the urban river sediment. International Journal of Sediment Research 29 (3), 368–377.

Stead-Dexter, K. and Ward, N.I. (2004) Mobility of heavy metals within freshwater sediments affected by motorway stormwater. Science of the Total Environment 334-335, 271–277.

Tejeda, S., Zarazúa-Ortega, G., Ávila-Pérez, P., García-Mejía, A., Carapia-Morales, L. and Díaz-Delgado, C. (2006) Major and trace elements in sediments of the upper course of Lerma River. J. Radioanal. Nucl. Chem. 270, 9–14.

Thevenot, D.R., Moilleron, R., Lestel, L., Gromaire, M.C., Rocher, V., Cambier, P., Bonte, P., Colin, J.L., de Ponteves, C. and Meybeck, M. (2007) Critical budget of metal sources and pathways in the Seine river basin (1994–2003) for Cd, Cr, Cu, Hg, Ni, Pb and Zn, Sci. Total Environ., 375, 180–203.

Turner, J.N., Brewer, P.A. and Macklin, M.G. (2008) Fluvial-controlled metal and As mobilisation dispersal and storage in the Río Guadiamar SW Spain and its implications for long-term contaminant fluxes to the Doñana wetlands. Science of The Total Environment, Volume 394, Issue 1, 1 May 2008, pp. 144–161.

Valls, M. and de Lorenzo, V. (2002) Exploiting the genetic and biochemical capacities of bacteria for the remediation of heavy metal pollution. FEMS Microbiology Reviews 26 (4), 3.

Walling, D. E., Owens, P. N., Cartera, J., Leeks, G. J. L., Lewis, S., Meharg A. A. and J. Wright (2003). Storage of sediment-associated nutrients and contaminants in river channel and floodplain systems. Applied Geochemistry 18 (2), 195–220, 27–338.

WERF (2005) Post-Project Monitoring of BMPS/SUDS to Determine Performance and Whole-Life Costs. Water Environment Federation, Alexandria, Virginia, USA.

Sustainable Drainage Systems: Delivering Multiple Benefits for People and Wildlife

Andy Graham

7.1 Introduction

For too long rainwater has been perceived as a waste product, hastened into a pipe, discharged into watercourses and then down to the sea as quickly as possible, with often disastrous implications for people and wildlife. With more widespread use of sustainable drainage systems (SuDS), there is an opportunity to change society's relationship with rainfall, particularly in urban areas, and to see it as a precious resource capable of transforming the environment and improving lives.

Increasing urbanisation and reduced opportunities for rainfall to infiltrate into the ground means that conventional drainage systems can struggle to cope with sudden, intense cloudbursts. Drains can be quickly overwhelmed, leading to floods and polluted streams and rivers. Thus, the impacts of conventional drainage systems on the chemistry, hydrology, biology and ecology of receiving watercourses and associated habitats are wide-ranging and severe. Species richness declines, those sensitive to pollutants decline or become locally absent, and streams and rivers lose their amenity value. Conversely, because aquifers beneath urban areas receive less rainfall through infiltration, rivers and streams are also adversely affected due to low flows, while high nitrate and phosphate levels cause algal blooms.

Conventional drainage systems cannot deliver the full range of benefits that a natural catchment or SuDS can offer. In general, a pipe cannot create wildlife habitats, protect water quality in streams and ponds, promote pollination, support climate change adaptation, reduce energy use, create attractive open spaces for people or contribute to a healthier, more cohesive community – a well designed and managed SuDS can do all of these things.

The drivers for changing the way rainfall is managed are clear; the challenge is to do so in a way that uses limited resources to manage surface water to alleviate flooding and protect water quality but also to capture more of the ecosystem services outlined above,

Sustainable Surface Water Management: A Handbook for SuDS, First Edition.
Edited by Susanne M. Charlesworth and Colin A. Booth.
© 2017 John Wiley & Sons, Ltd. Published 2017 by John Wiley & Sons, Ltd.

delivering more broadly for people and for wildlife. Intelligent, multi-functional use of urban spaces with high quality design that supports and enhances the environment is the requirement, and SuDS can help meet this challenge. They provide an opportunity to bring urban wetlands and other wildlife-friendly spaces into towns and cities, and to link these with existing habitats, creating blue and green corridors where people and wildlife can thrive. Designed and located appropriately, SuDS will protect downstream waters (particularly relevant here is their ability to support the delivery of the Water Framework Directive (WFD) objectives) and provide additional high quality habitat as well as meeting many other objectives.

While there are many good examples of SuDS, which blend creative thinking, high quality urban design and community benefits alongside flood attenuation and water quality protection (see Section 7.7), there is still a long way to go before SuDS fulfil their potential to transform living spaces and environment. Many early SuDS schemes focused on managing large volumes of surface water in a single, deep feature, at the expense of adopting measures to improve water quality and amenity. These drainage systems can seem more like bomb craters and are frequently ringed with security fences and keep out signs – a clear indication that amenity and wildlife value are not top priorities.

7.2   Getting Better SuDS

7.2.1   Delivering Policy and Strategy Objectives with SuDS

Using SuDS to bring wide-ranging benefits for people and wildlife is supported by national and international policy and legislation. Undoubtedly, these instruments will change and develop over time but central to them are arguments for the intelligent, multi-functional use of space; SuDS are a key means of delivering these objectives. For example, the National Planning Policy Framework (2012) (NPPF) makes it clear that local authority plans should seek to protect and enhance the natural environment and should adopt policies that recognise the wider benefits of ecosystem services, capturing the multiple benefits from land use including for wildlife, recreation, mitigation of flood risk and carbon storage; clearly, SuDS have a central role to play in delivering these multiple benefits.

SuDS will benefit a range of habitats and species, such as those which are targeted for specific conservation action in the UK (and elsewhere), for example, reed-beds and water voles along with many more species associated with wetlands and other habitats. They will be able to contribute to meeting many of the objectives for priority habitats and species contained in each of the four national biodiversity strategies (UK post 2012 Biodiversity Framework at http://jncc.defra.gov.uk/). Specifically, 'Biodiversity 2020: a strategy for England's wildlife and ecosystem services' aims to halt overall biodiversity loss, support healthy, well-functioning ecosystems and establish coherent ecological networks with more and better places for nature for the benefit of wildlife and people. Again, SuDS in urban areas will deliver these objectives.

The WFD is a key driver in promoting sustainable water management and aims to improve the chemical and ecological status of water bodies throughout the EU. In the UK, SuDS are identified as one of the key measures needed to achieve this grand objective. In river basin management planning, SuDS are seen as valuable tools in achieving compliance with WFD objectives.

7.2.2 Sustainable Development and Liveability – Supporting Quality of Life and Wellbeing Through Well Designed SuDS

The contribution of biodiversity to sustainable development is fully recognised in national, regional and local policies and programmes dealing with urban areas and development, including the NPPF and the localism agenda. The Localism Act introduced Neighbourhood Development Plans, a new and voluntary planning process as well as a Local Green Space designation. These new tools can drive the use of SuDS to create wildlife-rich developments.

Natural England's 'Nature Nearby' Accessible Natural Greenspace Guidance (2010) specifically identifies SuDS as an opportunity for creating new green space in urban areas and states that when incorporated into site master plans alongside new footpaths, greenways and woodlands, they deliver a range of benefits to people. These can include provision of places for recreation and relaxation, play areas for children, urban regeneration, education and improved health. For more details on how green space, including SuDS, can bring health benefits in England see http://www.naturalengland.org.uk/ourwork/enjoying/linkingpeople/health/default.aspx, and see http://www.snh.gov.uk/docs/A265734.pdf for information about the experience in Scotland.

7.2.3 Green Infrastructure (GI) and Blue Corridors

SuDS should not be seen as isolated features in the urban environment and care should be taken at the design stage to locate them in existing or future networks of habitats. They can act as linking habitats, stepping stones, or as part of a corridor. They are particularly useful in urban areas, allowing wildlife to move through and out into rural environments, as well as being urban habitats in their own right. They will help maintain and build ecological function in urban areas, particularly as part of a network of such sites. Integrating surface water management into Green Infrastructure strategies ((http://www.greeninfrastructurenw.co.uk/) at the local scale brings enormous benefits to local people. Similarly, Defra's (2001) 'blue corridors' (http://randd.defra.gov.uk/) scoping study shows how, with creative thinking and good planning, SuDS can play a significant part in delivering better places to live.

7.3 SuDS and How They Support Biodiversity

SuDS seek to manage rainfall in ways that mimic natural processes by using the landscape to control the flow and volume of surface water, to prevent or reduce pollution downstream of development and to promote recharging of groundwater resources. By keeping water at the surface and slowing its progress to the nearest watercourse there is the opportunity to use it creatively to produce a range of habitats, including wetlands (Figure 7.1). These habitats support biodiversity but also link existing habitats in towns and cities, and between urban and rural areas, permitting migration of species, and with it, exchange of genetic resources. They can be retrofitted to existing buildings and open spaces, and should always be part of any new development. There are few situations where wildlife-rich SuDS cannot be created and they are easily incorporated into buildings, open spaces both private and public, such as parks, gardens, road traffic islands, allotments, farms, designated nature reserves, existing blue/green spaces, orchards, canals, rivers, streams,

Figure 7.1   Wetland and swale in a housing estate, N. Hamilton Leicestershire, UK.

lakes, transport networks and, in fact, anywhere where surface water can be conveyed and stored safely in the built and rural environment.

Specifically, using SuDS will achieve the following:

Reduce pollution – Surface water runoff is often polluted with silt, oil and other contaminants that, when discharged into rivers, can harm wildlife and contaminate drinking water sources. Combined sewer overflows also discharge during periods of heavy rainfall when sewers are surcharged. SuDS can trap, store and treat these pollutants through natural processes and produce a supply of cleaner water for downstream use.

Attenuate flooding – Traditional piped drainage networks convey water far more quickly than natural processes. Rivers respond quickly and violently to rainfall, exacerbating downstream flooding. Flooding also occurs where housing and other urban development, such as the paving of gardens and the building of extensions (a process often referred to as 'urban creep'), increase the volume of surface water entering drains. SuDS contain water safely at the surface and slow its flow downstream into receiving drains and watercourses, thus attenuating flooding.

Support low flows in streams and rivers – Piped drainage prevents natural percolation of rainfall into groundwater that can support rivers and wetland flows, but keeping water on the surface in features that mimic natural wetland processes allows rainfall to recharge aquifers and supports flows in streams and rivers.

7.3.1   Wildlife Benefits

Many benefits are derived from SuDS, which manage water at the surface, but the greatest opportunities for habitat creation are by also allowing the discharge of rainwater into the ground, which supports aquifer recharge and thus existing rivers and wetlands. This should be an important consideration in design. Most SuDS features will permit some infiltration into the ground, which is desirable, but water should also be kept at the surface to generate wetland habitats and provide direct biodiversity benefits.

Not all SuDS features are permanently wet; some storage capacity needs to be available to safely store water after heavy rainfall. The need to include dry habitats that become inundated after rain brings additional opportunities for habitat development. SuDS, which feature a mix of habitats from permanently wet to only occasionally inundated, will offer the most opportunities for wildlife to colonise. Wet woodland, reed-beds, marsh, unimproved wet and dry grassland, scrub and open water are all valuable in terms of SuDS function and habitat provision. However, the value of individual features or habitats depends to an extent on their position in the system (e.g. a pond at the top of the system receives more polluted influent and is likely to be less biodiverse than a pond or other wetland downstream). SuDS features included as part of a management train improve water quality incrementally as it progresses downstream.

SuDS habitats, as outlined above, will benefit plants, invertebrates, reptiles, amphibians, mammals (e.g. bats and water voles), both in the devices themselves and also indirectly in receiving watercourses through improvements to water quality and by controlled discharge of surface water. Realising these benefits depends on adherence to some simple design principles, covered later in this chapter.

7.4   Involving People

Working with local authorities and other stakeholders will greatly assist in identifying the habitats and species most in need of conservation action, and this means that SuDS can be designed with the local context in mind, to maximise biodiversity benefits. This needs to be done at the design stage or earlier, at the masterplanning stage. It is strongly recommended that appropriate ecological advice be sought to inform the design and to secure a positive outcome for wildlife. This will help enhance awareness of the value and importance of the natural environment in local communities where SuDS exist.

7.4.1   Community Involvement and Participation

There are many ways in which those responsible for developing and managing SuDS can involve local communities. In fact, community SuDS management is likely to be one of the most straightforward ways of getting people involved in their local environment. With good design and an effective participation strategy, as well as expert ecological guidance, SuDS can readily become a focus for community life, where people are willing to get involved with local activities; for example, the retrofitting and creation of rain gardens (www.raingardens.info), seeding of community meadows or planting of wetlands.

Appropriate management of SuDS can provide many opportunities for learning, informal recreation, supported play and other community programmes. This has many social and health benefits and gives people a sense of pride, responsibility and ownership of their environment. Active interpretation, volunteering opportunities, guided walks and other forms of engagement provide ways in which people can become involved in decision-making and management of SuDS. This in turn can engender public support for SuDS, leading to increased awareness of wetlands and the natural environment.

Key principles include:

- involve local communities in 'masterplanning' their SuDS environment at the earliest stage
- involve them in the detailed design and management of SuDS
- establish amenity and biodiversity as high priorities in all SuDS, both new and retrofit schemes
- allocate adequate resources for design and long-term management; develop a SuDS management plan with the local community
- consider establishing community management of SuDS and related wildlife habitats.

7.5   Designing SuDS for People and Wildlife

7.5.1   The SuDS Management Train

The SuDS management train is the fundamental principle underpinning all SuDS design. It comprises a series of stages in a journey starting when rain falls onto a roof or other hard surface and then flows to its destination, normally a wetland, stream, river or aquifer. SuDS seek to mimic natural hydrological processes in order to incrementally reduce pollution, flow rates and volumes.

As rainfall flows from hard surfaces, it carries with it silt-size particles, organic debris and pollution. The most important component of this runoff is the silt, to which pollutants adhere. The management train aims to use enough treatment stages to clean runoff and improve water quality as it moves downstream. SuDS features, such as green roofs and permeable paving, trap polluted material at the beginning of the sequence, allowing natural biological and chemical processes in water, plants and soil to deal with it, a process called bioremediation. Water is therefore cleaned, ready for discharge to the receiving watercourse and able to support wildlife.

7.5.2   Designing for People and Wildlife

SuDS are constructed features designed to deal with polluted water, and so careful consideration of how to maintain a ready supply of high quality water is essential. Adhering to some basic principles will allow the creation of a functioning system containing a network of habitats linked to the wider ecological context that benefits people and wildlife. It should also be borne in mind that depending on geographical location, soils and design, SuDS will exhibit a range of hydrological conditions that will impact on planting plans and choices. Some SuDS features will be dry most of the time and only inundated during and immediately after rainfall (e.g. rain gardens, swales).

7.5.2.1 Design Principles

- Sufficient treatment stages are included, appropriate to the strength and volume of the influent; this is essential if SuDS are to generate sufficient high quality water for amenity and ecological uses.
- Existing biodiversity or other designations are understood and taken into account in the design.
- Appropriate biological surveys are undertaken to support the design process.
- Existing habitats are retained, such as hedgerows and ponds, and incorporated into landscape planning and management.
- Natural re-colonisation of locally occurring native species is encouraged by locating new SuDS close to semi-natural habitats, wherever possible.
- A mosaic of habitats is created with a variety of eco-hydrological conditions (e.g. create a number of connected, smaller detention basins that vary in depth and design) rather than fewer, larger features.
- Shallow(-sided) temporarily inundated habitats can be richer in wildlife than permanent ponds; a mix of temporary and permanently wet habitats will maximise habitat provision.
- Aquatic habitats, include areas of wet grassland and wet scrub/woodland (a scarce habitat), as well as drier habitats (such as wildflower meadows) where space allows.
- Use design and planting plans to ensure a diversity of light and shade around pools.
- Consideration should be given to how the site can be managed sustainably; small areas may be managed by local authority personnel, or even a local community group, and larger sites may even be grazed, which can often be more cost-effective and certainly will produce a more natural outcome.
- SuDS must slow and contain surface water flows, to facilitate gradual infiltration into the soil, allow bioremediation of pollutants and provide safe storage in extreme rainfall events.
- A controlled flow of clean water is critical for the development of SuDS with high wildlife and amenity value; poor water quality seriously reduces the likelihood of creating valuable wildlife habitats.
- Maximising the benefits of SuDS for wildlife and people will require expert input from ecologists, planners and landscape architects to generate an appropriate design; professional advice should be sought.
- Stay legal – legally protected species may be present in SuDS habitats, so expert advice should be sought.
- A management plan should be created for the site, which protects and enhances all wildlife including legally protected species (e.g. bats, birds in the breeding season, water voles, great crested newts) and integrates the views and needs of local people.

7.5.2.2 Other Key Points to Remember

- SuDS should be designed to facilitate easy maintenance and include features to contain and manage accidental spillages of contaminants.
- Direct pipe connections should be avoided beyond the first treatment stage wherever possible.
- Management plans are essential for the delivery of wildlife and people benefits.

- Training and supervision of contractors and other practical staff involved in SuDS management is essential.
- Where a supply of clean water is not guaranteed, existing wetlands *should not* be incorporated into SuDS. For example, while it might be desirable to channel clean roof water into a wildlife pond without prior treatment, using untreated road runoff would not be appropriate.

7.5.2.3 Planting Recommendations/Principles

- Designs and planting plans for new SuDS should meet local and national biodiversity conservation objectives.
- Ensure that plants are native and of local provenance and suited to local soils and hydrology.
- Never introduce invasive non-native species, such as water fern (*Azolla filiculoides*) or floating pennywort (*Hydrocotyle ranunculoides*).
- Seek expert advice if required.

7.6   SuDS Management Trains and Their Wildlife Benefits

7.6.1   Source Control

Managing rainfall at source (the point at which it reaches the ground or a building) is a fundamental SuDS concept. It reduces the chances of silt and contaminants entering the system and controls the flow and quality of water for use further downstream. A supply of clean water is the first requirement for generating wildlife and people benefits, and all SuDS should provide this; containing the water at – or close to – the source will provide this.

Green roofs need suitable roofs to be readily available to host them, but these roofs can be critically important in urban areas where wildlife habitat is limited (they are a valuable stepping stone habitat. They intercept rainfall and are particularly effective in short duration, intense summer downpours. Water is held in the soil and taken up by plants and so runoff rates are reduced. They trap air-borne pollutants and provide high quality invertebrate habitat and foraging areas for birds (Figure 7.2).

Rain gardens are shallow, free-draining depressions, located in the private or public realm, receive rainfall via downpipes or impermeable areas (although not car parks). They should be planted with species that tolerate brief inundation but have high wildlife value. They provide invertebrate habitat and foraging for birds (Figure 7.3).

Filter strips are broad, flat or gently sloping areas of vegetated land that directly intercept rainfall or receive it from an adjacent impermeable area (e.g. car park), normally as overland sheet flow. They intercept silt and other pollutants, passing on a cleaner supply of water for downstream features. Although these features may be turfed with amenity grasses they can be used to create wildflower-rich meadows of great value to invertebrates, reptiles and amphibians. Tussocky grasslands or even scrub add wildlife value, while maintaining SuDS functions.

Bioretention areas are formal or informal landscaped shallow depressions that receive polluted runoff from roads and car parks (Figure 7.4). Plants and soils bioremediate these

Figure 7.2   Green roof at Coventry University.

Figure 7.3   Plantings in the Northampton rain gardens (John Brewington STW).

Figure 7.4   Formal engineered bioretention area, N. Hamilton, Leicestershire.

pollutants allowing water to infiltrate to ground, evapotranspire and/or move downstream. They provide invertebrate habitat, as well as foraging for birds, and can be used as traffic-calming measure to improve the local street scene.

7.6.2   Site Control

Site control describes SuDS features in or at the edge of developments, which provide a second or third treatment stage, including storage for runoff that has been conveyed from source control structures, such as from a green roof through a water butt to a rain garden). They are located downstream of source control features.

Detention basins are vegetated depressions that only hold water temporarily, allowing it to flow slowly downstream to another SuDS feature and/or permitting infiltration to the ground (Figure 7.5). Detaining flows also allows removal of pollutants through bioremediation. Where water quality is good and habitat is diverse (wildflower grasslands, temporary pools, scrub, wet woodland) these features can benefit invertebrates, amphibians and birds. These areas are of great amenity benefit and are often multi-functional, providing informal or formal recreational opportunities.

7.6.3   Regional Control

This provides the final water treatment stage before discharge to a receiving watercourse and into the wider catchment. When storage of runoff cannot easily be accommodated within the development, it may be possible to convey these excess volumes out of the development itself into public open space. Here, there is great potential for maximising both wildlife and amenity benefits.

Regional control features are detention basins (using natural or created shallow basins to temporarily store large volumes of clean water – Figure 7.5), permanent open water

Figure 7.5 Detention basin, known as 'The Dip', in Coventry, UK. Local children use it as a football field and general play area.

(retention basins, including lakes) and associated other wetlands, including seasonally flooded woodland and grassland habitats, wet fens, reed-beds and marshes. The extended retention period constitutes a final 'polishing' of water quality before release into the wider catchment. They link smaller SuDS features and existing areas of habitat with the wider landscape. Regional controls are generally larger and feature a mix of habitats, including permanent water; these can therefore benefit a greater number of species, including plants, invertebrates, reptiles, amphibians, fish, bats and other mammals. Conservation objectives should address local and regional species and habitats of concern.

7.6.4   Conveyance Features

Swales are the most common vegetated conveyance features used in SuDS and bring ecological and amenity benefits to a site. Water may also be conveyed in a number of other interesting ways using hard landscape features (Figure 7.6), which in turn may also be enhanced with appropriate planting.

They are wide shallow grassed features that slow down runoff, trap sediments and allow some infiltration. They may also contain small check dams to hold water back in a series of shallow pools, offering potential for wetland plants to colonise (Figure 7.7). They can be under-drained where a dry surface is needed, providing additional filtering in the under-drain, or become permanently wet to create a linear wetland habitat rich in plants and invertebrates (although care should be taken that the conveyance function is not jeopardised, as surface flows must be able to pass downstream at a controlled rate).

Figure 7.6   (left) Engineered feature leading down to (right) small wetland area, swale and filter strips, N. Hamilton, Leicestershire, UK.

Figure 7.7   (left) Swale with leaky stone dam creating downstream wetland area, Oxfordshire, UK. (right) Wood and metal dam across a swale, N. Hamilton, Leicestershire.

Urban design uses many forms of architectural channels, cascades, rills and canals to convey water in the landscape. Where silt and major pollutant loads have been removed, they provide visual interest, are easily understood by observers and are easily maintained features. Although somewhat architectural in nature, there are still many opportunities to provide valuable urban wildlife habitat.

7.7   Community Managed and Wildlife-Rich SuDS – a Case Study of Springhill Cohousing, Stroud, Gloucestershire

7.7.1   Overview

A cohousing company developed the site to provide environmentally friendly housing in a supported community, centred on a community house and shared social space. The SuDS were developed with local community stakeholders participating fully with landscape designers, ecologists and planners in the design of the system. They also manage the system themselves and to date there have been no instances of flooding in the SuDS catchment. Anecdotal evidence suggests that flooding of the community downstream has been reduced. Wildlife value is high with SuDS habitats throughout the development bringing people closer to wildlife at every stage. Further details of this development are hosted on the Susdrain website at: http://tinyurl.com/h9yywer or the designer's website at: http://robertbrayassociates.co.uk/projects/springhill-cohousing/(see also The Architecture Centre, 2010 and Graham *et al.*, 2012).

7.7.2   SuDS Design

Surface SuDS features used on site include:

- permeable pavement
- short under-drained swale
- surface cascade
- planted grass swales
- open channels and rills (in residents' front gardens)
- raised ornamental pool (rich in wildlife, and a communal meeting place)
- grassed detention basin (also used as a children's play area).

Flows run overland from the upper terrace, down to the lower level and along the pedestrian street to an outfall, where a natural spring emerges at the SE corner of the site. The access and car parking court was identified as a primary pollution risk. Permeable paving collects and stores runoff underneath in a tank. Water leaving the tank is joined by un-attenuated roof runoff that flows to the lower level down a tile-hung cascade on the retaining wall. A swale allows most of the cleaned runoff to soak into the ground with excess flows conveyed to a pool in front of the community house. Runoff from the tarmac road surface and adjacent roofs flow to the rill that follows most of the lower side of the pedestrian street. Additional overflows from the rill and the pond are directed to a detention 'play basin', which is used for recreation and play most of the time but has contained up to 300 mm depth of stormwater during and immediately after heavy rainfall.

7.7.3   Management

To facilitate better care and maintenance, wherever possible, SuDS features should deal with water on the surface. The community maintain all surface features and there have been no reports of system failure. Indeed, when major flooding occurred nearby, no impacts were felt at this site with approximately 150 mm of water stored safely in the final detention/'play' basin.

7.7.4   Amenity Value and Use

Use of permeable paving and integrated underground storage demonstrates full use of the space to collect, clean, store and release controlled flows of clean water for amenity and biodiversity. The surface flow of water through the site begins with a T-piece terracotta pipe inlet to a tile-hung cascade – a cost-effective and visually spectacular alternative to the traditional drop manhole. A short, planted swale and channel links to a raised pool in front of the community building, contributing visually to the social space. The rill system provides both a collection and flow route for water, and separates private and public space. Each householder understands how it works and takes responsibility for managing their section. The community understands that occasionally the play/detention basin will hold water after heavy rainfall but that it will again be available as a public play area within a short period.

7.7.5   Biodiversity Value

Clean water is the essential ingredient for aquatic biodiversity, and this is assured at Springhill by source control treatment of the main risk areas, including a permeable pavement in the upper car court, and a clear management train. The vegetated swales, rill and channels provide connectivity of habitats for wildlife, and the pond is of significant value, as well as being a great feature of interest. The community is careful to enhance biodiversity and maintain the SuDS in a relaxed way to allow maximum opportunities for wildlife on this relatively dense urban development. Species found in the SuDS include frogs, newts, dragonflies, damselflies, other aquatic invertebrates and a range of native wetland plants and birds.

References

Graham, A., Day, J., Bray, R. and Mackenzie, S. (2012) *Sustainable Drainage Systems – Maximising the potential for people and wildlife: A guide for local authorities and developers.* WWT and RSPB. Available at: http://tinyurl.com/hrp8hsb

The Architecture Centre (2010) *Springhill Co-housing Stroud, Gloucestershire.* Available at: http://www.architecturecentre.co.uk/assets/files/case-studies/Springhill_March_2010.pdf

Amenity: Delivering Value for Society

Stella Apostolaki and Alison Duffy

'A thing is right when it tends to preserve the integrity, stability and beauty of the biotic community. It is wrong when it tends otherwise.'

Aldo Leopold, A Sand County Almanac, 1968

8.1   Emergence of the Amenity Concept

Today's landscape architecture, a contemporary art according to many (e.g. Rendell, 2006), embraces various trends, bringing together current ideas of design, development and infrastructure. When new landscapes are designed, several functional societal needs have to be met, namely ecological, historical, emotional and visual. Open urban spaces have a role in transforming the aesthetics of a culture by establishing strong links between ecology, beauty, culture, geography and local topography. Designers now include materials that express the natural landscape of the region and try to restore natural function to the environment. With water being one of the fundamental elements and values for life, rainwater management is evolving and moving towards the replacement of traditional forms of engineering in drainage, with innovative stormwater management systems, SuDS, that clearly address the amenity element.

Current trends and efforts, aimed at utilising natural functions and promoting the amenity value of developments, are also at the heart of the sustainability concept with the replacement of damaging technologies by more beneficial ones (Johnson, 1997). Such landscape design and planning also seeks to reduce the impacts of industrialisation, thus minimising energy consumption and the depletion of natural resources (Laurie, 1997).

Amenity in contemporary thinking and design is closely associated with services provided, aesthetics, nature and biodiversity preservation, recreation, leisure and pleasure and attractiveness. Having to recreate natural and sustainable environments is synonymous

Sustainable Surface Water Management: A Handbook for SuDS, First Edition.
Edited by Susanne M. Charlesworth and Colin A. Booth.

with creating an aesthetic that brings beauty to a city. In the natural environment the call for conservation and natural function is a call for aesthetics (Ministry for the Environment of New Zealand, 1991). Even though natural landscapes are not always as clean and tidy as they could be, designers have to find ways of making them beautiful so that people will value and protect them. This ensures that the public will not only accept sustainable solutions in their cities but will also realise their importance and seek to include them in urban design.

For an area to be considered an area of high amenity value, a sense of balance and interaction between people and nature is vital (Taylor, 1998). Sustainable design has to build a strong relationship between nature and urban dwellers. Although highly anthropocentric landscapes continue to exist, the creation of landscapes that promote stewardship for the environment and support environmental consciousness should be encouraged. A way to achieve this goal is to promote local community participation in landscape creation and maintenance (Dalton, 1994).

One of the main issues with amenity landscapes is to produce a living and liveable environment that people like and accept (Taylor, 1998). Perceptions of amenity and ideas of landscape, however, differ based on personal interests and aesthetics, cultural heritage and contemporary trends. To many, wild landscapes have the highest amenity value, while others prefer landscapes to be more controlled and organised. For the latter constructed landscape scenario, often a complete change of perception is required in order to see the beauty of nature in its pure form and with the least possible human intervention. On the other hand, wilderness supporters reject any kind of constructed landscapes as worthless, including ornamental gardens or exotic species introduced into the area, which for many years were used successfully in the quest for nature in urban areas (Jamieson, 2001). The idea of wilderness in support of the preservation of nature and its aesthetics can be considered, to some extent, to be embedded in the idea of amenity. According to the Wilderness Act (1964), an area of wilderness is an area untouched by society and its works, an area retaining its primeval character, affected only by the forces of nature where there is no noticeable human impact. The idea of wilderness can involve anything 'natural' or resembling 'natural' features such as those that have survived as primeval memories in people's minds ever since land was discovered and mapped. Nowadays, the idea of 'natural' is, however, more a matter of what is *perceived* as natural rather than what really is untouched by society. People are influenced by their cultural, educational and ideological background. Nature, while often fragile in practice, is durable in the imagination. The social constructiveness of the term 'natural' is clearly depicted in the example of Niagara Falls which, under the influence of the 'Romantic' movement and since the 1860s, has been repeatedly reconstructed to restore the natural wonder lost by the water diversions used for power and industry (Cronon, 1995) and to regain the element of amenity. Although it is known that Niagara Falls have undergone many changes in order to preserve the notion of wilderness and to attract visitors, for many it has become the epitome of wilderness, a powerful force of nature with a high amenity value. For others, it is a historic landmark, and a source of energy generation (Cronon, 1995).

The *Romanticism* movement, which clearly depicts a movement towards a respect for nature (Jamieson, 2001) and the prioritisation of amenity as an important component of nature preservation in urban landscapes, inspired the planners and architects of urban open spaces in the 19th century. The preference towards ornamental gardens in Britain and the 'garden city movement' are two distinct examples of this trend (Parsons and Schuyler, 2004).

The garden city movement was based on the ideas of Ebenezer Howard (see TCPA (2011) for further details) and reflected a new form of 'townscape', which was intended to bring together the economic and cultural advantages of both city and country living, with land ownership vested in the community, while at the same time discouraging metropolitan sprawl and industrial centralisation. 'Garden cities' were built in a circular fashion around a central park with a prominent presence of water, while green areas would be integrated into residential neighbourhoods (Hall and Ward, 1998). Howard's ideas had a phenomenal impact on British planning doctrine with its most impressive application being the plan for Greater London in 1944 and the creation of a ring of new towns, the London Greenbelt, following the passing of the New Towns Act, 1946. Loch Katrine in Glasgow, Scotland, is another example of a 19th century public works initiative, which has gradually been developed into an amenity feature for nature lovers and romantics. The water supply works at the Loch have always prioritised nature preservation and amenity value by gradually restoring natural woodlands through tree planting and the introduction of embankments to minimise noise and visual impact (Water Technology Net, 2006; Scottish Water, 2008).

8.2   Amenity, Recreation and Biodiversity in the Built Environment

8.2.1   Need for Public Open Spaces in Cities

Modern urban landscapes are capable of addressing natural processes, and at the same time are able to provide wildlife habitat, stormwater retention benefits and water quality improvements (Bradshaw and Chadwick, 1980). This approach can create beautiful and liveable environments in urban centres, places that possess a sense of pride, place, history, safety, good housing, friendly parks and open spaces. Urban landscapes can therefore be a collective of expectations, responses and remembrances (Johnson, 1997).

In recent years, ideas of green networks, protecting or enhancing natural resources, and creating open urban spaces of high recreational value are gaining ground in architectural design. Urban forestry and community forests – for example the Chicago Urban Forest Initiative (Nowak *et al.*, 2009) – as well as providing improvements to urban landscapes are becoming more common in urban development and are making significant contributions to the practical application of sustainable development (Blowers, 1993). Current concerns regarding environmental degradation has influenced this perception of urban design. Planning for sustainable development includes land use that enhances landscapes and protects natural environments and wildlife habitats. It is therefore necessary to view environmental planning as an integrated process moving around the idea of sustainability. A great number of people are drawn to parks and open green spaces due to an interest in sustaining ecosystem integrity and also to admire beautiful landscapes.

Currently, landscape architectural practices and ecological protection are promoted and safeguarded in 'urban green spaces', while ecological restoration has a central role in urban landscape design. In that context, the development of the 'new model village', a popular architectural trend among planners, recognises the need for urban citizens to live closer to nature, resulting in planners' trending towards the development of 'sustainable communities' (Kim and Kaplan, 2004). The new model village is usually placed in suburban areas and attempts to recreate rural scenarios, with the inclusion of small ponds (often associated with SuDS) or streams with rich tree and flower vegetation; this can be linked to the garden city movement of the 19th century (Parsons and Daniel, 2002). Such developments can serve the practical needs of a community, such as drainage and provision of space for recreation.

8.2.2   Urban Open Space Design

The new approach to city design, which incorporates green open spaces in the urban environment, proposes an ethic of nature that pays special attention to moral concepts such as respect, sympathy, care, concern, compassion, gratitude, friendship and responsibility (Van De Veer, 1998; Jamieson, 2001). Based on these ideas, the city provides the source of energy for the world around it, with people facilitating the change from natural resources to fabricated product. Traditional and contemporary ethical theories could also be extended to include animals, plants and inanimate natural features (Gunn and Vesilind, 1986). The modern citizen shops for 'natural products' in the artificially 'natural-looking' shopping mall or open market, while walking in 'nature' is synonymous with urban or suburban parks, with ponds, streams or rivers, and existing vegetation and wildlife, which are often alien to the area and could by no means be considered as 'natural' (Van De Veer, 1998).

To enhance public acceptability of new artificial open water features, including different types of drainage devices, such as ponds or swales in the urban environment, an attempt to make them resemble natural landscapes is vital. The concept of sustainability can include these features as amenity lakes with great importance for wildlife. For SuDS approaches, including the 'daylighting' or opening up of urban rivers, the presence of vegetation to support wildlife is fundamental, although there is a balance to be made between functionality and amenity. The modification of water bodies according to topographical variation and characteristics of the area is recommended as well as the creation of islands in these features, and the support of natural treatment processes for biodegradation. The introduction of plant species, for example in association with reed-beds, can biodegrade organic matter, utilise hydrocarbons as a nutrient source, and trap polluted particulates, thus providing biological water treatment (see Chapter 9 of this volume; Heal, 2014).

Wetland vegetation and wildlife are crucial for urban water body design and establishment in urban areas. The notion that native plants encourage wildlife and are more easily established than exotic ones is shared by most landscape architects (Watkins and Charlesworth, 2014). However, the focus should be on places where human control over ecosystems is minimal and where plants and wildlife (both native and exotic), once introduced, are free to interact.

The human impact on urban landscapes is overwhelming, a fact that results in the formation of specific plant and animal communities. The majority of the plant and animal species present in urban landscapes are alien, and are often ignored or undervalued by many ecologists and landscape architects. In fact, it has been estimated that 60–70% of urban vegetation has been deliberately introduced, with landscape styles varying from semi-natural to exotic (Fitzgerald, 2003; Forbes and Kendle, 2013). However unnatural the urban flora and fauna may be, it is usually welcomed by local residents, who lack the prejudices of professional ecologists, and are open and often enthusiastic in accepting colourful exotic plants and attractive wildlife species in their residential area. The climatic and geologic conditions of the site play very important roles in the establishment and support of alien or 'exotic' species in urban open spaces.

Forbes and Kendle (2013) describe the characteristics of naturalistic or ecological styles of landscape design as: low-cost and sustainable; of high intrinsic value for the area; in contrast to ornamental design style; requiring minimum maintenance; having high conservation, educational and recreational value.

A very sensitive point for naturalistic landscape architecture is the fact that most artificial urban landscapes are modelled on relatively immature communities, whereas landscape development should be considered in terms of good long-term management: SuDS in new housing developments and/or river restoration schemes. One of the main advantages of such artificial landscapes, in comparison to natural ones, is that the possibilities and opportunities for enhancing human contact with nature can be maximised through proper design and appropriate locations where there is easy public access and social interaction. It is generally known that people prefer to access countryside close to where they live for recreation and wish such 'beauty spots' to be protected (Cullingworth and Nadin, 2006). However, public attitudes towards landscapes and green open space design differ between countries. For example, in Britain people appreciate landscapes of panoramic view, hills, woods, lakes, while in Japan people take pleasure in noticing the details of trees and flowers (Anderson and Meaton, 2000).

8.2.3   Multi-Functionality of Urban Open Space and SuDS

A general attitude towards public landscapes is that people from various cultures prefer natural to built environments or those where the human impact is very obvious. The stress-releasing ability of natural landscapes supports evolutionary theories on landscape preferences (Ulrich, 1993) which are influenced by several parameters such as age (Balling and Falk, 1982; Lyons, 1983; Zube *et al.*, 1983), educational level and occupational interests (Yu, 1995).

The reasons why people express a preference for specific landscapes and wish to protect them varies; landscapes can serve human needs and can also have high ecosystem value. In combination, these two factors can urge people to protect the landscape or can influence their preference towards those that are more natural-looking (Kaltenborn and Bjerke, 2002).

The regeneration of ecologically degraded urban landscapes into more natural-looking areas also serves a number of functions:

■ conceptual – establishes a link between urbanism and nature
■ cultural – heritage plays a very important role in landscape design and in the way people value places
■ ecological – reintroduces nature into the urban landscapes
■ social – promotes social interaction, with urban parks being the places where democracy is worked out 'on the ground'
■ psychological – improves quality of life and provides a relaxed environment with a direct and positive impact on people's wellbeing
■ aesthetic – creates pleasant and beautiful environments that are highly valued by the public (Tress *et al.*, 2001).

The psychological function served by contact with nature and amenity features in an urban setting is underlined by Wells and Evans (2003), whereby symptoms of psychological distress were significantly reduced in children with higher exposure to amenity features. Groenewegen *et al.* (2006) referred to this access to green space having a positive effect on human wellbeing such as stress reduction, as well as better mental and physical health as being the effects of 'vitamin G'.

There has been an extensive debate between 'scenic' and 'ecological' aesthetics in landscape design, with ecological aesthetics becoming more and more important for landscape planners. According to ecological aesthetics, ecosystem principles such as biodiversity and sustainability are the main values that have to be taken into account when designing new urban landscapes. In the case of scenic aesthetics, the main importance is given to aesthetic

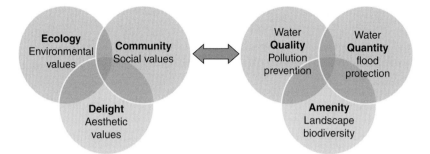

Figure 8.1 Conceptual overlaps between the three value areas: ecology, community, delight (after Thompson, 2002); and quality, quantity, amenity of SuDS (after Charlesworth, 2010).

preferences undermining the importance of ecological benefits, which for many results in the creation of superficial landscapes with no concern for wildlife. For the advocates of ecological aesthetics, scenic aesthetics are morally inferior (Parsons and Daniel, 2002).

The notion of ecological landscapes, however, is not just influenced by aesthetics and the senses but also by indirect knowledge of ecosystem health and sustainability. The conflict between scenic romantic landscapes and ecologically raw and unrestrained ones has always been an issue for human society (Appleton, 1975).

There is a congruent split between 'urbanists' and 'ruralists', but then aesthetic and amenity issues tend to be overshadowed by larger environmental concerns. Beauty and use are now linked to the idea of sustainability. The ecological approach that became popular in the 1970s and 1980s and claims that humans would benefit in psychological and social terms from contact with naturalistic landscapes of rich biodiversity, is becoming increasingly popular, especially in the concept of sustainability. This trend is clearly depicted in Figure 8.1, which demonstrates the interrelationship between environmental, social and aesthetic values, and is in accordance with the SuDS triangle interlinking the three sustainability components of: quality (environmental component, e.g. pollution control), quantity (economic component, e.g. flood mitigation) and amenity (social component, e.g. leisure and wildlife benefits) of sustainable water features.

This association clearly underlines the multi-functional role of SuDS in providing:

1. environmental benefits through vegetation supported in and around them
2. the attraction of wildlife
3. community benefits such as the provision of areas for gathering, socialisation and recreational use of the surrounding area
4. aesthetic benefits for the area and the local residents through the provision of natural-looking environments, which are generally highly appreciated by the public.

8.3   SuDS Amenity and Sustainable Development

Sustainable development aims to provide a holistic approach for all interrelated sectors of life and wellbeing. It pursues certain goals including protection of natural resources, a sustainable built environment, social equity between locations and generations and political participation to change values, attitudes and behaviours (Blowers, 1993).

Along the lines of sustainability, greening of policy is essential to reflect public attitudes and the public need for information and participation. Public education and community involvement are essential ingredients in any environmental strategy. For example, SuDS ponds and river restoration are both potentially sustainable, as they provide obvious environmental and social benefits, while their high self-maintenance ability is a component of economic efficiency. Although there is a big debate over the sustainability element of SuDS (e.g. Heal *et al.*, 2004), there is general agreement that the application of SuDS, even though it may not be entirely sustainable, is still a more 'sustainable' approach, an approach that encompasses the concept and elements of sustainable development more than traditional drainage. SuDS are considered as beneficial for the environment and society, while at the same time they are cost efficient, with construction costs comparable to or less than those of traditional drainage, reduced operational costs and with minimal cost of maintenance (Gordon-Walker *et al.*, 2007; Bray, 2015).

SuDS provide stormwater management to attenuate flows and thereby reduce the risk of flooding, while at the same time they serve the environmental, biodiversity and amenity needs of the city and urban citizens. SuDS are often considered for drainage solutions in new developments and represent part of a new trend in landscape architecture and urban design (Watkins and Charlesworth, 2014).

8.4   Reviewing the Public Perception of the Concept of Amenity and SuDS

In considering amenity, the major difficulty in its study is how it can be measured. Amenity encompasses ideas related to greening urban landscapes, returning to nature, providing useful or pleasant services to the public, encouraging leisure activities, social interaction and democracy (Woods Ballard *et al.*, 2015). However, their assessment is problematic; the main difficulty is in the classification of individual green spaces or SuDS device in terms of the 'quality' of their amenity value. One way to assess amenity is via regular inspections to identify the performance of SuDS in matters of maintenance, aesthetics, biodiversity present and use by the public. The results of these types of inspections, however, are not objective as they are often subject to individual bias, the budget available for maintenance and the effort made by the authority in charge. Another way to identify the amenity of SuDS is by collecting and analysing public views, rendering assessment of amenity more qualitative rather than quantitative.

Public perception surveys of SuDS ponds and wetlands in the UK (Apostolaki *et al.*, 2001, 2002) identified the need for the design of urban water bodies in the landscape to have amenity in mind. It was further found that where open water schemes and natural-looking sites were designed to enhance recreation and leisure, their presence in the landscape was highly appreciated by public and professionals alike. Specifically, participants in these surveys indicated their preference for SuDS of high aesthetic value that attracted biodiversity. They linked amenity of aesthetically pleasing SuDS with naturalness and attractiveness of the site, provision of recreation and leisure services and improvement in terms of everyday life via flood protection. In accordance with a survey conducted in Athens (Greece), the term 'amenity' seemed to include ideas such as visual and habitat enhancement; wildlife and biodiversity benefits; provision of an urban park environment for recreation, relaxation and leisure; stress relieving service and other social services; educational benefits; and high aesthetic value. In well-designed ponds that incorporated amenity, even safety concerns, which are frequently associated with open water bodies located in urban environments, were reduced. On the contrary, safety concerns seemed to be higher in sites of reduced perceived amenity value (Apostolaki and Jefferies, 2009). A follow-up survey, which took place in the Dunfermline Eastern Expansion, found similar results, but it demonstrated slightly reduced safety concerns as the SuDS ponds were better established and their perceived amenity value was higher (Shan Quek, 2010). Perceptions of SuDS seem to depend on where the balance of the argument is, flood prevention, amenity or water quality improvements. Design of SuDS is therefore seen as a key component when influencing attitudes towards amenity. This can be explained by the idea that amenity in landscape design is often perceived by the public to reduce the impacts of urbanisation in the city, introducing sustainable design, replacing damaging technologies with more beneficial and sustainable ones. In this way, it is possible to minimise the consumption of natural resources and maximise environmental conservation (Johnson, 1997; Laurie, 1997).

8.5   Conclusions

The transformation of SuDS sites into amenity and recreational features is increasingly important for local communities. The current trend is towards 'new model villages' or 'sustainable communities' and housing developments that are located around a pond or close to other types of watercourses which have high amenity value.

The need for preservation of the amenity value of SuDS, which is also closely linked to biodiversity benefits, is highly valued by the public. Amenity, associated with sustainability, encompasses societal and environmental benefits, while schemes of high amenity value can reduce the loss of natural functions in new construction, and improve urban design.

One concern that arises when referring to the design of SuDS is how *amenity* itself can be better designed. The public perception surveys and research undertaken identified several improvements for SuDS in this regard. Most of them are related to improvement of aesthetics and naturalness of the systems including: increased vegetation, introduction and support of existing wildlife, introduction of natural barriers as safety measures, introduction of features for leisure pursuits such as benches, picnic tables, children playgrounds, walkways and pathways that could link the SuDS features, introduction of green roofs and gardens on buildings, designing for safety, designing for social integrity and the promotion of municipal awareness raising and education. Most importantly it should not be forgotten that designing for amenity is *designing for humans*.

References

Anderson, M. and Meaton, J. (2000) Confrontation of consensus? A new methodology for public participation in land use planning. In: Miller D. and de Roo G. *Resolving Urban Environmental and Spatial Conflicts*. Groningen, GeoPress.

Apostolaki, S. and Jefferies, C. (2009) The social dimension of stormwater management practices in urban areas – Application of Sustainable Urban Drainage Systems (SuDS) and River Management options. VDM Verlag Dr. Müller e.K. ISBN 978-3-639-17692-6, Paperback, 224 pp.

Apostolaki, S., Jefferies, C., Smith, M. and Woods Ballard, B. (2002) Social acceptability of Sustainable Urban Drainage Systems. Fifth Symposium of the International Urban Planning and Environmental Association on *Creating Sustainable Urban Environments: Future Forms for City Living*, Christ Church, Oxford.

Apostolaki, S., Jefferies, C. and Souter, N. (2001) Assessing the public perception of SuDS at two locations in Eastern Scotland. *First National Conference on SuDS*, Coventry University, June 2001.

Appleton, I. (1975) *The experience of landscape*. John Wiley and Sons, New York.

Balling, J.D. and Falk, J.H. (1982) Development of visual preference for natural environments. *Environment and Behavior*. 14, 1, 5–28.

Blowers, A. (1993) '*Planning for a sustainable environment: A report by the Town and County Planning Association*', London, Earthscan.

Bradshaw, A.D. and Chadwick, M.J. (1980) *The Restoration of Land: The Ecology and Reclamation of Derelict and Degraded Land*. Oxford, Blackwell Scientific Publications.

Bray, B. (2015) SuDS behaving passively – designing for nominal maintenance. *SuDSnet International Conference*, Coventry University, UK. http://tinyurl.com/hcfcezd

Charlesworth, S. (2010) A review of the adaptation and mitigation of Global Climate Change using Sustainable Drainage in cities. *J. Water and Climate Change*. 1, 3. 165–180.

Cronon, W. (1995) *Uncommon Ground – Toward Reinventing Nature*. New York, W.W. Norton Company Ltd.

Cullingworth, B. and Nadin, V. (2006) *Town and Country Planning in the UK*. Routledge. 624 pp.

Dalton, R.J. (1994) *The Green Rainbow – Environmental Groups in Western Europe*', New Haven and London, Yale University Press.

Fitzgerald, F. (2003) Plant and Animal Communities in Urban Green Spaces. *Design Center for American Urban Landscape Design Brief, No 5*. http://www.china-up.com:8080/international/case/case/287.pdf

Forbes, S. and Kendle, T. (2013) *Urban Nature Conservation: Landscape Management in the Urban Countryside*. Taylor and Francis, Architecture, 368 pp.

Gordon-Walker, S., Harle, T. and Naismith, I. (2007) *Cost-benefit of SuDS retrofit in urban areas*. Science Report – SC060024. Environment Agency.

Groenewegen, P.P., van den Berg, A.E., de Vries, S. and Verheij, R.A. (2006) Vitamin G: effects of green space on health, wellbeing, and social safety. *BMC Public Health* 2006, 6:149. http://tinyurl.com/mlw9vbg

Gunn, A.S. and Vesilind, P.A. (1986) *Environmental ethics for engineers*. Lewis Publishers – Business and Economics – 153 pp.

Hall, P. and Ward, C. (1998) *Sociable Cities: the Legacy of Ebenezer Howard*. Wiley, Chichester, 229 pp.

Heal, K. (2014) Constructed wetlands for wastewater management. In: Booth, C. and Charlesworth S.M. (eds) (2014) *Water Resources in the Built Environment – Management Issues and Solutions*. Wiley Blackwell Publishing.

Heal, K., McLean, N. and D'Arcy, B. (2004) SuDS and Sustainability. *Proceedings 26th Meeting of the Standing Conf. on Stormwater Source Control*, Dunfermline, September 2004. At: http://tinyurl.com/or6ww2j

Jamieson, D. (2001) *A companion to Environmental Philosophy*. Blackwell Publishers, Oxford.

Johnson, M. (1997) Ecology and the urban aesthetic. In: Thompson G.F. and Steiner F.R. *Ecological Design and Planning*. Canada, Wiley and Sons Inc.

Kaltenborn, B.T. and Bjerke, T. (2002) Associations between environmental value orientations and landscape preferences. *Landscape and Urban Planning*, 59, 1, 1–11.

Kim, J. and Kaplan, R. (2004) Physical and psychological factors in sense of community: new urbanist Kentlands and nearby Orchard Village. *Environment and Behavior*. 36, 3, 313–340.

Laurie, M. (1997) Landscape architecture and the changing city. In: Thompson G.F. and Steiner F.R. *Ecological Design and Planning*. Canada, Wiley and Sons Inc.

Leopold, A. (1968) The Land Ethic. In: *A Sand County Almanac, Essay on Conservation from Round River*. New York, Oxford University Press. 286 pp.

Lyons, E. (1983) Demographic correlates of landscape preference. *Environmental Behaviour*, 15, 4, 487–511.

Ministry of the Environment of New Zealand. (1991) '*Resource Management Act*'. http://www.legislation.govt.nz/act/public/1991/0069/latest/DLM230265.html

Nowak, D.J., Hoehn III, R.E., Crane, D.E., Stevens, J.C. and Fisher, S.L. (2009) Assessing Urban Forests, Effects and Values: Chicago's Urban Forest. Northern Research Station, Resource Bulletin NRS-37 *United States Department of Agriculture Forest Service*. At: http://tinyurl.com/pqranyu

Parsons, R. and Daniel, T.C. (2002) Good looking: in defence of scenic landscape aesthetics. *Landscape and Urban Planning*, 60, 1, 43–56.

Parsons, K.C. and Schuyler, D. (2004) From Garden City to Green City. The Legacy of Ebenezer Howard. *Utopian Studies*. 15, 2, 265–270.

Rendell, J. (2006) *Art and Architecture: A Place Between*. I.B. Tauris, London. 212pp.

Scottish Water (2008) *Katrine Water Project*. Water Treatment and Supply. http://tinyurl.com/zuwvxk2

Shan Quek, B. (2010) *Public perception and amenity of SuDS retention ponds: a case study at Dunfermline Eastern Expansion area, Fife, Scotland*. Unpublished MSc thesis, University of Abertay.

Taylor, N. (1998) *Urban planning theory since 1945*. Sage Publications Ltd, London.

Thompson, I.H. (2002) Ecology, community and delight: a trivalent approach to landscape education. *Landscape and Urban Planning*. 60, 81–93.

Town and Country Planning Association (TCPA) (2011) *Re-imagining garden cities for the 21st century: benefits and lessons in bringing forward comprehensively planned new communities*. http://tinyurl.com/qeboobw

Tress, B., Tress, G., Décamps, H. and d'Hauteserre, A.-M. (2001) Bridging human and natural sciences in landscape research. *Landscape and Urban Planning*, 57, 137–141.

Ulrich, R.S. (1993) Biophilia, biophobia, and natural landscapes. In: Kellert, S.R., Wilson, E.O. (Eds) *The Biophilia Hypothesis*. Washington, DC, Island Press.

Van De Veer, D. (1998) *The Environmental Ethics and Policy Book*. 2nd edition, Wadsworth Publishing Company.

Water Technology Net (2006) *Katrine, water treatment project, Glasgow, United Kingdom*. http://www.water-technology.net/projects/katrine/

Watkins, S. and Charlesworth, S.M. (2014) Sustainable Drainage Design. In: *Water Resources in the Built Environment – Management Issues and Solutions*. Booth, C.A. and Charlesworth S.M. (eds) Wiley Blackwell.

Wells, N. and Evans, G. (2003) Nearby Nature: A Buffer of Life Stress among Rural Children. *Environment and Behaviour*. 35, 3, 311–330.

Wilderness Act (1964) *Public Law 88–577* (16 U.S. C. 1131–1136) 88th Congress, Second Session September 3, 1964.

Woods Ballard, B., Wilson, S., Udale-Clarke, H., Illman, S., Scott, T., Ashley, R. and Kellagher, R. (2015) The SuDS Manual. CIRIA, London, 968pp.

Yu, K. (1995) Cultural variations in landscape preference: comparisons among Chinese sub-groups and Western design experts. *Landscape and Urban Planning*. 32, 2, 107–126.

Zube, E.H., Pitt, D.G. and Evans, G.W. (1983) A lifespan developmental study of landscape assessment. *Journal of Environmental Psychology*. 3, 2, 115–128.

9

Biodegradation in Green Infrastructure

Alan P. Newman and Stephen J. Coupe

Alan P. Newman and Stephen J. Coupe

9.1 Introduction

There is a general expectation that a sustainable drainage device will be capable of producing an effluent where, for key pollutants, the quality of that discharged is better than the stormwater entering it. While this is not always the case, if sustainable drainage systems (SuDS) are to fulfil their planned lifetimes and continue to achieve the pollution reduction benefits that are expected, there is the regularly posed question of whether the pollutants in question will accumulate over time. The concern is that they will saturate the system, leading to unacceptable pollutant losses later in the life of the asset, clogging the system (leading to hydraulic failure) or producing conditions in the device that detract from amenity, ecological value or both. For organic pollutants the generally held belief is that the pollutants biodegrade in the system, but in many SuDS systems the role of biodegradation has, to some extent, been assumed rather than studied.

It is perhaps an artificial division to classify SuDS into 'green SuDS' and 'hard SuDS' not least because the best approach to the issue of environmentally sound drainage is to develop a treatment train, where the effluent from one device feeds into another to offer several levels of treatment. In such situations, hard SuDS, such as pervious pavements or filter drains, feed green SuDS such as swales, which together provide initial stormwater treatment, and thus offer some protection to a downstream device such as a wetland or a (normally dry) infiltration basin. This is the principle of source control, treating the bulk of the pollutant as close to the source as possible. Biodegradation occurs both in most green SuDS (balancing ponds, swales, rain gardens etc) and also in some hard SuDS (particularly pervious pavements). While the underlying microbiology and chemistry of biodegradation are essentially identical in green and hard SuDS, there is often a greater opportunity in the latter for the designer to attempt to optimise the biodegradation processes. Hard SuDS are covered in more detail in Chapter 10 of this volume.

Sustainable Surface Water Management: A Handbook for SuDS, First Edition.
Edited by Susanne M. Charlesworth and Colin A. Booth.
© 2017 John Wiley & Sons, Ltd. Published 2017 by John Wiley & Sons, Ltd.

This chapter starts with an introduction to biodegradation processes in general and SuDS in particular. It will conclude with a discussion on the role of biodegradation and nutrient dynamics in green SuDS, the research efforts that have been made in this area and the particular issues that have made the research into biodegradation in green SuDS more challenging than in the more easily controlled hard SuDS environment. Some methodological developments that may be new and innovative when applied to SuDS are discussed, and suggestions made as to how to integrate new research methods into SuDS as a discipline and how to provide further multi-disciplinary links to a research field that is still emerging and developing.

While there is little doubt that most SuDS will contribute to the biodegradation of organic pollutants to some extent, it has been shown that there can be sufficient build-up of petroleum derived hydrocarbons in sediments in the components of SuDS treatment trains, to make the disposal of sludge removed during maintenance problematic (e.g. Durand et al., 2004). This does not mean that biodegradation is not taking place, but it does imply that it is not fast enough to prevent an unacceptable accumulation of pollutants, either during the design life, or within the required maintenance intervals of the system. The extent of this issue varies with the pollutant, the device and the surrounding environment. Some SuDS devices provide conditions that are very amenable to biodegradation. Some have been applied to surface water drainage after initially being developed and validated as important components for tertiary stages of wastewater treatment. The best examples are constructed wetlands that may have the superficial appearance of natural ecosystems but are just as much a product of engineering and design as hard surfaced devices.

9.2   Environmental Conditions and Requirements for Biodegradation

Biodegradation results in the biologically mediated breakdown of pollutants, to simpler materials, and if biodegradation is considered to be complete and irreversible, then the final product should include a gaseous component of a nutrient cycle. Thus, if the pollutant is an organic compound, the aim should ideally be that the products should include carbon dioxide and water. The only other type of pollutant where this process leads to a gaseous by-product is in the transformation of organic or inorganic nitrogen compounds, which can, ultimately, be removed from the system as N_2, but almost always includes less desirable oxides of nitrogen, particularly N_2O, a highly potent greenhouse gas (Richardson et al., 2009). For the purposes of this chapter, such processes are included in the definition of biodegradation. Among these processes, filtration is identified as occurring on the plant stems and leaves in vegetated systems such as swales and grass filter strips, adsorption onto the solid surfaces of the biofilms coating filter media and also precipitation processes. These may involve biologically mediated chemical changes that do not result in elimination of the pollutant via the gas phase. The conditions that are necessary to bring about these changes are often influenced by the environment in the SuDS device, and occur as a result of the biological degradation process taking place in the system. For example, in engineered wetlands, the precipitation of metals as sulphides relies on anaerobic conditions produced by the biodegradation of either trapped organic matter, which might originally be part of the influent pollution load, or consist of dead plant material generated in the wetland. While biodegradation itself must be aimed at degradable substances, these other processes can help remove both inorganic pollutants and organic materials that are not readily amenable to biodegradation. In the latter case, either the biodegradation process leaves behind so-called 'dead end metabolites' (e.g. Andersson and Henrysson, 1996) or, where the pollutant is a mixture of related compounds (e.g. hydrocarbon fractions), persistence of the compounds that are most difficult to degrade.

As mentioned previously, SuDS devices are often organised into a treatment train where the downstream device adds to the effectiveness of the overall pollution attenuation process, but downstream devices often depend on retention upstream to ensure that their, often more delicate, ecological systems are not overloaded. The same can often be applied to single devices which have multiple layers of treatment where entrapment in the upstream parts of the system is used to protect the more biologically active downstream parts. Good examples are the trash filters and oil separators often integrated into the design of stormwater sand filters (USEPA, 2014).

The pollutants in stormwater that challenge the biodegradation capabilities of SuDS systems, range from highly complex compounds and mixtures (e.g. PAHs, lubricating oils and fuels) to some very simple compounds such as methanol and glycol. The contaminants might also include nitrogen-containing compounds ranging from proteins and nucleic acids, through to simple organic compounds, such as urea and uric acid, to ammonia and nitrate or nitrite.

The main requirements for biodegradation are commonly listed as:

- microorganisms (suitably adapted for both the target material and the environment)
- a source of carbon and energy (which may or may not be the target material)
- inorganic nutrients (and sometimes organic micronutrients)
- suitable water relations
- a suitable environment with respect to oxygen (or another electron acceptor).

In studies of biodegradation (Coupe *et al.*, 2003), a suitable substrate, or platform, upon which the microorganisms can grow and make contact with the target materials, is often included. However, a factor that is rarely highlighted sufficiently in the SuDS literature is the need for sufficient contact time between the target material and the microorganisms. This implies that for successful biodegradation, the pollutants themselves will need to be retained in preference to the water. This is easier to achieve where the pollutants form a separate phase and rapid physical processes can form the first step in a biodegradation process, but less so when contaminants are in solution. For example, hydrocarbons arising from motor vehicles, such as fuels and lubricating oils, when present as a separate phase, generally biodegrade quite slowly (having half lives of months rather than days) mainly because of their low availability through limited solubility, but also because of the toxicity of some of the components. This slow degradation rate is often compounded by the fact that the temperature or water availability may severely limit biodegradation for parts of the year.

A biodegrading SuDS system must be capable of accumulating the pollutants throughout the year and, at times, when conditions are optimal, biodegrading the pollutant sufficiently quickly that, within the design life of the system, it does not become overloaded. The fact that 'real-time' biodegradation is less important than the entrapment, storage and biodegradation model for many organic pollutants in SuDS, also points to the need to take into account remobilisation when conditions change. For example, Ellis *et al.* (2003) stress strongly that in constructed wetlands, negative removal efficiencies, including organic pollutants, could turn the wetlands into pollution sources when sediment-associated contaminants are flushed out during periods of intense flow or after prolonged dry periods. This implies that in such systems, the biodegradation rate does not exceed the accumulation rate and that eventually the maintenance that will be required to deal with inorganic deposits, such as silts, will be complicated by the accumulation of undegraded organic pollutants. Designers and operators of all SuDS should take this into account. It must be recognised that biodegradation is often aiming to extend the lifespan of the asset, or the interval between maintenance activities for a SuDS system, rather than allowing complete degradation of the pollutant.

9.3   Biofilms: What They Are, What They Do and How They Work

One of the most important aspects of both biodegradation and detention of pollutants in a green SuDS system is the formation of a biofilm (Figure 9.1). Biofilms are assemblages of either single species or numerous types of microbial population that are attached to surfaces through secretions known as extracellular polymeric substances (EPS). In biodegrading environments there is often a mixture of bacteria, fungi, protozoa (Charlesworth *et al.*, 2012) and, where light can penetrate sufficiently, algae (Singh *et al.*, 2006). The role of the predators of the bacteria and fungi that perform biodegradation, including multi-cellular organisms, either as part of a biofilm or free-living in the system, should not be ignored.

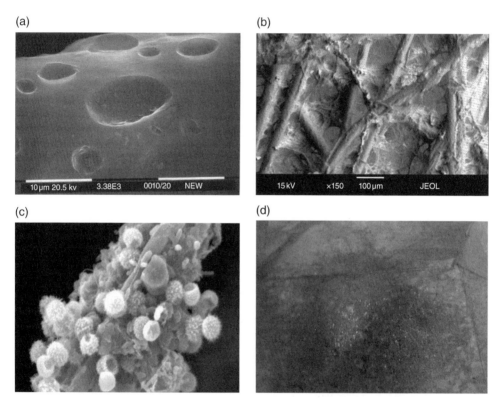

(a)

(b)

(c)

(d)

Figure 9.1   Biofilm development: (a) the surface of a geotextile fibre showing the pits in which biofilm begins to grow (b) vertical geotextile fibres with interwoven biofilm (c) a single geotextile fibre with microorgansims attached (d) sheet of geotextile with biofilm growing on the surface.

The grazing eukaryotes (including protozoa) play an important role in maintaining and controlling the biofilm, keeping water passages open and in recycling processes (Griffiths, 1995).

Singh *et al.* (2006) state that surface-attached biofilm cells express their genes differently from those in the planktonic stages of the same organism, and that this switch in gene expression is required for the regulation of biofilm formation and development. Biofilms have been described as 'cities built by micro-organisms' (Lewandowski, 2000) and it is well known that a biofilm works faster to biologically transform materials than the equivalent number of planktonic organisms (de Beer *et al.*, 1993). Bioremediation is also facilitated by enhanced gene transfer among biofilm organisms and by the increased bioavailability of pollutants for degradation as a result of chemotaxis (Singh *et al.*, 2006). The major components of the EPS include polysaccharides, proteins and in some cases lipids, with minor amounts of nucleic acids and other biopolymers (Flemming and Wingender, 2001). Flemming and Wingender (2001) went as far as to state that biofilm organisms can establish stable arrangements and function multi-cellularly as synergistic microconsortia. The matrix facilitates the retention and recycling of many complex substances cells invest greatly in forming, including exoenzymes, cellular debris and genetic material. The EPS also encourages materials, which are in short supply, to be recycled and would otherwise be lost from the system. Importantly, Flemming and Wingender (2001) point out that the ESP sequesters materials from solution and this is key in the detention of dissolved phase pollutants in a flow through system.

In the absence of surfaces on which to form, biofilm organisms often form flocs, or planktonic clumps of biofilm (Güde, 1982). The formation of flocs in this way is the next best thing to forming an attached biofilm, and indicates that for most organisms, life as a biofilm is the norm and the planktonic phase is simply the mode of growth that organisms are forced into when a biofilm cannot be achieved. While this can be advantageous to the organisms, it may have negative consequences for the performance of the SuDS device, due to the blockages and water flow restrictions that may result. Thus, in the design of SuDS, where biodegradation is to be encouraged, conditions should be provided in which attached biofilm development is encouraged and where materials for degradation are concentrated where the biofilm is most easily maintained.

9.4   Biodegradation in Green SuDS

Unlike hard SuDS, green SuDS do not lend themselves readily to studies of the degradation of individual compounds or related groups of compounds using laboratory-based microcosms. Compared with hard SuDS, such as pervious pavements, there are greater difficulties – although not unsurmountable ones – in creating the systems in the laboratory that sufficiently reflect the field situation. In part, this is a matter of scale, since they are often extensive, which plays a major role in their function.

Maintenance of the ecosystem in laboratory models of green SuDS involves providing light and a suitable diurnal rhythm as well as a means of removing any heat that builds up. However, the greatest problem is that it is not possible to study the biodegradation of a specific organic pollutant by ensuring that it is, essentially, the sole source of energy and carbon. Green SuDS inevitably contain a lot of potentially degradable carbon, making study of their activity by simply monitoring carbon dioxide production much more complicated, particularly in the establishment of a 'baseline' activity level. Photosynthesis adds complications as significant oxidation of non-target material, either previously generated by photosynthesis or naturally present in the growth medium, will release a high background of carbon dioxide. Studying biodegradation in green SuDS, in both field and laboratory conditions, has involved indirect methods, such as identification and quantification of species of microorganisms known to degrade the target compounds of interest (Kämpfer *et al.*, 1993; Zarda *et al.*, 1998). Also useful are molecular methods to identify specific genes (Rakoczy *et al.*, 2011). Green SuDS have also been the subject of much study in terms of the overall balance between the input and output of pollutants (including volatilisation), but the most successful have been recent investigations using isotopes.

9.4.1   Constructed Wetlands

Among the most widely studied green SuDS elements are constructed wetlands. Scholz (2010) gives a useful introduction to their role in controlling urban runoff. Studies of the degradation of specific compounds in constructed wetlands have rarely been associated with a SuDS treatment train but have been more targeted at those in wastewater and groundwater treatment systems (e.g. Wallace and Davis, 2009). Some studies targeted systems operated in such a way that they would be unlikely to be included in a treatment train; for instance, where they are augmented by forced aeration (e.g. Wallace and Kadlec, 2005). Thus, while these studies can provide useful information, care must be exercised in translating results from one situation to another.

Distinguishing between the removal of pollutants by biodegradation, volatilisation and other attenuation processes is often an important issue in wetland studies. Isotope-based studies are relatively recent, but more traditional approaches that seek to balance inputs, outputs and accumulation have been more widely used. For example, Eke and Scholz (2008) used benzene as a model to study the removal of low molecular weight petroleum components in vertical-flow experimental wetlands (six indoors and six outdoors), constructed using different layered structures. Findings indicated that removal efficiencies were higher for the indoor rigs (controlled environment) than those outdoors. Outside rigs achieved mean removal efficiencies for benzene of 85% and chemical oxygen demand (COD) of 70%, with indoors rigs at 95% and 80%, respectively. However, from a separate experiment it was found that benzene removal was predominantly due to volatilisation. This conclusion contrasts with the work by Rakoczy *et al.* (2011) who used a combination of isotope studies and molecular techniques (aimed at detecting functional genes) to study the biodegradation of benzene in a model constructed wetland over a period of 370 days. They found that, despite low measured dissolved oxygen in the aqueous system, aerobic conditions seemed to dominate in the system, citing the complete oxidation of ammonium to nitrate and ferrous iron as evidence of efficient oxygen transfer into the sediment by plants, with benzene removal highly efficient after day 231 (>98% removal). Compound specific isotope analysis was used to study in-situ benzene degradation, revealing that it was degraded aerobically, mainly via the monohydroxylation pathway. This was additionally supported by the detection of the BTEX mono-oxygenase gene tmoA in sediment and root samples. They were able to calculate the extent of biodegradation from isotope signatures and demonstrated that at least 85% of benzene was degraded in this way, and thus only a small fraction was removed without microbiological involvement. The difference between their conclusions and those of Eke and Scholtz (2008) may be due to the extended period that the system was given to develop before any attempt was made to determine the extent of biodegradation taking place; this is certainly something to bear in mind. A similar isotopic approach is underway by Watzinger *et al.* (2014) in Austria, focusing on the degradation of diesel hydrocarbons, in this case contaminated groundwater. However, from the minimal information to date it appears that isotopic studies have been confined to very small scale microcosms.

Within the microbial communities of constructed wetlands, the relationships between microorganisms, substrates and plants is highly complex; their contribution to the conduction of oxygen into the root zone is neither universally supported nor well understood. Faulwetter *et al.* (2009) reviewed microbial processes influencing the performance of treatment wetlands, finding that there were questionable assumptions made about the influence of microbial processes, based primarily on inferences from known processes in wastewater treatment systems and natural wetlands. Circumstantial evidence was used to corroborate these basic assumptions, based on changes in water chemistry with the underlying biological processes being poorly understood.

Until recently, there was a lack of direct evidence of specific microbial consortia at work, but molecular methods have been used to study processes in constructed wetlands, improving the quality of evidence significantly. Faulwetter *et al.* (2009) and Truu *et al.* (2009) both gave overviews of the methods available to study microbial biomass, its activity and community composition in constructed wetlands. These included biochemical measures of diversity based on fatty acid methyl esters (FAMES) extracted from organisms in wetland communities (Weavera *et al.*, 2012) and molecular fingerprinting techniques. The molecular methods have included 'terminal restriction fragment length polymorphism' (TRF) (Sleytr *et al.*, 2009) and denaturing gradient gel electrophoresis (DGGE). These techniques have commonly been followed by cloning, to produce bacterial libraries, which can be studied in more detail by techniques such as amplified ribosomal DNA restriction analysis (ARDRA), for example Ibekwe *et al.* (2007), who investigated the relationship between plant cover and diversity. It was concluded by Ibekwe *et al.* (2007) that there were consistently higher microbial diversity indices in samples with 50% plant cover than those with 100% cover and that the bacteria in the dominant group of their clone library belonged to unclassified taxa.

Using microbial diversity to study biodegradation in constructed wetlands has not translated into the study of wetlands used to treat urban surface waters. However, Ancion *et al.* (2014) did investigate the efficacy of a stormwater treatment train, which included rain gardens, grassy swales, a stormwater filter and a wetland, by monitoring changes in the composition of the biofilm bacterial community at multiple locations throughout the system. These changes were assessed by automated ribosomal intergenic spacer analysis (ARISA), which detected significant differences in bacterial community composition as it gradually changed to that similar to the receiving stream; the discharge of treated stormwater appeared to have minimal effect on communities found in the receiving stream.

9.4.2 Biodegradation in Other SuDS Devices

Biodegradation studies in other green SuDS are less common and have not used isotopic methods. However, Charlesworth *et al.* (2012) investigated the use of compost to enhance, among other things, the oil retention and biodegradation performance of swales. Insights into improved swale design and more optimised materials to be used in the construction of the swales, particularly the necessary characteristics of the compost (added to the swales as organic matter to provide pollutant treatment and biological seeding), were key outcomes from the study. As part of this investigation the more traditional techniques of culturing, isolation, microscopy/staining and biochemical/physiological testing were used to study the diversity and numbers of microorganisms, including oil-degrading bacteria and fungi. Only species capable of growing in aerobic conditions were considered. The limitation of traditional approaches is that they only identify contributions to microbial diversity of organisms that are culturable under laboratory conditions. Appropriate use of a molecular biology approach can overcome this difficulty. For example, LeFevre et al. (2012) studied both the microbiology (using molecular methods) and the total petroleum hydrocarbon content of 58 rain gardens of differing ages and catchments, concluding that oil degrading microbial populations were higher in raingardens planted with deep rooting species than those planted with turf grass. In a separate study of raingarden soils LeFevre et al., 2012 showed that initial degradation rates of naphthalene could be correlated with the number of copies of bacterial 16S rRNA genes present.

9.5　Nitrogen in Green SuDS

Nitrogen can enter SuDS in a variety of forms, including both reduced (NH_4^+), oxidised inorganic nitrogen (NO_3^- and NO_2^-) and both dissolved and particulate organic N. The relative importance of inorganic versus organic forms can vary as a function of land use (Collins et al., 2010). Flint and Davis (2007) proposed that organic N can be abundant in the first flush of stormwater runoff from highly urbanised areas. As previously stated, organic nitrogen can exist in the form of complex macromolecules, such as proteins or nucleic acids, but also as simpler compounds including urea, uric acids and simple amines.

In an aerobic environment, NH_4^+ can be nitrified to NO_3^- in a two-step microbial process, and organic N can also undergo ammonification to NH_4^+ and nitrification to NO_3^-. These microbial processes transform N, but unless the pH is very high, they do not remove it, leading to the loss of ammonia to the atmosphere, an undesirable outcome, associated as it is with local and global environmental problems. Thus, as stormwater runoff passes through SuDS, nitrogen can be removed via uptake by plants and microorganisms, by adsorption and microbiologically mediated denitrification. Nitrogen uptake and adsorption only temporarily remove N and are therefore not biodegradation processes but, since microbial denitrification results in the permanent removal of nitrate from the system in gaseous forms, such as N_2O or N_2, they can fit the definition of biodegradation previously discussed. In the final step of denitrification, microorganisms use nitrate as a substitute for oxygen as the final electron acceptor, and thus require relatively anoxic conditions. The ammonium must be oxidised to provide the nitrate used in the denitrification step and thus an aerobic phase must be present either in the SuDS device itself (separated either in time or space from the anaerobic step) or upstream in the treatment train.

Research has shown that changes to the design of green SuDS can sometimes improve N removal, but hydrologic residence time has to be taken into account as it can be a critical step in denitrification (Kaushal et al., 2008; Klocker et al., 2009). Collins et al. (2010) caution against any possible negative consequences of these changes, which may produce loss of amenity or ecological value. For example, to achieve the anoxic stage, changes can include water retention, slowing of infiltration and drainage times, all of which require additional storage to allow the system to cope volumetrically, either in the device or at upstream points in the treatment train. If the proposed change includes electron donor amendments, such as to organic matter and elemental sulphur (Sutton-Grier et al., 2011), there is the potential for reduction in other environmental aims such as biological oxygen demand removal and even release of methane, a highly potent greenhouse gas. There can even be other unexpected consequences, such as release of phosphorus and heavy metals. Another important factor is the extent to which the system favours nitrogen release as N_2 rather than the very potent greenhouse gas N_2O.

In research into nitrogen removal in SuDS, it is critical to distinguish between denitrification and immobilisation, as mechanisms for N removal, since denitrification is the only process representing the permanent removal of N from the system. In the cases of both plant uptake and immobilisation by microorganisms, N is still present in the organism, and has the potential to be exported from the system in the future in a variety of organic and inorganic forms. To address these concerns, research must go beyond the approach of analysing N inputs and outputs to quantifying N cycling and removal processes in SuDS, and to do so across several seasons and types of installation. There are established methods for quantifying N mineralisation and nitrification in terrestrial systems (e.g. Hart *et al.*, 1994; Robertson *et al.*, 1999). Denitrification is more complicated because, as pointed out by Groffman *et al.* (2005), accurate methods for quantifying denitrification in terrestrial systems, which fluctuate between aerobic and anaerobic conditions, are not fully established, and all methods have significant disadvantages at different times and locations and with different conditions.

Measuring the denitrification potential of green SuDS has been approached by examining specific denitrification enzyme activity (Bettez and Groffman, 2012). They compared potential denitrification in this way in five types of SuDS system (wet ponds, dry detention ponds, dry extended detention, infiltration basin and filtering systems) and forested and herbaceous riparian areas in Baltimore, Maryland, USA. Denitrification activity was found to be about three times higher in the SuDS than in the riparian areas.

Stable isotope techniques will probably become the method of choice to quantify both denitrification and assimilation. One way forward, as carried out by Payne *et al.* (2014), was to use column-type microcosms to study the partitioning of $^{15}NO_3$ between biotic assimilation and denitrification. They found that, contrary to expectations, assimilation was the dominant means of NO_3^- removal under typical stormwater concentrations (1–2 mg N/l), contributing an average 89–99% of $^{15}NO_3$ turnover in columns containing the most effective plant species, while only 0–3% was denitrified and 0–8% remained in the pore water. Denitrification played a greater role for columns containing less effective species, processing up to 8% of $^{15}NO_3^-$, and increased with nitrate loading. Payne *et al.*, (2014) pointed out that their results required validation under field conditions with extension by tracing nitrogen fate over longer periods. Seasonal effects, particularly in temperate zones, would also be an important subject for future research. Nevertheless, the findings form an initial step in identifying and quantifying nitrogen processes in green SuDS and, thus, 'represent an important advance on the predominantly black-box approach of studies to date' Payne *et al.*, (2014).

9.6   Conclusions

Recent research into nutrient relations in green SuDS takes its place alongside parallel efforts to improve the understanding of nutrient and pollutant dynamics in natural soils and water. Better environmental protection principles through the use of green SuDS will only be achieved by determination of the attenuation and release of compounds of interest over appropriate time lengths, in field and laboratory trials. The combination of newly applied methods to SuDS, such as molecular techniques and stable isotopes, are shedding new light on the organisms implicated in transformation processes and the resulting partitioning of nutrients in water, plants, microbes and soil.

An opportunity has been presented for both SuDS laboratory and field trials to contribute knowledge to some of the most important questions in soil processes and environmental pollution. This is because it is possible to generate new, specifically designed experimental SuDS systems as part of a drainage scheme, to control the inputs in terms of flow rates and volumes, to rapidly observe and record pollution episodes and to install monitoring equipment in the drainage device. Additional advantages of using SuDS devices to answer fundamental research questions are that the ownership of such systems is relatively simple to establish, and gaining permission to monitor a site can be relatively straightforward. As stated in this chapter, care must be taken when assuming equivalence in the comparison of processes and overall effectiveness between, for example, natural and constructed wetlands, but many of the underlying scientific principles are similar, and SuDS devices should not be automatically assumed to be non-representative of the wider environment.

The detailed analysis of biological processes in the soils and water of green SuDS are prime candidates for further study, particularly given the known link between plants, soil microbes (bacteria, protozoa and metazoa such as nematodes) and soil nutrient flow. As stated, the tendency of microbes to associate in biofilms, the abundance of microbes in the rhizosphere and the association between plants and soil aeration, nutrient uptake, the partitioning and cycling of compounds are all mechanisms that contribute vast amounts to the effectiveness of green SuDS. These must be understood more fully, as links in a chain of processes and part of a functioning whole, in order to achieve the difficult goal of understanding such complex dynamic systems and to be able to unlock the black box.

References

Ancion, P.Y., Lear, G., Neale, M., Roberts, K. and Lewis, G.D. (2014) Using biofilm as a novel approach to assess stormwater treatment efficacy. *Water Research*, 49, 406–15.

Andersson, B.E. and Henrysson, T. (1996) Accumulation and degradation of dead-end metabolites during treatment of soil contaminated with polyaromatic hydrocarbons with five strains of white rot fungi. *Applied Microbiology and Biotechnology*, 46, 647–652.

Bettez, N.D. and Groffman, P.M. (2012) Denitrification Potential in Stormwater Control Structures and Natural Riparian Zones in an Urban Landscape. *Environmental Science and Technology*, 46, 10909–10917. Available at: http://pubs.acs.org/doi/abs/10.1021/es301409z

Charlesworth, S.M., Nnadi, E.O., Oyelola, O., Bennett, J., Warwick, F., Jackson, R.H. and Lawson, D. (2012) Laboratory based experiments to assess the use of green and food based compost to improve water quality in a sustainable drainage (SUDS) device such as a swale. *Science of the Total Environment*, 424, 337–343.

Collins, K.A., Lawrence, T.J., Stander, E.K., Jontos, R.J., Kaushal, S.S., Newcomer, T.A., Grimm, N.B. and Ekberg, M.L.C. (2010) Opportunities and challenges for managing nitrogen in urban stormwater: A review and synthesis. Ecological Engineering, 36: 11, 1507–1519.

Coupe, S.J., Smith, H.G., Newman, A.P. and Puehmeier, T. (2003) Biodegradation and microbial diversity within permeable pavements. *European Journal of Protistology* 39: 495–498.

de Beer, D., van den Heuvel, J.C. and Ottengraf, S.P.P. (1993) Microelectrode measurements of the activity distribution in nitrifying bacterial aggregates. *Applied and Environmental Microbiology*, 59, 573–579.

Durand, C., Ruban, V. and Oudot, J. (2004) Characterization of the organic matter of sludge: determination of lipids, hydrocarbons and PAHs from road retention/infiltration ponds in France. *Environmental Pollution*, 132, 375–384.

Eke, P.E. and Scholz, M. (2008) Benzene removal with vertical-flow constructed treatment wetlands. *Journal of Chemical Technology and Biotechnology*, 83: 1, 55–63.

Ellis, J.B., Shutes, B.E.R. and Revitt, M.D. (2003) *Constructed wetlands and links with sustainable drainage systems*. Environment Agency. ISBN 1857059182. Available at: https://eprints.mdx.ac.uk/6076/1/SP2–159-TR1-e-p.pdf

Faulwetter, J.L., Gagnon, B.V., Sundberg, C., Chazaren, C.F., Burr, M.D., Brisson, B.J., Camper, A.K. and Steina, O.R. (2009) Microbial processes influencing performance of treatment wetlands: A review. *Ecological Engineering*, 35, 987–1004.

Flemming, H.C. and Wingender, J. (2001) Relevance of microbial extracellular polymeric substances (EPSs) – Part I: Structural and ecological aspects. *Water Science and Technology*, 43, 1–8.

Flint, K.R. and Davis, A.P. (2007) Pollutant mass flushing characterization of highway stormwater runoff from an ultra-urban area. *Journal of Environmental Engineering*, 133, 616–626. doi:10.1061/(ASCE)0733-9372(2007)133:6(616).

Groffman, P.M., Dorsey, A.M. and Mayer, P.M. (2005) Nitrogen processing within geomorphic features in urban streams. *Journal of the North American Benthological Society*, 24, 613–625.

Griffiths, B.S. (1995) *Soil nutrient flow*. In: Darbyshire, J.F. (ed.), Soil Protozoa. CAB International, Guildford.

Güde, H. (1982) Interactions between floc-forming and non floc-forming bacterial populations from activated sludge. *Current Microbiology*, 7: 6, 347–350.

Hart, S.C., Stark, J.M., Davidson, E.A. and Firestone, M.K. (1994) *Nitrogen mineralisation, immobilisation and nitrification* In: Weaver R.W., Angle L.S., Bottomley, P.J., Bezdicek D.F., Smith (eds) Methods of soil analysis Part 2 Microbiological and chemical processes. Soil Science Society of America. Madison, Wisconsin, 986–1018.

Ibekwe, A.M., Lyon, S.R., Leddy, M. and and Jacobson-Meyers, M. (2007) Impact of plant density and microbial composition on water quality from a free water surface constructed wetland. *Journal of Applied Microbiology*, 102: 4, 921–36.

Kämpfer, P., Steiof, M., Becker, P.M. and Dott, W. (1993) Characterisation of chemoheterotrophic bacteria associated with the in situ bioremediation of a waste oil contaminated site. *Microbial Ecology*, 26, 161–188.

Kaushal, S.S., Groffman, P.M., Mayer, P.M., Striz, E., Doheny, E.J. and Gold, A.J. (2008) Effects of stream restoration on denitrification in an urbanizing watershed. *Ecological Applications*, 18, 789–804.

Klocker, C.A., Kaushal, S.S., Groffman, P.M., Mayer, P.M. and Morgan, R.P. (2009) Nitrogen uptake and denitrification in restored and unrestored streams in urban Maryland, USA. *Aquatic Sciences*, 71, 411–424.

LeFevre, G.H., Hozalski, R.M. and Novak, P.J. (2012) The role of biodegradation in limiting the accumulation of petroleum hydrocarbons in raingarden soils. *Water Research*, 46: 20, 6753–6762.

Lewandowski, Z. (2000) *Structure and Function of Biofilms*. In: Biofilms: Recent Advances in Their Study and Control, L.V. Evans (ed.), 2000 Harwood Academic Publishers. pp 1–17. ISBN 3-093-7.

Payne, E.G.I., Fletcher, T.D., Russell, D.G., Grace, M.R., Cavagnaro, T.R., Evrard, V., Deletic, A., Hatt, B.E. and Perran, L.M. (2014) Temporary Storage or Permanent Removal? The Division of Nitrogen between Biotic Assimilation and Denitrification in: Stormwater Biofiltration Systems. *PLoS ONE* 9: 3. e90890. doi:10.1371/journal.pone.0090890.

Rakoczy, J., Remy, B., Vogt, C. and Richnow, H.H. (2011) A bench-scale constructed wetland as a model to characterize benzene biodegradation processes in freshwater wetlands. *Environmental Science and Technology*. 45: 23, 10036–44. doi: 10.1021/es2026196.

Richardson, D., Felgate, H., Watmough, N., Thomson, A. and Baggs, E. (2009) Mitigating release of the potent greenhouse gas N_2O from the nitrogen cycle – could enzymic regulation hold the key? *Trends in Biotechnology*, 27: 7, 388–97.

Robertson, G.P., Wedin, D., Groffman, P.M., Blair, J.M., Holland, E.A., Nadelhoffer, K.J. and Harris, D. (1999) *Soil carbon and nitrogen availability: nitrogen mineralization, nitrification and carbon turnover*. In: Robertson, G.P., *et al.* (eds) *Standard Soil Methods for Long-Term Ecological Research*. Oxford University Press, New York. 258–271. Available at: http://tinyurl.com/ogjbukd.

Scholz, M. (2010) *Wetland Systems.* Springer. ISBN-13: 9781849964586.

Singh, R., Paul, D. and Jain, R.K. (2006) Biofilms: implications in bioremediation. *Trends in Microbiology*, 14: 9, 389–397.

Sleytr, K., Tietz, A., Langergraber, G., Haberl, R. and Sessitsch, A. (2009) Diversity of abundant bacteria in subsurface vertical flowconstructed wetland. *Ecological Engineering*, 35, 1021–1025.

Sutton-Grier, A.E., Keller, J.K., Koch, R., Gilmour, G.J. and Megonigal, P. (2011) Electron donors and acceptors influence anaerobic soil organic matter mineralization in tidal marshes. *Soil Biology and Biochemistry*, 43: 7, 1576–1583.

Truu, M., Truu, J. and Heinsoo, K. (2009) Changes in soil microbial community under willow coppice: the effect of irrigation with secondary-treated municipal wastewater. *Ecological Engineering*, 35, 1011–1020.

USEPA (2014) *Sand and Organic Filters.* USEPA. Available at: http://www.epa.gov/npdes.

Wallace, S. and Davis, B.M. (2009) *Engineered Wetland Design and Applications for On-Site Bioremediation of PHC Groundwater and Wastewater.* Society of Petroleum Engineers. 4, 01, 1–8. doi:10.2118/111515-PA.

Wallace, S. and Kadlec, R. (2005) BTEX degradation in a cold-climate wetland system. *Water Science and Technology*, 51: 9, 165–171.

Watzinger, A., Kinner, P., Hager, M., Gorfer, M. and Reichenauer, T.G. (2014) Removal of diesel hydrocarbons by constructed wetlands – Isotope methods to describe degradation. In-situ Remediation '14 Conference, London, UK, 2nd-4th September 2014. Available at: http://theadvocateproject.eu/conference/presentations.html.

Weavera, M.A., Zablotowicza, R.M., Jason-Krutza, L.J., Brysona, C.T. and Lockeb, M.A. (2012) Microbial and vegetative changes associated with development of a constructed wetland. *Ecological Indicators*, 13, 37–45.

Zarda, B., Mattison, G., Hess, A., Di Hahn, D., Höhener, P. and Zeyer, J. (1998) Analysis of bacterial and protozoan communities in an aquifer contaminated with monoaromatic hydrocarbons. *FEMS Microbiology Ecology*, 27, 141–152.

Hydrocarbon Biodegradation in Hard Infrastructure

Stephen J. Coupe, Alan P. Newman and Luis Angel Sañudo Fontaneda

Stephen J. Coupe, Alan P. Newman and Luis Angel Sañudo Fontaneda

10.1 Introduction

This chapter considers 'hard' SuDS infrastructure, as opposed to vegetated, green or 'soft'; the structure and function of the latter and the associated biofilm are covered in the previous chapter. Biodegradation is an important function of sustainable drainage devices, and the designer has an opportunity to optimise the system for enhancing this process. Essentially, hard devices are located in the subsurface and have one or more layers upon which a biofilm can form, and which include such structures as filter drains, pervious pavement systems (PPS) and similar devices. Filter drains run parallel with major roads and highways in the UK and in Europe and, as their name suggests, they drain excess surface water from impermeable pavements and also filter the water through their structure before conveying it to the receiving watercourse. They have an important role in the safety of road users in flood prevention and also protection of the pavement by removal of stormwater. PPS also remove excess water by means of infiltration, but are normally installed in lightly trafficked areas, or used for car parking and pedestrianised areas. More details of the structure and function of PPS can be found in Charlesworth *et al.* (2014).

The pollutants considered in this chapter will be confined to hydrocarbon fractions including fuels and lubricating oils, since these are utilised as a source of the nutrients by the microorganisms living in the biofilm. It has long been known that such hydrocarbon pollutants are common in the urban environment (Whipple and Hunter, 1979) and are associated with increased urbanisation, traffic and industrialisation.

Sustainable Surface Water Management: A Handbook for SuDS, First Edition.
Edited by Susanne M. Charlesworth and Colin A. Booth.

10.2　Hard SuDS Structure, Design and Related Technologies

Compared with green SuDS it is relatively easy to construct a laboratory scale model of a PPS or filter drain, which can include the determination of respiration rates, and hence biofilm activity, by the monitoring of carbon dioxide production. There are no macrophytes that require light in hard SuDS and it is reasonably straightforward to provide a hydrocarbon pollutant as the only significant carbon source. This 'black box' approach is, thus, more reasonable for laboratory experiments than would be the case for green SuDS, but for fully-sized outdoor models and studies on live installations the problems of measuring carbon dioxide flux would be considerable. These difficulties would be particularly great if a mass balance was used to determine the ratio of pollutant attenuation by biodegradation relative to attenuation mechanisms that are simply entrapment based. In a live situation, there is also the problem of measuring the pollution input onto the surface. The authors are not aware of research involving the use of stable isotopes in hard SuDS that has proven to be as successful as research on constructed wetlands, as discussed in Chapter 9.

It is worthwhile here to counter a popular misconception. The type of PPS with a stone sub-base is commonly compared with the trickling filter used in sewage treatment, particularly by manufacturers and trade bodies. While the trade bodies and commercial organisations are trying to simplify the story for communication purposes, the same misconception can also be found in the research community. While there is a similarity between the PPS and a trickling filter, particularly with respect to hydrocarbon degradation, this is purely superficial. Unlike a trickling filter, the PPS is only intermittently supplied with both water and the hydrocarbon pollutant. Thus, unless the system is designed to retain the hydrocarbon for a long period, there would be less opportunity for an oil-degrading biofilm to develop. Also unlike a trickling filter, the conditions with respect to temperature and pollutant loading are highly variable. Then, a trickling filter receives settled sewage, whereas hard SuDS are relatively low in nitrogen and phosphorus (although their nutrient removal capability is often considered an issue). Equally significant is the fact that the greatest mass of hydrocarbons enters a PPS in the form of a separate phase. This is normally as a thin iridescent film but it can, during catastrophic events, be in the form of virtually neat fuel or lubricating oil. In a wastewater treatment plant the operator will make every effort to exclude floatables, including free phase oil and grease, which will be separated effectively in primary sedimentation (or before). Finally, unlike a trickling filter, where the water inflow rate is controlled to ensure that the system does not flood uncontrollably, the void spaces of a PPS are designed as a means of storage, such that variable amounts of water saturation will be achieved for most significant rain events. Full water saturation, up to the base of the laying course, might be expected during the pavement's maximum design storm, and flooding to the surface may be seen if the design storm is exceeded; this is very different from the conditions in a trickling filter where every effort is made to maintain a very thin layer of liquid in intimate contact with the granular medium. Thus the environment in a PPS is very different, and the idea that significant biodegradation takes place as the water trickles through the stone bed during a rain event must be in doubt for any but the most readily biodegradable compounds. Physical entrapment and sorption, followed by subsequent biodegradation *between rain events* must be the dominant mechanism. An important factor in the performance of the PPS is the ability of the biofilm to reduce the concentration of dissolved phase hydrocarbons in the water trapped in the pavement between rain events, thus reducing the tendency to release dissolved phase hydrocarbons during a first flush. The importance of this becomes

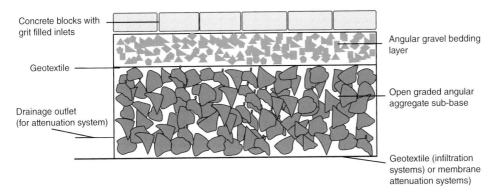

Concrete blocks with grit filled inlets

Geotextile

Drainage outlet (for attenuation system)

Angular gravel bedding layer

Open graded angular aggregate sub-base

Geotextile (infiltration systems) or membrane attenuation systems)

Figure 10.1   The layers in an idealised pervious pavement system. The overall depth would typically be 450 mm.

insignificant however, if free product release is occurring, and this will be critically dependent on the design of the PPS.

The cross-section of the most popular pervious pavement design is shown (Figure 10.1) and deserves closer inspection to identify where the conditions are most amenable to biodegradation. While the surface course of concrete blocks readily absorbs oil, they will dry too quickly for significant biodegradation for much of the time and, in the heat of the sun, the high surface temperatures can be sufficiently high to kill microorganisms. This means that any oil-degrading biofilm at the block surface would need to re-establish each time the conditions are right, meaning that the degree of biodegradation here is less important than in other parts of the system. However, photodegradation and evaporation may play important roles in pollutant removal and induce changes to the chemistry of the trapped pollutants. Also important is storage of oil on or in the material of the blocks for release, at a rate that does not overcome the biodegradation capacity of the lower layers. Figure 10.2 shows oil staining the blocks of an experimental car park several months after loading with oil at a rate of 10 ml.m^{-2}. The infiltration gaps between the blocks can form very good sites for biodegradation. The lower reaches of the infiltration channels remain relatively damp for long periods, as does the gravel laying course.

The tendency of the infiltration channels and grit-filled gaps between the blocks to support weed growth indicates that water relations can be very good and that the accumulation of finer particulates can aid in the retention of inorganic nutrients entering from the surface. However, it has been shown that where a suitable geotextile overlays a 10 mm laying course, the geotextile can provide the highest contribution to the initial detention of oil (Bond, 1999). The role of geotextiles in PPS is covered in more detail in Chapter 11.

While biodegradation certainly occurs in the sub-base, the majority of separate phase oil is trapped higher up in the structure, and the relatively large void ratio would provide less of a barrier to oil moving through the system as a sheen than would the 10 mm laying course. This would be particularly relevant in heavy rain events. However, the stone surfaces do stay damp for long periods, and generally aerobic conditions have been shown to be maintained (Pratt et al., 1999). It is highly likely that this is the zone where the oil-degrading microorganisms survive longest during long-term drought periods, and the sub-base may form a reservoir of organisms capable of recolonising areas that have been killed by drought.

Figure 10.2   A contaminated field car park showing oil staining on permeable blocks and on the gravel bedding layer (reproduced by permission of Tim Puehmeier).

10.3   Evidence of Biodegradation in Hard SuDS

Although the potential for biodegradation in porous pavements was proposed as early as the 1970s by Thelen and Howe (1978) in their work on porous asphalt, this appears to be largely based on assumptions. The first controlled laboratory study on the retention and biodegradation of realistic regular additions of oil in pervious pavement models (simulating something close to reasonable non-catastrophic inputs) was carried out by Bond (1999). The oil-loading rate used was estimated at about ten times the normal daily loading rate onto a UK car park. Model rigs were inoculated with a commercially available mixture of microorganisms and artificially applied inorganic nutrients, on the basis that in a laboratory situation the inputs of these nutrients from the environment would be zero. Thus, Osmocote slow release fertiliser granules were applied, and it was shown (Pratt *et al.*, 1999) that elevated carbon dioxide levels could be maintained for over 400 days without any further additions of the slow release nutrients (Figure 10.3).

Smaller test rigs were used to produce an approximate mass balance, study the kinetics of the system and to demonstrate that the degradation of the one-off application of oil appeared to be following first-order kinetics. It was found that the degradation rates were not particularly high but appreciable, with the best half-lives achieved being around six months. In the context of a low-tech, low-input, long-term solution to water management and pollution (Bond, 1999) was an encouraging development in the establishment of hard SuDS as a multiple benefit technology.

Newman *et al.* (2002a) reported that the system continued to perform well, despite oil application for a further four years, and while the nutrient and water conditions were

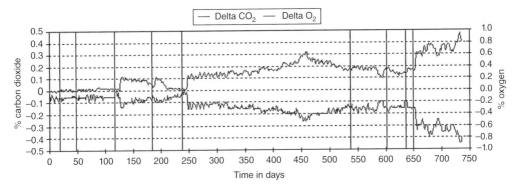

Figure 10.3   Carbon dioxide evolution and oxygen consumption in a model PPS over 750 days (Source: Bond, 1999).

maintained, biodegradation continued to occur. The model was also subjected to stress by gradually allowing nutrients to reduce for 200 days; this produced a decline in respiration rate, but was quickly reversed within 48 hours, once fresh inorganic nutrients were applied, and a similarly quick recovery was seen after a 75-day period of imposed drought.

After establishing the long-term capability of the PPS to maintain an oil-degrading biofilm, one of the important questions that needed to be answered was whether the biodegradation process required initiation by the addition of an oil-degrading microbial population at the outset, or whether microorganisms could, in a short timeframe, be recruited either from the pavement materials themselves or from external environmental inputs. It was also not known whether the commercial mixtures previously used by Bond (1999), were optimal for the PPS environment (Newman *et al.*, 2002b). In order to investigate this, denaturing gradient gel electrophoresis (DGGE) was used in conjunction with polymerase chain reaction (PCR) to target part of the 16S ribosomal RNA genes from cells collected from test rig effluent and the original inoculum (for more details of these techniques see Chapter 9). This showed that organisms extracted from the long-term PPS model and from the initial inoculum produced significantly different banding patterns, which confirmed that over time the population in the porous pavement changes and the initial inoculum appears to be out-competed by organisms from the environment (Newman *et al.*, 2002b).

A parallel study by Coupe (2004) monitoring the microbial activity in rig effluent as indicated by the rate of hydrolysis of fluorescein diacetate (Lundgren, 1981; Schnurer and Rosswall, 1982) showed that after about 20 weeks the inoculated and un-inoculated rigs displayed the same activity (Figure 10.4).

The effluent's viable bacterial count was around 10^4ml^{-1} in both inoculated and un-inoculated systems at the end of the study, with a similar amount of oil retained on the rig materials. Mineral oil was the only carbon source in the system, and in both treatments a diverse and abundant protozoan and metazoan community had been established. From Figure 10.4, it is notable that initially the activity in the inoculated rig was almost double that of the un-inoculated one, illustrating the potential benefit of a suitable inoculum early in the life of the PPS, advantageous for rapidly establishing a biofilm in studies where there are time limitations. There was later evidence, however, of increased numbers of protist predators in response to the inoculation of adapted assemblages, particularly large suspension feeding ciliates. This partly explains the disappearance of inoculum organisms in DGGE results, which could also be involved in initial differences in FDA activity patterns.

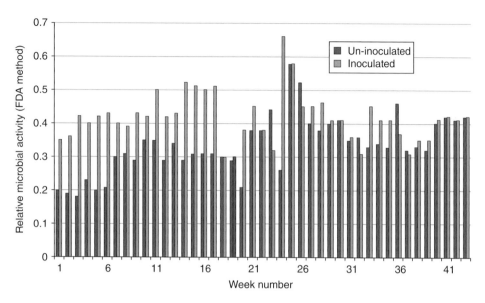

Figure 10.4   Results from a fluorescein diacetate assay (FDA) to determine PPS system activity, comparing inoculated and non-inoculated models (modified from Coupe, 2004).

10.4   Hard SuDS Microbiology and Biofilms

Hydrocarbon degrading organisms have been studied for many years, largely in relation to significant oil spills in both marine and terrestrial environments. Before the 1950s over 100 species of hydrocarbon degrading bacteria, yeasts and moulds had been discovered (Puehmeier 2009). Since then many other species have been reported. The most commonly isolated bacteria are, in decreasing order, members of the genera *Pseudomonas, Achromobacter, Flavobacterium, Nocardia, Rhodococcus, Arthrobacter, Corynebacterium, Acienebacter, Bacillus, Micrococcous, Brevibacterium, Mycobacterium, Alcaligenes and Aeromonas* (Atlas, 1978, 1981, 1995; Leahy and Colwell, 1990; Kämpfer *et al.*, 1991; Atlas and Bartha, 1992; Riser-Roberts, 1992, 1998; Arino *et al.*, 1996; Singh and Ward, 2004). Many of the previously documented species have been detected in oil contaminated PPS, and numerous DNA sequences have been detected for other unnamed organisms previously identified at contaminated sites (Puehmeier, 2009).

Hydrocarbon-degrading fungi (e.g. *Aspergillus, Penicillium, Trichoderma*) have also been regularly found and reported (Leahy and Colwell, 1990; Riser-Roberts, 1992, 1998; Andersson and Henrysson, 1996), although their relative contributions vary. Song *et al.* (1986) reported that 82% of n-hexadecane mineralisation in sandy loam soil was attributed to bacteria while only 13% was attributed to fungi, but this is likely to vary from system to system. Certainly fungal hyphae and fruiting bodies are regularly observed in oil-degrading pervious pavements and can be detected by both molecular methods (Puehmeier, 2009) and biochemical markers (phospholipid fatty acids), but they seem to be absent under some conditions, the reasons for which have not really been established. One theory for the absence of fungi in some oil-degrading systems could simply be time: fungi typically take a much longer time to grow significant colonies than bacteria, which can form biofilms covering surfaces in a matter of days. Also, fungi are less readily discharged

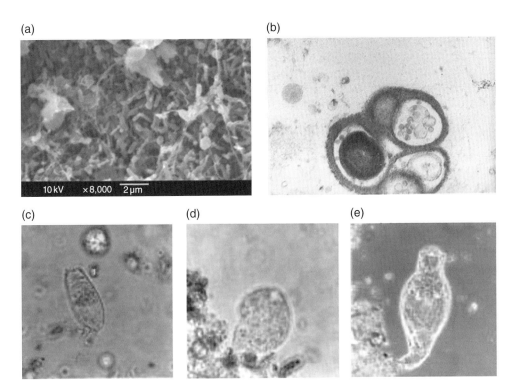

Figure 10.5   (a) SEM biofilm (b) TEM bacteria from effluent (c) cilliate of genus *Colpoda* (d) testate amoeba genus *Euglypha* (e) rotifer (reproduced from Newman, A.P., Pratt, C.J., Coupe, S.J. and Cresswell, N. (2002) Oil bio-degradation in permeable pavements by microbial communities *Water Science & Technology* 45: 7. 51–56, with permission from the copyright holders, IWA Publishing).

and eluted from flow systems than other microbial groups which may explain their lower numbers on some sampling occasions. For a general introduction to the mechanisms of hydrocarbon biodegradation, Singh and Ward (2004) is recommended.

Developing an understanding of the structure and arrangement of the oil-degrading biofilm in PPS was also important, and both optical and electron microscopy have been used to monitor biofilm development.

Figure 10.5a shows an image of a biofilm using scanning electron microscopy (SEM) and Figure 10.5b is a transmission electron micrograph (TEM) image of a bacterium from a heavily oiled PPS geotextile. Clearly visible is the presence of oil droplets in enlarged cell vacuoles, which were similar to structures observed by Cameotra and Singh (2009) in an oil-exposed *Pseudomonas* species. They found that cells were connected to each other by means of numerous fibrous projections concentrated in areas of the network formed by an extracellular secretion. All these structures were absent when they were grown on glucose as the carbon source. It was suggested that the fibre-like network could be a form of alkane and surfactant complex, and the complex could be a way by which hydrocarbon is transported to the cell surface for uptake.

While the protozoa and other non-oil-degrading eukaryotes (Figure 10.5c,d,e) are considered to be very important to the health of the oil-degrading ecosystem, they are not oil-degrading in themselves but depend on grazing the microorganisms which utilise the oil as their source of carbon and energy. It is believed that the regulating effect of predators on the bacterial and fungal biofilm is particularly important. Protozoa are also important in keeping water flow paths open in granular systems (Mattison and Harayama, 2002). It is possible that, where the protozoan and metazoan communities form a dense, species-rich population, they may stimulate the biodegradation process. It has been shown (Coupe *et al.*, 2003) that biodegradation is facilitated to a similar degree by bacteria or fungi, as demonstrated by inhibition with appropriate antibiotics while a full microbial community which included protozoa, but without inhibition by antibiotics, degraded the greatest mass of oil. It was suggested that recycling of inorganic nutrients by protozoan predation was an important factor, as indicated earlier by Kahlert and Baunsgaard (1999).

10.5   Design and Diversification from Standard Hard SuDS

In PPS based on plastic sub-base replacement units rather than stone, the available surface area for biofilm formation and for mechanical retention of the oil by smearing on surfaces is much less than in a stone sub-base system. In an attempt to compensate for this, a floating mat device, which provided a surface upon which oil could be absorbed and upon which the microorganisms could grow was designed by Puehmeier *et al.* (2005). In the prototype, floatation was originally provided by attaching geotextile to a grid made from polypropylene, incorporating a blowing agent to produce bubbles in the structure. A later development incorporated the buoyancy element directly into the geotextile. Electron microscopy showed that a very dense and highly structured biofilm had formed on the geotextile layer (Figure 10.6).

Initial experiments were carried out in small-scale chambers (Puehmeier *et al.*, 2005) with considerable differences found in both the structure and density of the biofilm growing on the mats when high and low nutrient conditions were compared. In the low nutrient condition, without the addition of inorganic fertiliser, an almost continuous biofilm still resulted, which could only be obtaining energy from the degradation of oil. Carbon dioxide

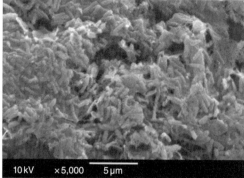

Figure 10.6   Scanning electron micrographs of bacterial biofilm growing on floating mats with oil contamination in PPS.

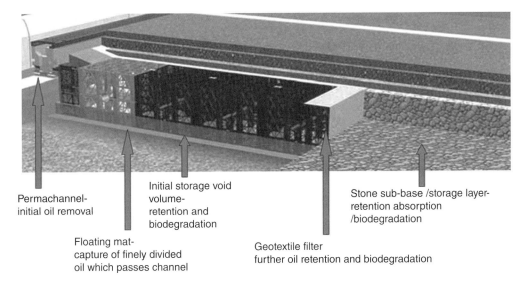

Permachannel-
initial oil removal

Initial storage void
volume-
retention and
biodegradation

Stone sub-base /storage layer-
retention absorption
/biodegradation

Floating mat-
capture of finely divided
oil which passes channel

Geotextile filter
further oil retention and biodegradation

Figure 10.7   A section through a macro-pervious pavement showing interception, storage and treatment areas (reproduced by permission of SEL Environmental).

measurements in the headspace above the water indicated that, under low nutrient conditions, the oil-degradation rate was enhanced on the floating mat in comparison with oil floating free on the water body (Puehmeier *et al.*, 2005). Molecular methods (DGGE) were applied to a study of the biofilm populations, which showed a more diverse bacterial community in the low nutrient-floating mat chambers compared to all other chambers. Under high nutrient conditions the species present on the floating biofilm were very similar to those in the water body, while under low nutrient conditions there was a significant difference.

These floating mats were also applied to macro-pervious pavements which have discrete oil-separating infiltration points to direct the stormwater into underground pervious sub-bases (Newman *et al.*, 2013) close to the point where water enters the structure (Figure 10.7).

In these devices, surfaces for biofilm formation were available on the floating mats, the vertical geotextile filter and the stone sub-base, but the majority of hydrocarbons were retained in the channel collector, which was found to contain hydrocarbons at concentrations thousands of times greater than that found in the effluent (Newman *et al.*, 2013).

10.6   Other Hard SuDS Biodegradation Studies

In Spain, Bayón *et al.* (2005) used electron microscopy to study biofilm on the geotextile, and were able to show that a novel geotextile designed to reduce evaporation from the sub-base was as good at maintaining an oil-degrading biofilm as the two most commonly used at the time (Gomez-Ullate *et al.*, 2010). They later extended their work to field studies on PPS car parks in the north of Spain where they were able to detect biodegradation by means of increased carbon dioxide in the sub-base atmosphere (Sañudo-Fontaneda *et al.*, 2014).

Scholz and Grabiowiecki (2009) and Grabiowiecki (2010) proposed the use of PPS to recover energy using ground source heat pumps and have demonstrated that, using gully pot liquor as a source of an organic pollutant load (presumably containing some mineral oils), it was possible to detect biodegradation in the system despite the pavement being used as a heat exchanger. Tota-Maharaj *et al.* (2010) also applied DGGE and other molecular methods to their studies.

10.7   Design Optimisation for Catastrophic Pollution Events

For biodegradation to be successful in PPS, there are two important aspects needing to be optimised: improving retention under very heavy oil loading and providing inorganic nutrients. It had been shown that horticultural slow release fertiliser was an effective substance for long-term provision of the inorganic nutrients required (Bond, 1999; Pratt *et al.*, 1999). However concerns relating to excess nutrient release have subsequently been confirmed by Nnadi *et al.* (2013) who showed the potential for eutrophication to be caused to receiving water bodies. Some work has been done on incorporating slow-release nutrient additives into geotextile fibres located where the microorganisms need it (Spicer *et al.*, 2006; Newman *et al.*, 2011). However, there is, as yet, no commercial uptake of these geotextiles and the supply of inorganic nutrients over the life of the PPS remains an area where great improvements could be made to biodegradation rates.

However, the most important factor associated with enhancing biodegradation in PPS, as has been discussed here, is retaining the oil in the system long enough for biodegradation to provide an effective remedy, particularly after catastrophic losses. While Bond (1999) was the first to study biodegradation in PPS systems subjected to regular small oil additions, Brownstein (1997) investigated levels of oil contamination simulating one-off releases of oil from a simulated car sump failure. The retention of oil in the system was significant (up to 90%) (Pratt *et al.*, 1996), but Jones *et al.* (2008) found that percentage retention of a pollutant in a SuDS device is of limited use if the output concentrations still exceed acceptable environmental standards. In Brownstein's (1997) experiment, losses of oil were sufficient to produce measurable thicknesses of free product in the effluent. An important lesson was learnt, in that once the various layers of material in the PPS were contaminated with oil beyond their holding capacity, release of hydrocarbons from the structure far exceeded the biodegradation rate.

The rapid loss of hydrocarbons from a PPS was further illustrated by a very dramatic field experiment, in which Newman *et al.* (2004a) used an experimental car park to simulate the total loss of oil from the sump of a large car, followed by two simulated 13 mm rain events. After about 22 hours, the concentration of oil in the effluent had exceeded 8000 mg/l. Clearly, the oil retaining capabilities of the PPS had been overcome and the case was clearly established for work on a means of retaining the oil that could be released in a major pollution event. Two approaches to modify the structure of the PPS itself have been utilised, the first of which was to incorporate additional retentive capacity in the upper layers of the system by means of improved geotextiles (Puehmeier and Newman 2008) or to incorporate natural materials such as compost (Bentarzi *et al.* 2010, 2013) or artificial interception media such as open cell foam (Lowe, 2006).

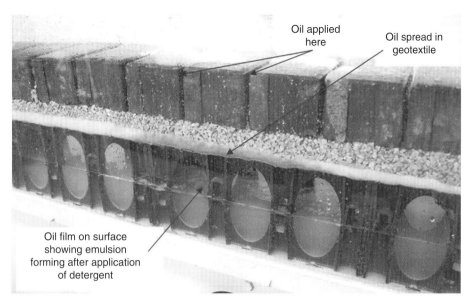

Figure 10.8   Oil in PPS with sub-base replacements following high oil application and mobilisation of oil by deter-gents (Wilson, S., Newman, A.P., Puehmeier, T. and Shuttleworth, A. (2003) Performance of an oil interceptor incorporated into a pervious pavement, ICE Proc: *Engineering Sustainability*, 156, (ES 1), pp. 51–58. Figure 7. ICE Publishing).

The second approach has been to deal with large hydrocarbon losses by means of shallow gravity separators incorporated into the pavement. These can be constructed using traditional stone sub-bases or sub-base replacement units. The principle is that any free product that passes the primary retention layer will be stilled by the low velocity of water in the system and will float on the permanent pool of water. This system that was equipped with a sub-base replacement is illustrated in Figure 10.8 (Wilson *et al.*, 2003) and was able to show that the system was limited when exposed to detergents used in the cleaning of motor vehicles. This has important consequences for the management of these systems, particularly following major oil releases. The ability of the system, when constructed with sub-base replacement boxes, to allow the rapid evacuation of almost the entire liquid contents with, for example, a gully sucker, following a major oil spill is an important advantage. Such a process would then leave a manageable body of residual contamination to be dealt with by biodegradation.

The microbiology of a laboratory PPS model was studied over a 4-month period, and the numbers of oil-degrading bacteria were found to increase from 10^4 to $10^{12}\,ml^{-1}$ after adding inorganic nutrients (Figure 10.9). Sequencing of amplified DNA from the bacterial groups found in samples collected from the model was indicative of oil-degradation (Newman *et al.*, 2004b). As with traditional PPS, most of these groups had been isolated in other studies from samples derived from sites contaminated by pollutants such as petroleum, coal tar and polycyclic aromatic hydrocarbons. Removing most of the separate phase oil followed by blocking the outlet and raising the water level would deposit the small amount of remaining residual free oil back onto the geotextile and granular laying course where biodegradation would be favoured.

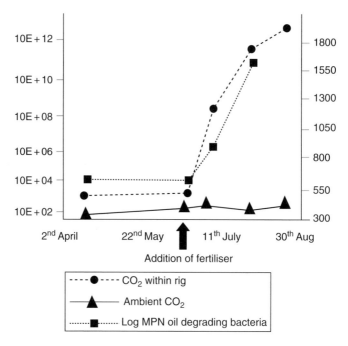

Figure 10.9   The impact of acute oil contamination on bacterial numbers and CO_2 evolution in a field-based PPS (redrawn from Newman *et al.*, 2004b).

10.8   Conclusions

Hard SuDS have been shown, by many years of research, to be colonised extensively by microoganisms recruited from the surrounding environment. The surfaces of the structures provide an ideal platform for the deposition of pollutants, allowing the necessary time and conditions for appreciable and maintenance-free biodegradation and decontamination of water prior to discharge. The manipulation of the physical structure of hard SuDS, to meet the specific needs of a site, allows the design to fit the required purpose and can trap and treat chronic low level pollution and point source spills that may threaten to overwhelm systems with a lower capacity for attenuation. The key therefore is the immobilisation of pollution, to provide the necessary time for aerobic microbes to ultimately turn organic pollution into glucose, carbon dioxide and water.

Pervious paving is an established technology for water management and pollution prevention. In-depth study of environmental engineering, chemistry and biology has increased the knowledge base on the performance and optimisation of hard SuDS for environmental protection. Research should continue into the fundamental science underpinning the technology, particularly in the areas of biological treatment, resource protection (e.g. rainwater harvesting) and new and emerging pollutants. Hard SuDS are now a feature in the urban landscape and are well placed to provide resilience against extreme weather and to protect downstream environments in a future that is increasingly variable and full of uncertainty.

References

Andersson, B.E. and Henrysson, T. (1996) Accumulation and degradation of dead-end metabolites during treatment of soil contaminated with polyaromatic hydrocarbons with five strains of white rot fungi. *Applied Microbiology and Biotechnology*. 46: 5-6. 647–652.

Arino, S., Marchal, R. and Vandecasteele, J.P. (1996) Identification and production of a rhamnolipidic biosurfactant by a *Pseudomonas* species. *Applied Microbiology and Biotechnology*. 45: 162–168.

Atlas, R.M. (1978) Microorganisms and petroleum pollutants. *BioScience*. 28: 387–391.

Atlas, R.M. (1981) Microbial degradation of petroleum hydrocarbons: An environmental perspective. *Microbiology Reviews*. 45: 180–209.

Atlas, R.M. (1995) *Bioremediation*. *Chemical and Engineering News*. 73: 32–42.

Atlas, R.M. and Bartha, R. (1992) Hydrocarbon biodegradation and oil spill bioremediation. *Adv. Microbial Ecology*, 12: 287–338.

Bayón, J.R., Castro, D., Moreno-Ventas, X., Coupe, S.J. and Newman, A.P. (2005) *Pervious pavement research in Spain: Hydrocarbon degrading microorganisms*. Proc. 10th International Conference on Urban Drainage, Copenhagen/Denmark, 21–26 August 2005. Available at: http://tinyurl.com/zfpkh5y

Bentarzi, Y., Ghenaim, A., Terfous, A., Wanko, A., Hlawka, F. and Poulet, J.B. (2010) New material for permeable and purifying pavement in the urban areas: estimation of hydrodynamic characteristics (in French). *8th International Conference on Sustainable Techniques and Strategies in Urban Water Management*. Novatech, Lyon, France. Available at: http://tinyurl.com/zjynz6g

Bentarzi, Y., Ghenaim, A., Terfous, A., Wanko, A., Hlawka, F. and Poulet, J.B. (2013) Hydrodynamic characteristics of a new permeable pavement material produced from recycled concrete and organic matter. *Urban Water Journal*. 10: 4. 260–267.

Bond, P.C. (1999) *Mineral Oil Biodegradation within Permeable Pavements: Long-Term Observations*. Unpublished PhD thesis, Coventry University, UK.

Brownstein, J. (1997) *An investigation of the potential for the bio-degradation of motor oil within a model permeable pavement structure*. Unpublished PhD thesis, Coventry University, UK.

Cameotra, S.S. and Singh, P. (2009) Synthesis of rhamnolipid biosurfactant and mode of hexadecane uptake by *Pseudomonas* species, *Microbial Cell Factories*. 8, 16, 1–7. Available at: http://tinyurl.com/z429fto

Charlesworth, S.M., Lashford, C. and Mbanaso, F. (2014) Hard SUDS Infrastructure. Review of Current Knowledge, Foundation for Water Research.

Coupe, S.J. (2004) *Oil Biodegradation and Microbial Ecology within Permeable Pavements*. Unpublished PhD Thesis. Coventry University, UK.

Coupe, S.J., Smith, H.G., Newman, A.P. and Puehmeier, T. (2003) Biodegradation and Microbial Diversity within Permeable Pavements, *European Journal of Protistology*. 39. 1–4.

Gomez-Ullate, E., Bayón, J.R., Coupe, S. and Castro-Fresno, D. (2010) Performance of pervious pavement parking bays storing rainwater in the north of Spain. *Water Science and Technology*. 62: 3. 615–621.

Grabiowiecki, P. (2010) Combined Permeable Pavement and Ground Source Heat Pump Systems. Unpublished PhD Thesis. University of Edinburgh. Available at: http://tinyurl.com/hverbt4

Jones, J., Clary, J., Strecker, E. and Quigley, M. (2008) 15 reasons you should think twice before using percent removal to assess BMP performance. *Stormwater Magazine* Jan/Feb 2008. p. 10.

Kahlert, M. and Baunsgaard, M.T. (1999) Nutrient recycling – a strategy of a grazer community to overcome nutrient limitation. *Journal of the North American Benthological. Society*. 18: 363–369.

Kämpfer, P., Steiof, M. and Dott, W. (1991) Microbiological characterization of a fuel-oil contaminated site including numerical identification of heterotrophic water and soil bacteria. *Microbial Ecology*. 21. 227–251.

Leahy, J.G. and Colwell, R.R. (1990) Microbial degradation of hydrocarbons in the environment. *Microbiology Review*. 54: 3. 305–315.

Lowe, T.R. (2006) Paving System, UK Patent Application No. CA2595539 A1 (US Application Number US8104990 B2: Published 2012).

Lundgren, B. (1981) Fluorescein Diacetate as a Stain of Metabolically Active Bacteria in Soil. *Oikos* 36: 1. 17–22.

Mattison, R.G.H. and Harayama, S. (2002) The Bacterivorous Soil Flagellate *Heteromita globosa* reduces bacterial clogging under denitrifying conditions in sand-filled aquifer columns. *Applied Environmental Microbiology.* 68: 9. 4539–4545.

Newman, A.P., Pratt, C.J. and Coupe, S. (2002a) Mineral oil bio-degradation within a permeable pavement: microbiological mechanisms. *Water Science and Technology.* 45: 7. 51–56.

Newman, A.P., Coupe, S., Puehmeier, T., Morgan, J.A., Henderson, J. and Pratt, C.J. (2002b) *Microbial ecology of oil degrading porous pavement structures; global solutions for urban drainage.* Proceedings of the 9th International Conference on Urban Drainage, Portland OR, USA, 8–13 Sept 2002.

Newman, A.P., Duckers, L., Nnadi, E.O., and Cobley, A.J. (2011) Self fertilising geotextiles for use in pervious pavements: A review of progress and further developments. *Water Science & Technology,* 64 (6) 1333–1339.

Newman, A.P., Puehmeier, T., Kwok, V., Lam, M., Coupe, S.J., Shuttleworth, A. and Pratt, C.J. (2004a) Protecting groundwater with oil retaining pervious pavements: historical perspectives, limitations and recent developments, *Quarterly Journal of Engineering Geology.* 37: 4. 283–291.

Newman, A.P., Puehmeier, T., Schwermer, C., Shuttleworth, A., Wilson, S., Todorovic, Z. and Baker, R. (2004b) The next generation of oil trapping porous pavement systems. 5th int. conf. sustainable techniques and strategies in urban water management, Lyon, France, Groupe de Recherche Rhone-Alpes sur les Infrastructures et l'Eau. Novatech, 2004, pp. 803–810.

Newman, A.P., Aitken, D. and Antizar-Ladislao, B. (2013) Stormwater quality performance of a macro-pervious pavement car park installation equipped with channel drain based oil and silt retention devices. *Water Research.* 47: 2. 7327–36.

Nnadi, E.O., Newman, A.P. and Coupe, S.J. (2013) Geotextile incorporated permeable pavement system as potential source of irrigation water: effects of re-used water on the soil, plant growth and development. *CLEAN – Soil Air Water.* 42, 2, 125–132.

Pratt, C.J., Newman, A.P. and Bond, P.C. (1999) Mineral oil bio-degradation within a permeable pavement: long term observations. *Water Science and Technology,* 29: 2. 103–109.

Pratt, C.J., Newman, A.P. and Brownstein, J. (1996) Use of porous pavements to reduce pollution from car parking surfaces – some preliminary observations. *Proceedings 7th International Conference on Urban Storm Drainage.* Hanover, Germany. Available at: http://tinyurl.com/no2ljeu

Puehmeier, T. (2009) *Understanding and optimising pervious pavement systems and source control devices used in sustainable urban drainage.* Unpublished PhD. Coventry University, UK.

Puehmeier, T. and Newman, A.P. (2008) Oil retaining and treating geotextile for pavement applications, Proc. 11th International Conference on Urban Drainage, Edinburgh, UK. Available at: http://tinyurl.com/zauu7x6

Puehmeier, T., De Dreu, D., Morgan, J.A.W., Shuttleworth, A. and Newman, A.P. (2005) Enhancement of oil retention and biodegradation in stormwater infiltration systems, Proc.10th *International Conference on Urban Drainage,* Copenhagen/Denmark. Available at: http://tinyurl.com/nqwjdtn

Riser-Roberts, E. (1992) *Bioremediation of Petroleum Contaminated Sites.* CRC Press, Boca Raton, FL.

Riser-Roberts, E. (1998) *Remediation of Petroleum Contaminated Soil: Biological, Physical, and Chemical Processes.* Lewis Publishers, Boca.

Sañudo-Fontaneda, L.A., Charlesworth, S., Castro-Fresno, D., Andrés-Valeri, V.C.A. and Rodríguez-Hernández, J. (2014) Water quality and quantity assessment of pervious pavements performance in experimental car park areas. *Water Science and Technology* 69: 7. 1526–1533.

Schnurer, J. and Rosswall, T. (1982) Fluorescein diacetate hydrolysis as a measure of total microbial activity in soil and litter. *Applied and Environment Microbiology.* 43: 6. 1256–1261.

Scholz, M. and Grabiowiecki, P. (2009) Combined permeable pavement and ground source heat pump systems to treat urban runoff. *Journal of Chemical Technology and Biotechnology.* 84: 3. 405–413.

Singh, A. and Ward, O.P. (2004) *Biodegradation and Bioremediation.* Springer Science & Business Media.

Song, H.G., Pedersen, T.A. and Bartha, R. (1986) Hydrocarbon mineralization in soil: relative bacterial and fungal contribution. *Soil Biology and Biochemistry.* 18. 109–111.

Spicer G.E., Lynch D.E. and Coupe S.J. (2006) The development of geotextiles incorporating slow-release phosphate beads for the maintenance of oil degrading bacteria in permeable pavements. *Water Sci. Technol.* 54(6-7) 273–80.

Thelen, E. and Howe, L.F. (1978) *Porous Pavement*. Franklin Institute Press, Philadelphia.

Tota-Maharaj, K., Scholz, M., Ahmed, T., French, C. and Pagaling, E. (2010) The synergy of permeable pavements and geothermal heat pumps for stormwater treatment and reuse. *Environmental Technology*, 31:14. 1517–1531.

Wilson, S., Newman, A.P., Puehmeier, T. and Shuttleworth, A. (2003) Performance of an oil interceptor incorporated into a pervious pavement, *ICE Proc: Engineering Sustainability*, 156: 1. 51–58.

Whipple, Jr, W. and Hunter, J.V. (1979) Petroleum hydrocarbons in urban runoff. *Journal of the American Water Resources Association*, 15, 1096–1105.

Use of Geosynthetics for Sustainable Drainage

Luis Angel Sañudo Fontaneda, Elena Blanco-Fernández, Stephen J. Coupe, Jaime Carpio, Alan P. Newman and Daniel Castro-Fresno

11.1 Introduction to Geosynthetics

According to the international standard ISO 10318-1 (2006), a geosynthetic is a 'Generic term describing a product, at least one of whose components is made from a synthetic or natural polymer, in the form of a sheet, a strip or a three-dimensional structure, used in contact with soil and/or other materials in geotechnical and civil engineering'. This definition is so generic that it could include some construction materials that traditionally are not considered as geosynthetics, such as roofing felts. However, the huge variety of products that have been developed in this area requires a wide description, not a very precise definition. The term 'geotextile' (later modified to 'geosynthetic') was introduced by Giroud (1977), at the International Conference on the Use of Fabrics in Geotechnics, the first International Conference on Geosynthetics, and the foundation of the International Geosynthetics Society, which has worked to expand the use of these materials, as well as to educate engineers in their proper use.

Concerning their history, fibrous materials to reinforce the soil have been used for a long time. For applications such as the hydraulic and/or physical separation of soils or for drainage, the materials traditionally used were clays and gravel, respectively, and synthetic materials were not used before the first half of the 20th century. Some key milestones in the history of geosynthetics are, for example, the use of cotton fabrics to reinforce roads in South Carolina (USA) in 1926 (Koerner, 2012) and the use of fabrics and plastics for coastal protection in the Netherlands in the 1950s (Van Santvoort, 1994). Since then, factors such as:

- good quality of the products, which are very well controlled in factories
- good design techniques
- well established technical standardisation

Sustainable Surface Water Management: A Handbook for SuDS, First Edition.
Edited by Susanne M. Charlesworth and Colin A. Booth.
© 2017 John Wiley & Sons, Ltd. Published 2017 by John Wiley & Sons, Ltd.

- cost competitiveness and quick and relatively easy installation
- better environmental performance, due to lower use of natural resources and lower carbon footprint

have promoted an explosion of geosynthetic use, and in 2012, despite the world economic crisis, 3.4 billion m² of geosynthetic were used worldwide, suggesting more than $6 billion of economic value. The material is expected to surpass a demand of 5 billion m² in 2017 (Geosynthetics Magazine, 2014).

11.2   Classifications, Functions and Applications of Geosynthetics

As explained, there is an enormous variety of geosynthetics on the market, and there are various criteria to classify them. Figure 11.1 shows some examples of common types.

Geosynthetics are generally classified according to their permeability, thus dividing them into permeable, non-permeable and geocomposite, as described in Table 11.1.

Other classification criteria, some of them only applicable to specific geosynthetic groups, are (Koerner, 2012):

- polymeric material: this could be polyolefins, polyamides and polyesters, and in geomembranes some kind of rubber is also important
- fibre length used to produce the geosynthetic: short, long or continuous
- fabric style: woven, non-woven or knitted.

However, more important than classifying according to shape is the classification according to function, since it will define the final performance of each specific product under real

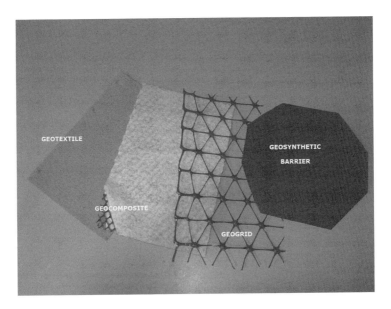

Figure 11.1 Examples of geosynthetics.

Table 11.1 Classification of geosynthetics according to permeability (ISO, 2006).

Permeable	Geotextiles: planar, permeable, polymeric (synthetic or natural) textile material used in contact with soil and/or other materials in geotechnical and civil engineering applications. They may be non-woven, knitted or woven. Geotextile-related products: planar, permeable, polymeric (synthetic or natural) material, which does not comply with the definition of a geotextile. They can be geogrids, geonets, geomats, geostrips, geocells, etc.
Non-permeable (Geosynthetic barriers)	Polymeric: factory-assembled structure of geosynthetic materials in the form of a sheet which acts as a barrier. The barrier function is created by a polymer. They are usually known as geomembranes. Clay barriers: factory-assembled structure of geosynthetic materials in the form of a sheet which acts as a barrier. The barrier function is created by clays. Bituminous: factory-assembled structure of geosynthetic materials in the form of a sheet which acts as a barrier. The barrier function is created by bitumen.
Geocomposites	Manufactured material using at least one geosynthetic product among the components. The most well-known example is the drainage geocomposite, composed of a geonet between two other geosynthetic, usually non-woven, geotextiles. Its use is now almost an essential in SuDS.

conditions. Furthermore, depending on the function required of each product, a specific design method is required. There are seven main functions recognised by the International Geosynthetics Society (Geosynthetic Society, 2015) and also in the CE marking regulations (European Commission, 1996):

- separation of two soil materials with different granulometry
- reinforcement of earth, soil or other granular materials
- erosion control: retention of soil particles, to avoid them washing away by water runoff, rivers, waves, etc.
- protection of soils or other geosynthetics: a geotextile can protect a geomembrane from puncturing by sharp aggregates in direct contact
- impermeable barriers: in order to avoid undesirable leakage
- filtration: geotextiles, especially, can allow the water to permeate, retaining fine aggregates
- drainage: redirecting the course of water, performing like a gravel layer or drainage pipe

In addition to these applications there are a number of specialist applications for geotextiles, which include their use to transmit water by capillary action, known as 'wicking geotextiles', and those with specific pollutant retention or degradation capabilities including enhanced hydrocarbon detention and biodegradation. Degradation enhancement is achieved by nutrient-loaded geotextiles and a buoyant device inserted into plastic void-forming structures used as alternatives to crushed stone sub-bases (Newman *et al.*, 2004; Puehmeier *et al.*, 2005).

In terms of application, geosynthetics are widely used in roads, railways, foundations, retaining walls, slopes, tunnels, channels, reservoirs, landfills, and they are, of course, used extensively in SuDS and in many kinds of drainage systems.

11.3   Application of Geotextiles in SuDS

There are many applications for geotextiles in SuDS, and many accounts relating to their use in pervious pavements systems (PPS) (Castro Fresno *et al.*, 2005; Gomez-Ullate *et al.*, 2011a; Sañudo-Fontaneda *et al.*, 2014c) and filter drains (FD) (Andrés-Valeri *et al.*, 2014; Coupe *et al.*, 2015). The main aims of geotextiles in SuDS are to serve as a separation layer between the base and sub-base aggregates, in the case of PPS, and as a filter layer for the surface runoff for both PPS (Coupe *et al.*, 2006; Gomez-Ullate *et al.*, 2010; Castro-Fresno *et al.*, 2013; Sañudo-Fontaneda *et al.*, 2013; Sañudo-Fontaneda *et al.*, 2014a) and FD (Andrés-Valeri *et al.*, 2013; Coupe *et al.*, 2015). There is also a third objective of a geotextile in SuDS, which consists of the reinforcement of the structural integrity of the drainage system that they are placed in (Pratt, 2003; Castro Fresno *et al.*, 2005). Monitoring of the geotextile has therefore been carried out to include engineering aspects such as hydraulic and structural performance. Sometimes the geotextile can be a wrapping layer in the case of a filter drain (National SUDS Working Group, 2003; Newman *et al.*, 2015) or an impermeable layer to allow water storage in PPS (Gomez-Ullate *et al.*, 2010, 2011; Castro-Fresno *et al.*, 2013; Sañudo-Fontaneda *et al.*, 2013; Sañudo Fontaneda, 2014).

11.3.1   The Role of Geotextiles in Improving Water Quality

In addition to the engineering properties explained above, geotextiles have an important role as a pollutant removal layer (see Chapter 9). It has been widely shown by research carried out across the world that the geotextile does make a difference in the water quality of the stored water in a PPS (Rodríguez *et al.*, 2005; Coupe *et al.*, 2006; Sañudo-Fontaneda *et al.*, 2014b). They have therefore been monitored for their water quality and microbiological properties, both in the laboratory and in the field. However, the geotextile layer has also been identified as the most likely layer to become clogged together with the surface layer in a PPS which may be related to its pollutant retention function (Legret *et al.*, 1996; Rommel *et al.*, 2001; Gomez-Ullate *et al.*, 2011; Sañudo-Fontaneda *et al.*, 2013; Sañudo Fontaneda, 2014; Sañudo-Fontaneda *et al.*, 2014a).

There have been many studies at the laboratory scale, but few have been able to validate these results in the field; an exception is the monitoring work at the Las Llamas car park at the University of Cantabria, Santander (Spain) by Gomez-Ullate Fuente, 2010; Gomez-Ullate *et al.*, 2010; Sañudo-Fontaneda *et al.*, 2014b. At 45 individual tanked car park bays, it was thought to be the biggest fully monitored PPS car park in the world when opened in 2006. The main aim of this project involved the study of the performance of several types of PPS with different surfaces (grass reinforced with concrete cells, grass reinforced with plastic cells, porous asphalt, porous concrete, and impervious concrete blocks with permeable joints) with and without a geotextile layer and the influence of different kinds of geotextiles when used in the PPS structure (Inbitex, One-Way, Composite, Polyfelt TS30 and Danofelt PY150). The project also included the development of tailored, optimised geotextile materials for biofilm growth, their impact on chemical and biological water quality of the infiltrating water and also their water harvesting potential at the field scale (Gomez-Ullate, 2010). The influence of a geotextile as a layer to support the biofilm that degrades hydrocarbons in the structure of the PPS was also studied, including evaporation performance (Gomez-Ullate Fuente, 2010; Gomez-Ullate *et al.*,

2010; Gomez-Ullate *et al.*, 2011a; Castro-Fresno *et al.*, 2013; Sañudo-Fontaneda *et al.*, 2014b). The main conclusions of this work are summarised as follows:

- Runoff pollution treatment: the PPS surfaces can be grouped into three main groups with similar properties: open, closed and green surfaces (Gomez-Ullate *et al.*, 2011a; Castro-Fresno *et al.*, 2013).
- Infiltration capacity: the influence of the type of surface was more significant than the type of geotextile, and the PPS demonstrated a high capacity to store water in the sub-base (Gomez-Ullate *et al.*, 2011b).
- Rainwater reuse : Gomez-Ullate *et al.* (2011b) demonstrated that after a year of storage, the water retained in the sub-base of the PPS had high enough quality, under the Spanish laws (España (Spain), 2007), to be reused in irrigation of green areas and/ or road cleaning.

Castro-Fresno *et al.* (2013) and Andrés-Valeri *et al.* (2014) carried out a study comparing the water quality performance of conventional drainage (a concrete ditch) with two sustainable linear drainage systems, a swale and a filter drain. Both of the sustainable linear systems had a geotextile in their profiles, acting as separation, filtration and treatment layers (Andrés-Valeri *et al.*, 2014). In order to complete this research, three stretches of 20 m length each were designed and constructed in a roadside car park outside the hotel El Castillo de La Zoreda near to the city of Oviedo (Asturias, Spain). After the analyses of the outflow water quality of each drainage system, Andrés-Valeri *et al.* (2014) concluded that the quality of the water from the two sustainable linear drainage systems was substantially better than the one from the conventional drainage system (concrete ditch). Of the two sustainable drainage systems, the filter drain presented the better performance in reducing TSS and turbidity (Andrés-Valeri *et al.*, 2014), enabling the effluent water to be reused in certain applications such as irrigation under the Spanish Royal Decree 1620/2007 (España (Spain), 2007).

Fernández Barrera (2009) explored a different approach in the application of a geotextile layer for the treatment of runoff pollutants from impervious surfaces. The concept of the system for catchment, pre-treatment and treatment (SCPT) was developed successfully and widely published in Castro-Fresno *et al.* (2009), Fernández-Barrera *et al.* (2010), Rodríguez-Hernández *et al.* (2010) and Fernández-Barrera *et al.* (2011). Also, Fernández Barrera (2009) established the operational behaviour of the SCPT in the long term with respect to oil degradation and hydraulic conductivity in the geotextile filter with biodegradation processes taking place inside the SCPT. The hydraulic conductivity of the geotextile decreased slowly with successive rainfall events (Fernández-Barrera *et al.*, 2011), while suspended solid and oil treatment efficiencies after 14 consecutive simulated rain events were high, reducing solids by 80% and oils by 90% (Rodríguez-Hernández *et al.*, 2010).

Later, a project 'Development of catchment of stored rainwater systems, using porous pavements in parking lots, for non-potable use with geothermal low-enthalpy energy' funded by the Spanish Ministry of Science and Innovation (BIA2009-08272) led to publications by Sañudo-Fontaneda *et al.* (2010, 2011 and 2012) and provided further confirmation of the use of a geotextile as an important layer in PPS structures in Spain, leading to the concept and philosophy of sustainable urban construction. It also addressed energy harvesting, exploring the possibility of using the stored water in low-enthalpy geothermal energy systems, and using the harvested rainwater for irrigation and cleaning purposes. Sañudo Fontaneda (2014), then highlighted the benefits of the geotextile layer in reducing peak flow in heavy rainfall events, providing the basis for current research on the hydraulic performance of SuDS structures with geotextile layers.

In the case of the use of geotextile as a hydrocarbon biodegrading layer, work by Newman *et al.* (2011) suggested that the most important factors in its encouragements are:

- availability of oxygen and water
- a suitable surface on which oil can be trapped and microorganisms can grow
- availability of suitable microorganisms and inorganic nutrients.

The suitability of polymeric geotextiles as surfaces on which microorganisms can grow is well illustrated by electron microscope studies carried out by Newman *et al.* (2002). This was later supported by Gomez-Ullate *et al.* (2011), Sañudo-Fontaneda *et al.* (2014b) and Bayón *et al.* (2015), and Puehmeier and Newman. (2008) on treated polyester fabrics complementing Newman *et al.*'s (2010, 2011) work relating to self-fertilising, or nutrient dosed geotextiles.

Coupe (2004) and Jenkins (2002) monitored an in-service car park in the UK, finding that the amount of nitrogen passing through the system was adequate for biofilm formation but that phosphorus was the limiting nutrient. It was later proposed by Newman *et al.* (2011) that animal excreta, leaf fall, materials brought in on car tyres and in the gas phase as oxides of nitrogen from car exhausts, contributed to total nitrogen in the system. Naturally-occurring nitrogen fixing bacteria (e.g. *Rhizobium*) could also have added to the total nitrogen in the PPS. When Osmocote slow release fertiliser pellets were subsequently applied by Jenkins (2002) to the outdoor car park, Newman *et al.* (2011) found that the amounts of inorganic nutrients released from the system were, worryingly, far higher than those reported by Bond (1999), who had shown that they produced elevated biodegradation rates for more than 12 months following initial application. It was first suggested by Jenkins (2002) that the pellets were crushed by traffic rocking the pervious blocks between which they had been brushed, causing unacceptable release rates. Later work by Nnadi (2009), however, clearly showed that this was incorrect, and probably simply an experimental artefact (Newman *et al.*, 2010). Thus, except in circumstances where the PPS effluent is be trapped and reused for irrigation, the use of slow release fertiliser pellets in PPS applications should be discouraged.

More recent research by Mbanaso *et al.* (2013), Charlesworth *et al.* (2013) and Mbanaso *et al.* (2014) has shown that application of herbicides onto PPS test rigs, such as those containing glyphosate (GCH), have substantial impacts on the water quality improvement benefits of the geotextile in PPS. It was found that the diversity of the microbial community in the biofilm was reduced, and thus their function impaired. Some hydrocarbon was released according to the concentration of GCH applied, and metals such as Pb, Cu and Zn were released in higher concentrations in comparison with rigs without herbicide application. In this way, fundamental long-term investigations of microbial diversity, taxonomy and ecology were applied in a polluted urban context.

A floating mat device was developed by Newman *et al.* (2003) and Puehmeier *et al.* (2005), whereby floatation was achieved by stitching the geotextile to a buoyant plastic grid that had been laser cut to fit into the load-bearing void formers and would therefore rise and fall due to incoming stormwater (Puehmeier *et al.*, 2005). The aim was to interact with any thin film of floating hydrocarbons and hold onto the film long enough for biodegradation to take place. It was originally designed to replace the large surface area available on a stone sub-base PPS when plastic void formers were used. In the commercially produced version, the buoyant plastic grid was dispensed with in favour of a single layer of die-cut geotextile, which consisted of a polymer that was treated before spinning to create buoyant bubbles in the fibre matrix. This system formed an integral part of a macro-pervious pavement which was the subject of an extensive field trial between 2012 and 2013 (Newman *et al.*, 2013). Retention of hydrocarbons well within acceptable ranges was observed over the entire period.

11.3.2 Addition of Nutrients to Geotextiles

Spicer *et al.* (2006) reported on other field studies in which the inorganic nutrient needs of oil-degrading microorganisms were satisfied by incorporating an additive directly into the geotextile in the PPS by means of organic micro-beads. While the beads were successful in providing the nutrients required, they were costly to implement and difficult to incorporate due to poor fibre mechanical properties (Newman *et al.*, 2010); subsequent work has developed a commercially available additive known as PM957 (AddMaster UK Ltd., Stafford, UK).

Figure 11.2 shows the results of an experiment reported by Newman *et al.* (2011) in which biodegradation in sealed model microcosms was monitored at two-weekly intervals. The models contained a 10 mm pea gravel laying course on top of a geotextile-supported plastic void forming unit, one set of three with the additive in the geotextile and one set of three without. Activity was measured using carbon dioxide production (replacing the model atmospheres with clean air after each measurement and is presented in Figure 11.2, as a time series with notable events highlighted as numbered points.

Both sets of models were inoculated with the effluent from another established pervious pavement model. Figure 11.2 thus shows that 'wild type' PPS microbes (i.e. not pre-adapted commercially available strains of bacteria and fungi) could be provided with inorganic nutrients by deploying a dosed geotextile in a realistic simulation. Between point 1 on the graph and point 2 (the discontinuation of the fortnightly 1.4 ml oil additions) the models with the nutrient enriched textile showed enhanced oil degradation performance compared with the untreated control. It would appear that the population of oil-degrading organisms grew almost exponentially between points 1 and 2 on the treated textile but a much slower

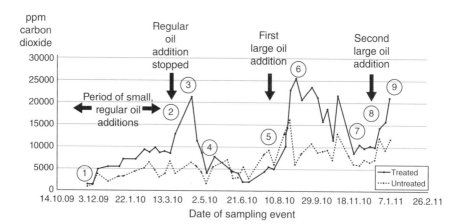

Figure 11.2 Time series for carbon dioxide monitoring of the two sets of models from Newman *et al.* (2011).

trend in carbon dioxide output growth was seen in the control. This was interpreted by Newman *et al.* (2011) as the growth of oil degraders being limited by the amount of available P, with any P being provided largely by the added oil and to a lesser extent by impurities in the simulated rainwater inputs.

After cessation of regular small oil inputs at point 2 until point 3 on the graph, there was a continuous increase in CO_2 concentration in the treated textile models as the oil degrading organisms continued to grow, using the oil remaining in the system. For the control textile there was still plenty of oil available as the previous degradation rate had been much lower than for the treated textile. Between points 3 and 4, the CO_2 output from the treated textile fell off rapidly, due to the most easily degraded fraction of the remaining oil being depleted and the organisms switching their metabolism to utilise the heavier, less degradable fractions. After point 4 and up to the first large (14 ml) oil addition at point 5, the organisms in both sets of models seemed to be operating in a low activity, semi-dormant state. After the 14 ml oil addition, the CO_2 production rate increased rapidly in the treated models with the control models also showing a similar initial response. However, in the latter, this was not sustained for long and by point 6 had fallen back considerably. Between points 6 and 7 the treated models demonstrated significant enhancement in performance over the controls, followed by a gradual decline due to the reduction in carbon as the oil was used up. The controls exhibited a more extended period of activity as their oil utilisation rate was much lower. The response to the second large pulse of oil at point 8 seemed to be following the same as for the earlier addition at point 5.

In contrast to the untreated textiles, the biofilm on the treated textile was visible and significantly populated by fungi. This was not previously observed to the same extent in textiles fertilised with Osmocote pellets, although both Coupe (2004) (via fatty acid methyl ester studies) and Puehmeier (2009) (via studies of microbial DNA) had previously reported that fungi were confirmed components of the microbial assemblage in model systems.

Figure 11.3 shows an electron micrograph of fruiting bodies on one of the fungal hyphae growing on the treated geotextile. This was put forward by Newman *et al.* (2011) as a possible explanation for the rapid response of the treated models, following an application of a large pulse of oil after starvation. It was suggested that the fungal spores would be easily distributed in the models, even in areas not previously contaminated with oil. Fungal spores were both relatively resistant to periods without the availability of a ready carbon source and quick to respond once that carbon source was re-established. Newman *et al.* (2011) also reports that for the models containing treated textiles, the fungal hyphae bridged between the textiles and those areas of the gravel bed that were visibly contaminated with oil but not in direct contact with the ready source of P.

In the previous chapter, the need for long-term retention of hydrocarbons to give time for biodegradation processes to take place was stressed. Obviously, the greater the retention capability the longer a pavement can be stressed with hydrocarbon contamination without the risk of breakthrough. Puehmeier and Newman (2008) described a modified geotextile which, under laboratory testing conditions, showed a hydrocarbon retention capability many times greater than standard non-woven geotextiles. The modified system was shown to be capable of holding back 600 ml of oil per m^2 without exceeding 6 mg/l hydrocarbons in the effluent when 50 mm/hour simulated rain events were applied, a great improvement on standard geotextiles which were producing over 100 mg/l under identical conditions. To date there has been no attempt to incorporate nutrient releasing capability with enhanced oil retention, although it seems logical that this could be easily achieved in a bilayer structure.

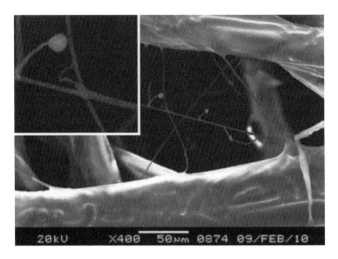

Figure 11.3 Growth of fungi on phosphate treated geotextile.

11.4   Secondary Uses for Urban Water

Previous sections have shown that geotextiles are a versatile tool that can be integrated into the SuDS approach and that can contribute to the environmental protection of downstream environments. While the initial functions of geotextiles in PPS were very similar to those required in geotechnical applications for separation, hydraulic control and the stabilisation of granular material, it was realised that the geotextile may provide additional hydraulic benefits such as the control of discharge rates, aiding evaporative losses where this was a site requirement and provide a degree of water quality improvement (e.g. Pratt *et al.*, 1995; Andersen *et al.*, 1999). It was found that, due to its hydrophobic properties, the geotextile encouraged the development of a shallow but important temporary storage area on the upper surface of the geotextile, establishing a hydraulic head, encouraging percolation and the partitioning of contaminants from water (Brownstein, 1999; Bond, 1999). It was shown that not only were geotextiles the prime site of hydrocarbon and sediment accumulation in PPS (Pratt *et al.*, 1999; Pratt, 2004) but that the stored material was immobilised in an environment with high humidity (partly due to the hydraulic head and the location below concrete blocks away from direct insolation) and with the potential for the addition of nutrients to degrade the trapped hydrocarbons. The interior of the PPS and the position of the geotextile were ideal conditions for aerobic biodegradation, and early studies by Brownstein (1999) and Bond (1999) validated this function of the PPS against a wide spectrum of simulated urban pollution events, explaining to what extent the geotextile worked as a retention structure and bioreactor, what abiotic conditions helped or hindered decontamination and how long the beneficial processes could be maintained. Using plant growth experiments Nnadi (2009) demonstrated that if the water was stored, it could be used for irrigation, instead of potable water, particularly appropriate in the developed world where mains water is routinely used for garden watering. Additional benefits such as growing fruits and vegetables in soil irrigated by stormwater without any risk of soil salinisation, gives further credibility to the 'trap and treat' processes for pollutants, supported by geotextiles and increasing the overall sustainability of hard SuDS and turning a waste into a valuable resource (Nnadi, 2009; Nnadi *et al.*, 2014, 2015).

In applications where water stored in SuDS devices is to be used directly to provide water to a granular substrate, whether the aim is to support plants, simply to evapotranspirate the water to reduce runoff (or a combination of both) or to keep the surface damp for other reasons, the use of geotextiles with wicking capabilities can be advantageous. Although not directly applied to SuDS, but dealing with applications that are very relevant, Azevedo and Zornberg (2013) reported studies on a number of geotextiles with wicking properties. Wicking geotextiles have been applied in SuDS most commonly in green roofs where water is partially stored in plastic void-forming units below the substrate (Voeten, 2014; Voeten *et al.*, 2016). In this case, the capillary action can be entirely provided by the geotextile or achieved using a combination of the geotextile and fibre-based capillary cylinders inserted into the hollow load-bearing columns of plastic void-forming units (Anon, 2014, 2015). An even more specialised application is where sports surfaces are integrated with SuDS principles (e.g. Wilson *et al.*, 2015). Of these applications, the most high profile example is the equestrian surface used at Greenwich Park during the London 2012 Olympics (Pennington, 2014). Built on a temporary steel platform to protect the historic park surface, the equestrian arena had to be designed with SuDS principles in mind and able to deal with a 100-year storm event without discharge to existing sewers. The system also had to deal with a limited available water supply. A combination of sub-platform water harvesting system wicking geotextile played an important role in achieving the water conservation aims.

11.5   Conclusions

Some of these developments in PPS work can be seen as a transition from asking **if** the structures work (e.g. are PPS and components materially and structurally adequate for geotechnical applications or can PPS and their components intercept and retain priority pollutants to the required standard) to questions of **how** they work. These **how** questions may include specific explanatory details of the movements of water through SuDS materials, subject to varying input volumes, velocities and suspended loads, or investigation of the microbiological biodiversity of PPS mesocosms and the detailed interactions between taxa (decomposition, competition, consumers, predators) and the flow of energy-rich decomposable material, such as hydrocarbon pollution, through a PPS.

It is important to note here that the distinction between 'if' or 'how' questions is not linked to specific disciplines or research fields, as it is entirely possible that microbiological studies could be focused on compliance issues, for example the microbiological safety of stored recycled rainwater if Legionnaires were suspected. Equally, a material design change in the PPS, including the specification of the geotextile, would require an exploration of its physical properties and performance, based on demonstrating the first principles underlying the observed processes. Similarly, capabilities that had been developed over many years in laboratory simulations of rainfall for SuDS and also chemical analyses of discharged effluent (Nnadi, 2009) were then taken further in determining the flood prevention properties and infiltration rates of intense rainfall through hard SuDS (Nnadi *et al.*, 2012). Longer-term holistic views of this kind of empirical research would suggest that emerging and challenging fields such as environmental and ecological engineering can comfortably entertain fundamental and applied research questions simultaneously and so provide better answers.

References

Andersen, C.T., Foster, I.D.L. and Pratt, C.J. (1999) Role of urban surfaces (permeable pavements) in regulating drainage and evaporation: Development of a laboratory simulation experiment. *Hydrological Processes.* 13 (4). 597–609.

Andrés-Valeri, V.C., Castro-Fresno, D., Sañudo-Fontaneda, L.A. and Rodríguez-Hernández, J. (2014) Comparative analysis of the outflow water quality of two sustainable linear drainage systems. *Water Science and Technology* 70 (8), 1341–1347.

Anon (2014) Permavoid System: Passive Capillary Irrigation, *Permavoid System Technical Bulletin* Issue No: 5. Available from : http://www.polypipe.com/cms/toolbox/PCL_14_224_Permavoid_Technical_Bulletin_Issue_5a_V2.pdf

Anon (2015) Permavoid in the Urban Environment – Urban Streetscapes. Available from: Permavoid System Technical Bulletin Issue No: 6 May-2015 p1 http://www.polypipe.com/cms/toolbox/PCL_15_293_Permavoid_Technical_Bulletin_Issue_6_V2_LR.pdf

Azevedo, M. and Zornberg, J.G. (2013) Capillary barrier dissipation by new wicking geotextile Advances in Unsaturated Soils – Proc. 1st Pan-American Conference on Unsaturated Soils organised in Cartagena de Indias, Colombia, February 2013. pp 559–566, Caicedo B, Murillo C, Hoyos L., Esteban Colmenares J., Rafael Berdugo I.(eds), CRC/Taylor & Francis Group, London, ISBN 978-0-415-62095-6.

Bayón, J.R., Jato-Espino, D., Blanco-Fernández, E. and Castro-Fresno, D. (2015) Behaviour of geotextiles designed for pervious pavements as a support for biofilm development. *Geotextiles and Geomembranes.* 43 (2). 139–147.

Bond, P.C. (1999) *Mineral Oil Biodegradation within Permeable Pavements: Long-Term Observations.* Unpublished PhD thesis, Coventry University, UK.

Brownstein, J. (1999) *An investigation of the potential for the biodegradation of motor oil within a model permeable pavement structure.* Unpublished PhD thesis, Coventry University, UK.

Castro-Fresno, D., Rodríguez-Bayón, J., Rodríguez-Hernández, J. and Ballester-Muñoz, F. (2005) Sistemas urbanos de drenaje sostenible (SUDS) (Sustainable urban drainage systems, SUDS), *Interciencia.* 30(5). 255–260.

Castro-Fresno, D., Rodríguez-Hernández, J., Fernández-Barrera, A.H. and Calzada-Pérez, M.A. (2009) Runoff pollution treatment using an up-flow equipment with limestone and geotextile filtration media. *WSEAS Transactions on Environment and Development.* 5 (4). 341–350.

Castro-Fresno, D., Andrés-Valeri, V.C., Sañudo-Fontaneda, L.A. and Rodríguez-Hernández, J. (2013) Sustainable drainage practices in Spain, specially focused on pervious pavements. *Water (Switzerland).* 5 (1). 67–93.

Charlesworth, S.M., Mbanaso, F.U., Coupe, S.J. and Nnadi, E.O. (2013) Utilization of glyphosate-containing herbicides on pervious paving systems: laboratory-based experiments to determine impacts on effluent water quality. *CLEAN – Soil, Air, Water.* 42 (2), 133–138.

Coupe, S.J. (2004) *Oil Biodegradation and Microbial Ecology within Permeable Pavements.* Unpublished: PhD Thesis, Coventry University, UK.

Coupe, S.J., Sañudo-Fontaneda, L.A., Charlesworth, S.M. and Rowlands, E.G. (2015) *Research on novel highway filter drain designs for the protection of downstream environments.* SUDSnet International Conference 2015. SUDSnet. September 2015. Coventry, UK.

Coupe, S.J., Newman, A.P., Davies, J.W. and Robinson, K. (2006) *Permeable pavements for water recycling and reuse: initial results and future prospects.* 8th International Conference on Concrete Block Paving. November 6–8, 2006. San Francisco, California, USA.

España (Spain) (2007) Real Decreto 1620/2007, de 7 de diciembre, por el que se establece el régimen jurídico de la reutilisación de las aguas depuradas [Royal Decree 1620/2007 of 7 December, that establish the legal status of the reutilisation of depurated water]. Boletín Oficial del Estado (BOE) 294, 08/12/2007, 50639-50661.

European Commission (1996) *Commission Decision 96/581/EC.* Available from: http://tinyurl.com/o8kcqdu.

Fernández-Barrera, A.H., Castro-Fresno, D., Rodríguez-Hernández, J. and Vega-Zamanillo, Á. (2011) Long-term analysis of clogging and oil bio-degradation in a system of catchment, pre-treatment and treatment (SCPT). *Journal of Hazardous Materials.* 185 (2–3). 1221–1227.

Fernández-Barrera, A.H., Rodríguez-Hernández, J., Castro-Fresno, D. and Vega-Zamanillo, A. (2010) Laboratory analysis of a system for catchment, pre-treatment and treatment (SCPT) of runoff from impervious pavements. *Water Science and Technology*. 61 (7). 1845–1852.

Fernández-Barrera, A.H. (2009) *Desarrollo de un sistema de tratamiento del agua de escorrentía superficial procedente de aparcamientos impermeables usando flujo ascendente y geotextiles*. Published PhD thesis. University of Cantabria. Spain. Available from: http://tinyurl.com/p25u3aq.

Geosynthetics Magazine (2014) *World geosynthetics demand to surpass 5 billion square meters in 2017*. Available from: http://tinyurl.com/zcydtws.

Geosynthetic Society (2015) *Geosynthetics Functions*. Available from: http://tinyurl.com/ov2b3ka.

Giroud, J.P. (1977) *Commentaires sur l'utilisation des geotextiles et les spécifications*. Proceedings of the International Conference on the Use of Fabrics in Geotechnics, Volume III. April 1977. Paris, France.

Gomez-Ullate Fuente, E. (2010) *Study of an experimental pervious pavement parking area to improve sustainable urban water management through the storage and re-use of rainwater*. Published PhD thesis. University of Cantabria, Spain. Available from: http://tinyurl.com/ozwsfwg.

Gomez-Ullate, E., Novo, A.V., Bayón, J.R., Rodríguez-Hernández, J. and Castro-Fresno, D. (2010) *Design and Construction of an Experimental Pervious Paved Parking Area to Harvest Reuseable Rainwater*. In Proceedings of the 7th International Conference on Sustainable Techniques and Strategies in Urban Water Management (Novatech 2010). Novatech. 27 June – 1 July 2010. Lyon, France.

Gomez-Ullate, E., Castillo-Lopez, E., Castro-Fresno, D. and Bayón, J.R. (2011a) Analysis and Contrast of Different Pervious Pavements for Management of Stormwater in a Parking Area in Northern Spain. *Water Resources Management*. 25 (6). 1525–1535.

Gomez-Ullate, E., Novo, A.V., Bayón, J.R., Hernández, J.R. and Castro-Fresno, D. (2011b) Design and construction of an experimental pervious paved parking area to harvest reusable rainwater. *Water Science and Technology*. 64. 1942–1950.

Koerner, R.M. (2012) *Designing with Geosynthetics*. 6th Edition. XLibris. USA. Chapter 1. Standard ISO 10318:2006. Geosynthetics – Part 1: Terms and definitions. General information.

Jenkins, M.S.B. (2002) *A study on the release of inorganic nutrients from an experimental permeable pavement structure and the development of a flow proportionate sampler for PPS runoff studies*, Unpublished BSc thesis, Coventry University, Coventry, UK.

Legret, M., Colandini, V. and Le Marc, C. (1996) Effects of a porous pavement with reservoir structure on the quality of runoff water and soil. *Science of the Total Environment* 189 (190). 335–340.

Mbanaso, F.U., Coupe, S.J., Charlesworth, S.M. and Nnadi, E.O. (2013) Laboratory-based experiments to investigate the impact of glyphosate-containing herbicide on pollution attenuation and biodegradation in a model pervious paving system. *Chemosphere*. 90, 737–746.

Mbanaso, F.U., Charlesworth, S.M., Coupe, S.J., Nnadi, E.O. and Ifelebuegu, A.O. (2014) Potential microbial toxicity and non-target impact of different concentrations of glyphosate-containing herbicide in a model pervious paving system. *Chemosphere*. 100 (34). 34–41.

National SUDS Working Group (2003) *Framework for Sustainable Drainage Systems (SUDS) in England and Wales*. TH-5/03-3k-C-BHEY.

Newman, A.P., Pratt, C.J., Coupe, S.J. and Cresswell, N. (2002) Oil bio-degradation in permeable pavements by microbial communities. *Water Science and Technology*. 45 (7). 51–56.

Newman, A.P. (2003) Liquid Storage Module with a Buoyant Element, Patent GB 29 399 567A.

Newman, A.P., Puehmeier, T., Kwok, V., Lam, M., Coupe, S.J., Shuttleworth, A. and Pratt, C.J. (2004) Protecting groundwater with oil retaining pervious pavements: historical perspectives, limitations and recent developments, *Quarterly Journal of Engineering Geology*. 37. (4) 283–291.

Newman, A.P., Nnadi, E.O., Duckers, L.J. and Cobley, A.J. (2010) Self fertilising geotextiles for use in pervious pavements: A review of progress and further developments, 8th Int. Conf. Sustainable Techniques and Strategies in Urban Water Management, Lyon, France, Groupe de Recherche Rhone-Alpes sur les Infrastructures et l'Eau. Novatech, 2010, CD-ROM.

Newman, A.P., Nnadi, E.O., Duckers, L.J. and Cobley, A.J. (2011) Further developments in self-fertilising geotextiles for use in pervious pavements. *Water Science and Technology*. 64. 1333–1339.

Newman, A.P., Aitken, D. and Antizar-Ladislao, B. (2013) Stormwater quality performance of a macro-pervious pavement car park installation equipped with channel drain based oil and silt retention devices. *Water Resources*. 47 (20) 7327–7336.

Newman, A., Nnadi, E.O. and Mbanaso, F.U. (2015) Evaluation of the effectiveness of wrapping filter drain pipes in geotextile for pollution prevention in response to relatively large oil releases, Proc. World Environmental and Water Resources Congress 2015, Floods Droughts and Ecosystems, May 17–21 2015 Austin, Texas. pp. 2014–2023.

Nnadi, E.O. (2009) An evaluation of modified pervious pavements for water harvesting for irrigation. Unpublished PhD thesis, Coventry University, UK.

Nnadi, E., Newman, A., Duckers, L., Coupe, S. and Charlesworth, S. (2012) Design and validation of a test rig to simulate high rainfall events for infiltration studies of permeable pavement systems. *Journal Irrigation and Drainage Engineering*. 138 (6) 553–557.

Nnadi, E.O., Coupe, S.J., Sañudo-Fontaneda, L.A. and Rodríguez-Hernández, J. (2014) An evaluation of an enhanced geotextile layer in permeable pavement design to improve water quantity and quality. *International Journal of Pavement Engineering* 15 (10) 925–932.

Nnadi, E.O., Newman, A.P., Coupe, S.J. and Mbanaso, F.U. (2015) Stormwater harvesting for irrigation purposes: An investigation of chemical quality of water recycled in pervious pavement systems. *Journal of Environmental Management*. 147, 246–256.

Pennington, P. (2014) London 2012 legacy: design and reuse of temporary equestrian platforms. *Proc. ICE – Civil Engineering*, Civil Engineering Special Issue, 167 (CE6) pp. 33–39.

Pratt, C.J. (2003) *Application of geosynthetics in sustainable drainage systems*. 1st UK National Geosynthetics Symposium. Geosynthetics Society. June 2003. Loughborough, UK.

Pratt, C.J. (2004) *A Review of Published Material on the Performance of Various SUDS Components*. UK Environment Agency report. Available from: http://tinyurl.com/qhx64jt

Pratt, C.J., Mantle, J.D.G. and Schofield, P.A. (1995) UK research into the performance of permeable pavement, reservoir structures in controlling stormwater discharge quantity and quality. *Water Science and Technology*. 32 (1) 63–69.

Pratt, C.J., Newman, A.P. and Bond, P.C. (1999) Mineral oil bio-degradation within a permeable pavement: long term observations. *Water Science and Technology*. 29 (2). 103–109.

Puehmeier, T., de Dreu, D., Morgan, J.A.W., Shuttleworth, A. and Newman, A.P. (2005) Enhancement of oil retention and biodegradation in stormwater infiltration systems, Proc.10th International Conference on Urban Drainage, Copenhagen/Denmark, 21–26 August 2005. CD-ROM.

Puehmeier, T. and Newman, A.P. (2008) Oil retaining and treating geotextile for pavement applications, *Proc. 11th International Conference on Urban Drainage*, Edinburgh, UK, 31 Aug – 5 Sept 2008, CD-ROM.

Puehmeier, T. (2009) *Understanding and Optimising Pervious Pavement Systems and Source Control Devices Used in Sustainable Urban Drainage*. Unpublished PhD thesis, Coventry University, UK.

Rodríguez, J., Castro, D., Calzada, M.A. and Davies, J.W. (2005) *Pervious pavement research in Spain: structural and hydraulic issues*. 10th International Conference on Urban Drainage (ICUD). August 2005. Copenhagen, Denmark.

Rodríguez-Hernández, J., Fernández-Barrera, A.H., Castro-Fresno, D. and Vega-Zamanillo, A. (2010) Long-term simulation of a system for catchment, pre-treatment, and treatment of polluted runoff water. *Journal of Environmental Engineering*. 136 (12) 1442–1446.

Rommel, M., Rus, M., Argue, J., Johnston, L. and Pezzaniti, D. (2001) *Car park with 1 to 1 (impervious/permeable) paving: performance of formpave blocks*. 4th International Conference Novatech: Sustainable techniques and strategies in urban water management. Novatech. June 2001. Lyon, France.

Sañudo-Fontaneda, L.A. (2014) *The analysis of rainwater infiltration into permeable pavements, with concrete blocks and porous mixtures, for the source control of flooding*. Published PhD Thesis. University of Cantabria, Spain. Available from: http://tinyurl.com/zf8rxmb.

Sañudo-Fontaneda, L.A., Rodríguez-Hernández, J., Calzada-Pérez, M.A. and Castro-Fresno, D. (2014a) Infiltration behaviour of polymer-modified porous concrete and porous asphalt surfaces used in SuDS techniques. *Clean – Soil, Air, Water*. 42 (2) 139–145.

Sañudo-Fontaneda, L.A., Charlesworth, S., Castro-Fresno, D., Andrés-Valeri, V.C.A. and Rodríguez-Hernández, J. (2014b) Water quality and quantity assessment of pervious pavements performance in experimental car park areas. *Water Science and Technology* 69 (7) 1526–1533.

Sañudo-Fontaneda, L.A., Andrés-Valeri, V.C.A., Rodríguez-Hernández, J. and Castro-Fresno, D. (2014c) Field study of the reduction of the infiltration capacity of porous mixtures surfaces tests. *Water (Switzerland)* 6 (3) 661–669.

Sañudo-Fontaneda, L.A., Rodríguez-Hernández, J., Vega-Zamanillo, A. and Castro-Fresno, D. (2013) Laboratory analysis of the infiltration capacity of interlocking concrete block pavements in car parks. *Water Science and Technology.* 67 (3) 675–681.

Sañudo-Fontaneda, L.A., Castro-Fresno, D. and Rodríguez-Hernández, J. (2012) *Investigación y desarrollo de firmes permeables para la mitigación de inundaciones y la 'Valorización Energética del Agua de lluvia (VEA)'*. VI Congreso Nacional de la Ingeniería Civil. Retos de la Ingeniería Civil. February 2012. Valencia, Spain.

Sañudo-Fontaneda, L.A., Castro-Fresno, D., Rodríguez-Hernández, J. and Borinaga-Treviño, R. (2011) *Comparison of the infiltration capacity of permeable surfaces for Rainwater Energy Valorisation*. 12th International Conference on Urban Drainage (ICUD). September 2011. Porto Alegre (Rio Grande do Sul), Brazil.

Sañudo-Fontaneda, L.A., Castro-Fresno, D., Rodríguez-Hernández, J. and Ballester-Muñoz, F. (2010) *Rainwater energy valorization through the use of permeable pavements in urban areas*. 37th IAHS World Congress on Housing Science. Design, Technology, Refurbishment and Management of Buildings. October 2010. Santander, Spain.

Spicer, G.E., Lynch, D.E. and Coupe, S.J. (2006) The development of geotextiles incorporating slow-release phosphate beads for the maintenance of oil degrading bacteria in permeable pavements. *Water Sci. Technol.* 54(6-7) 273–80.

Van Santvoort, G.P.T.M. (1994) *Geotextiles and Geomembranes in Civil Engineering*. 2nd Edition. AA. Balkema/Rotterdam/Brookfield. Rotterdam, The Netherlands. Cap. 3.

Voeten, J.G.W.F. (2014) Vertical capillary water transport from drainage to green roof, Proc. 12th Annual Green Roof and Green Wall Conference, Nashville, Nov 12–16 2014. Available from: http://www.greenroofs.org/index.php/component/content/category/8-mainmenupages (including audio/visual recording of all presentations).

Voeten, J.G.W.F., van de Werken, L. and Newman, A.P. (2016) Demonstrating the use of below-substrate water storage as a means of maintaining green roofs – Performance data and a novel approach to achieving public understanding. Paper accepted for presentation at World Environmental and Water Resources Congress, 2016, May 21–26, Palm Beach FL. USA.

Wilson, S., Culleton, P.D., Van Raam, C.H., Shuttleworth, A.B. and Andrews, D.G. (2015) Areas for equestrian activities using structural modules. Patent US 8657695 B2 CA2753344A1, EP2401435 A1,US20120040767,WO2010097579A1.

Section 4   Multiple Benefits of Sustainable Drainage Systems

Natural Flood Risk Management and its Role in Working with Natural Processes

Tom Lavers and Susanne M. Charlesworth

12.1 Introduction

This chapter considers the emerging research field of 'Natural Flood Risk Management' (NFRM), and the importance of developing and collating the existing evidence to meet national and international policy agendas in adapting to the impacts of climate change by working with natural processes. However, this innovative field is not without challenges when it comes to practical application and meeting wider stakeholder and financial support in a changing political and economic climate.

12.2 Defining NFRM

NFRM is defined here as the alteration, restoration or use of landscape features in order to work closely with catchment-based natural processes so as to alleviate flood risk (adapted from POST, 2011). The majority of research into NFRM has emerged in the past decade across England, Wales, Scotland and continental Europe.

NFRM works with natural processes to alleviate current and future flood risk. The Pitt (2008) report, which was undertaken after the UK flooding of the summer 2007, concluded that flooding from a range of sources could no longer be managed by building ever higher, lengthier and heavier defences in urban and rural areas. The review emphasised the need to 'work with natural processes' as part of integrated portfolios of responses to flooding and coastal erosion, as highlighted through recommendation 27. NFRM capitalises on this, principally by implementing measures in the rural environment, as part of the responses required in current flood risk management (FRM) (Doak, 2008).

NFRM is a philosophy of FRM (Freitag *et al.*, 2009), which considers catchment-wide flow regimes, along with developing a wider flood resilient community, defined by Thieken

Sustainable Surface Water Management: A Handbook for SuDS, First Edition.
Edited by Susanne M. Charlesworth and Colin A. Booth.
© 2017 John Wiley & Sons, Ltd. Published 2017 by John Wiley & Sons, Ltd.

Table 12.1   The two main approaches to natural flood risk management (WWF, 2007).

Broad methods	Description
Restoration	The process of returning the existing system to a more natural one (e.g. re-meandering, restoration of disconnected floodplains, uplands grid blocking, restoration of native catchment woodlands, reinstatement of riparian woodlands and coastal realignment).
Alteration (including enhancement)	Is the improvement to, or enhancement of, an existing function for the purpose of flood risk management, including partial restoration or natural processes and soft engineering, e.g. enhancing the capacity for floodplains to store water (washlands), increasing channel roughness.

et al. (2014) as a community aware of their current flood risk and enabled to adapt to future implications of their changing flood risk. This is often predominately associated with purely theoretical studies although with a growing body of support across relevant agencies and organisations. As an approach, NFRM considers a wide range of methods (Table 12.1) that are broadly associated with restoration and alteration (WWF, 2007).

While working with natural processes considers restoring or altering existing practice to emulate natural processes, the Environment Agency Working Group's response to the Pitt Review (Defra, 2009), recognised that working with natural processes to manage flood risk could involve considerable intervention and, as a result, be far from natural. However, NFRM is supposed to be as natural as possible, with minimal intervention (Johnstonova, 2009).

Both methods of restoration and alteration are not unique to NFRM, and have been included in the Environment Agency Rural SuDS Guidance and the Environment Agency Guidance for Working with Natural Processes (Environment Agency, 2014; Avery, 2012). Rural SuDS guidance (Avery, 2012) recognises the importance of 'working with natural hydrological and morphological processes', as does NFRM, yet the key difference is the method by which they are applied. NFRM's approach is unique in that it can integrate with rural SuDS as part of a holistic catchment-based approach to FRM. In order to achieve flood resilience using these approaches, knowledge of the contributing hydrological regimes is required, as is the likelihood of achieving two aims in terms of flood alleviation: 'upstream thinking' as well as 'flow desynchronisation' (delaying the flood peak) (POST, 2014).

Flow desynchronisation refers to the interruption of flood peaks across a catchment before they reach a settlement, while *upstream thinking* similarly considers FRM in the higher reaches of the fluvial system. This develops the source–pathway–receptor relationship discussed in the rural SuDS guidance (Avery, 2012) and information on runoff attenuation features (RAFs), by addressing water quantity at the source as well as the importance of measures that deal with all three elements on a much larger scale (Blanc *et al.*, 2012).

It is recognised internationally that many populations who are vulnerable to flood risk are located in the lower reaches of catchments; this will only worsen with further population increases, combined with regional impacts of more intense rainfall and sea level rise (Feyen *et al.*, 2012). Mauch and Zeller (2009) recognised the significance that fluvial systems play in the history of settlement, with the fertility of low-lying floodplains attracting early civilisations. Upland thinking therefore considers intercepting and retaining flows with the potential to flood before they impact vulnerable low-lying communities across catchments.

Upstream thinking, a term coined by the Westcountry Rivers Trust and Cornwall Wildlife Trust (2015), was derived from a project that installed NFRM measures across Cornish river catchments, principally the Taw, with the primary aim of improving the ecological water quality status of the catchment's watercourses (Couldrick *et al.*, 2014). A second example is that of the Pontbren, Wales, where a group of farmers who were initially concerned with improving land management practice for water retention found that measures to improve agricultural land in upland catchments had flood alleviating potential downstream (Wheater *et al.*, 2008). Most of the agricultural fields had been previously intensified, including the installation of subsurface drainage, and the area was largely used for sheep farming (Ballard *et al.*, 2010). However, tree shelter belts and sensitive farming practices that have been adopted, including suitable locations for access to water for cattle, are not directly for FRM benefits but, nonetheless, it was found that they provided flood resilience to downstream communities (Wheater *et al.*, 2008). While not all NFRM measures are designed for upstream reaches of the catchment, the ideology is still pertinent, in that measures can provide multiple benefits in areas that were not previously considered to provide any value in terms of land management change, especially where they were identified as contributing considerable runoff to flood generation (Morris and Wheater, 2006).

The role of flow desynchronisation develops this idea of addressing land-use contribution to flood generation in terms of hydraulic conductivity across a catchment. On a localised scale, hydraulic conductivity is the measure of how fast water will travel under a unit pressure gradient, usually used in the context of soils or other porous media (McIntyre and Thorne, 2013). Emerging research has begun to identify that land-management practices in the rural environment can impact hydraulic conductivity across sub-catchments. The Floods and Agriculture Risk Matrix (FARM) (Wilkinson *et al.*, 2013) accepts this contribution, and developed a decision-support tool to assist farmers on a local level to realise their impacts at the larger scale. Importantly, this process recognises the significance of successful engagement with land managers, further discussed in Sections 12.3.2 and 12.3.3.4.

The NFRM approach therefore considers that alternatives to 'hard engineering' are a more sustainable approach to FRM, in managing both the sources and pathways of floodwaters. Werritty (2006) perceives this to be a 'new paradigm' in which adaptation to climate change can be addressed, particularly in the context of the isolation and long-term displacement of rural communities when a large flood event does occur (Lane *et al.*, 2006). This was evident with the recent Cumbrian floods in the UK (December 2015), in which abnormally large quantities of intense rainfall cut off major transport links across the Lake District. While the existing defences proved effective to a certain level, it begged the question of whether an integrated 'catchment systems engineering' approach could contribute to climate change adaptation through flood alleviation (Wilkinson *et al.*, 2014).

Emerging theoretical research into ecosystem services shows that these measures can also provide wider benefits than just flood alleviation (e.g. Iacob *et al.*, 2012). This includes benefits to water quality, improving local recreational space and improvements in biodiversity. Nonetheless, many studies (outlined in Table 12.2) have been predominantly theoretical with limited evidence, based either on mapping and modelling research with limited monitoring of the effectiveness of such measures. This is discussed further in the following section.

12.3   Examples of NFRM Studies

The pilot studies and limited research that have been conducted into NFRM have identified the theoretical gains sought through changing land use practices in variable catchments, through either alteration or enhancement (Table 12.2). While this chapter refers to this

Table 12.2 Information of various NFRM studies, adapted from Iacob et al. (2012) and Environment Agency (2014) to consider catchment specifics (CEH 2009). With the exception of Poyo, Spain and Kamp, Austria as FEH catchment descriptors are only applicable in the UK.

Catchment and NFRM measure	Country	Area (km²)	FEH catchment descriptors				Approach and reference
			BFIHOST	FARL	SAAR (mm)	SPRHOST (%)	
Upland afforestation (riparian, floodplain and wider catchment)							
Poyo	Spain	380	—	—	—	—	1D Modelling – Francés et al. (2008)
Kamp	Austria	600	—	—	—	—	1D Modelling – Francés et al. (2008)
Parrett	England	1675	0.447	0.987	860	43.3	1D/2D Modelling – Park et al. (2006)
Pont Bren	Wales	12	0.464	0.966	1659	37.3	1D/2D Modelling – Wheater et al. (2008)
Pickering Beck	England	66	0.688	1.000	838	20.1	1D/2D Modelling – Odoni et al. (2010)
Upland drainage (RAFs and rural SuDS)							
Ripon	England	120	0.421	0.970	901	40.2	1D/2D Modelling – JBA (2007)
Blacklaw Moss	Scotland	0.07	0.348	1.000	1543	50.1	Monitoring – Robertson et al. (1968)
Llanbrynmair	Wales	3	0.456	1.000	1587	38.3	Monitoring – Leeks and Roberts (1987)
Belford	England	10	0.329	0.997	677	41.5	Opportunity Mapping, 1D/2D Modelling and Monitoring – Quinn et al. (2013)
River Darent	England	7	0.771	0.981	768	20.8	Opportunity Mapping and Engagement – Evans et al. (2014)
Wetlands and floodplain alteration							
Cherwell	England	135	0.386	0.958	669	43.7	1D Modelling – Acreman (1985)
Quaggy	England	1	0.520	1.000	646	30.6	Monitoring – Potter (2006)
Sinderland	England	2	0.662	1.000	822	22.7	Monitoring – Defra et al. (2010)
Combined Measures							
Eddleston Water	Scotland	37	0.501	0.994	900	36.2	Monitoring – SEPA (2011)
River Laver	England	79	0.420	0.982	912	39.9	1D/2D Modelling – Nisbet and Thomas (2008)

process as NFRM, and primarily discusses the flood risk potential in relation to NFRM examples, there are other benefits with equal significance in meeting other agendas, including reducing soil erosion, beneficial water quality, carbon storage and biodiversity. With regard to flooding, CEP (2010) discussed a number of options for land use change that may have local beneficial impacts on flood risk. Odoni (2014) found that measurable differences are often more easily monitored and determined at the smaller catchment scale (<50 km²), primarily due to the increased complexity of upscaling the benefits to larger catchment areas and accurately determining the benefits with a quantitative evidence base (Blanc et al., 2012).

Table 12.2 shows that monitoring studies are often conducted at smaller scales, due to the feasibility in data collection. However, POST (2011) states that the existing evidence base can determine general relationships between NFRM features and their varied effectiveness dependent on catchment specifics, as well as meeting the wider benefits of other ecosystem-services. This variable relationship across a catchment is illustrated in Figure 12.1. This includes land management and land-use change, which is most effective upstream on a large scale (spatially diffuse), as opposed to, for example, wetlands that are more suited in the lower reaches of a catchment.

While Figure 12.1 gives a broad overview, catchment specifics must be considered; generalisations between two catchments would be meaningless due to the variation in rainfall, topography, soil type, geology and land-use, all influencing the flow dynamics that NFRM measures would need to address (SEPA 2016). These are reflected in the Flood Estimation Handbook (FEH) catchment descriptors (CEH, 2009), which determine such values based on the principles shown in Table 12.3.

While the relationship in Figure 12.1 may seem simple, catchments are complex interfaces between rainfall and runoff. As the scale increases, so does the intricacy of both physical and social elements. Firstly, the area that intercepts the rainfall event becomes larger, and

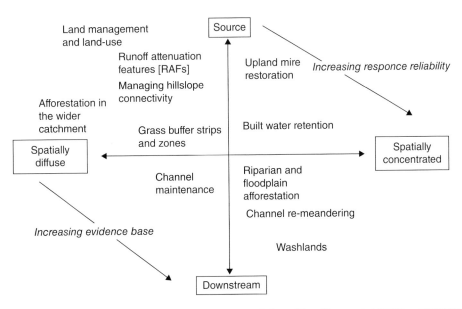

Figure 12.1   Catchment-scale classification of NFRM measures (adapted from Thorne et al. (2007) and POST (2011)).

Table 12.3   FEH catchment descriptors (NRFA, 2015).

Catchment descriptor	Definition
Area (km²)	The catchment or sub-catchment drainage area
Base flow index of hydrology of soil types (BFIHOST)	This base flow index is a measure of catchment responsiveness, derived from the 29-class hydrology of soil types (HOST) classification as determined from Boorman *et al.* (1995)
Flood attenuation by reservoirs and lakes (FARL)	Any reservoirs or lakes within a catchment are believed to have an effect on flood response, but it is those directly linked to the channel networks that are most likely to influence flood attenuation. Values close to unity (1.0) indicate the absence of attenuation due to lakes and reservoirs, whereas index values below 0.8 indicate a substantial influence on flood response.
Standard average annual rainfall (SAAR) (mm)	Average annual rainfall in the standard period (1961–1990) in millimetres.
Standard percentage runoff of hydrology of soil types (SPRHOST)	Standard percentage runoff (%) associated with each HOST soil class. This can be used to derive SPRHOST over a catchment. SPRHOST can be derived from channel flow data (if available).

therefore how the runoff responds has greater uncertainty. Larger catchments are also likely to have more contributing sub-catchments, which make determining the influential flow regimes more complex (Wainwright and Mulligan, 2004). Quinn *et al.* (2013) refer to this ethos as 'catchment systems engineering', recognising that a flood event is indicative of the storm event, illustrated through varied volumes and extents across the catchment.

The variable rainfall event itself also generates uncertainty, with no two storms identical, and a larger catchment is likely to be more impacted by variance in storms (Shaw *et al.*, 2011). Current monitoring research (evident in Table 12.3), illustrates the limited scope in data for analysis of how NFRM measures respond to great annual exceedance probability (AEP) events. But it is recognised that the scale of alleviating benefits from NFRM is likely to decrease with an increase in return interval annual exceedance probability (AEP) (Carter, 2014). However, long-term monitoring past a QMED return period event, termed the index flood (Lambe *et al.*, 2009), for a return period of every two years, would be needed to obtain greater accuracy. The pedological, hydrogeological and land-use variation across the catchment also influences the varied response at a larger scale, such as infiltration and runoff rates, producing different hydrological regimes (O'Connell *et al.*, 2004).

12.3.1   NFRM Measures

NFRM can be broadly aligned into three categories within both restoration and alteration, including:

1. upland afforestation, such as recognise riparian, floodplain and wider catchment planting, for alleviating flood risk in upland areas of a catchment (Sharp, 2014)
2. upland drainage alteration, including changes through runoff attenuation features and rural SuDS (Quinn *et al.*, 2013)
3. wetlands and floodplain alteration (in the lower reaches of the catchment) based on catchment-systems engineering and often supported by a combination of measures

(including 1 and 2), which are considered the most effective method of implementing NFRM at a catchment scale (Spense and Sisson, 2015).

The following section describes NFRM examples from the above approaches and explains their flood alleviation potential.

12.3.1.1 Upland Afforestation

Studies to date have shown that targeted afforestation outside the floodplain, in areas that intercept runoff, is likely to have an impact on flood risk at a very local scale (Chisholm, 2014), with infiltration likely to increase, along with interception, to as much as 60 times greater than areas of pastureland. Wheater *et al.* (2010) concluded that this kind of approach has three main influences:

1. The water would be intercepted and evapo-transpired by the trees, although this is unlikely to exceed 10% of the total volume during heavy rainfall events.
2. It would contribute to root system development, especially in mature trees, increasing infiltration capacity (Armbruster *et al.*, 2006).
3. Increasing the surface cover would increase Manning's 'n' value for surface runoff, reducing both discharge rates and suspended solids (hence, maintaining channel conveyance/capacity) and importantly desynchronising flood flows. This is highly dependent on a good management/maintenance plan to ensure that canopy growth does not prevent surface growth from developing. However, a lack of surface growth can be detrimental to the determined value of Manning's 'n' value for roughness and can decrease water quality through topsoil loss and increased discharge rates (Nisbet and Thomas, 2008).

There is, however, a lack of physical monitoring of the systems efficiency in order to quantify the benefits of addressing surface runoff from storm events (McIntyre and Thorne, 2013). Further, due to the complexity of larger catchments, modelling studies are predominately based on smaller ones (<50 km²). However, one exception was Broadmeadow *et al.* (2013), who used spatial datasets to develop a tool to locate areas for planting new woodland in the catchment to reduce rainfall runoff. This data encouraged the uptake of afforestation as part of a 'rural management train', in accordance with SuDS guidance (McBain *et al.*, 2010; Avery, 2012).

12.3.1.2 Upland Drainage Alteration

Measures involving upland drainage are based on catchment classification, not only of the characteristics, but also of an accurate understanding of base flow in order to know whether desynchronising flows through measures, such as drain blocking, could generate unwanted larger peaks elsewhere (JBA, 2007). Upland drainage is based on two principal methods: on-line and off-line storage. On-line refers to in-channel conveyance that is intercepted or stored by a particular NFRM measure; off-line is the diversion of water from the in-channel conveyance system (Wilkinson *et al.*, 2014).

An example of on-line storage includes large woody debris (LWD) dams, shown below in Figure 12.2 from Stroud, Gloucestershire. LWDs are in-channel measures that directly

Figure 12.2   LWDs near Stroud, Gloucestershire (TL), where the red arrow illustrates interruption of high-flow pathways.

interrupt high-flow pathways with the potential to hold back water up to 0.2–0.5 km upstream (Nisbet *et al.*, 2011). The term LWD can be applied to pieces of dead wood larger than 0.1 m in diameter and 1.0 m length, as well as rooted riparian woodland systems that accumulate flow pathways (Linstead and Gurnell, 1998; Thomas and Nisbet, 2012).

There are two methods of LWDs being introduced into flow pathways or fluvial systems: continual or episodic inputs (Faustini and Jones, 2003). Continual mechanisms include the regular introduction of wood as a result of natural tree mortality or gradual bank undercutting (Figure 12.2). This tends to add small amounts of wood at frequent intervals. In contrast, episodic inputs, such as severe flood events, occur more infrequently but can add large amounts of debris to the channel network. In contrast, off-line storage intercepts flow pathways and directs them out of the main in-channel conveyance systems, as shown in Figure 12.3 from Honeydale Farm, Evenlode Valley (Wilkinson *et al.*, 2014; Cotswolds Honeydale Farm, 2015).

These include measures such as off-line storage ponds that can be regulated to monitor discharge levels and released through leaky dam structures in order to not synchronise flood flows, alternatively known as 'flood spilling' (Wilkinson *et al.*, 2014). However, developing a quantitative evidence base for such measures is challenging because traditional modelling techniques are more suited to determining 1D in-channel conveyance (Thomas and Nisbet, 2012). This is because features outside of the channel must be assigned 2D values for Mannings 'n' roughness in order to represent frictional drag (Rose, 2011); this is further discussed in Section 12.3.3.2.

12.3.1.3 *Wetlands and Floodplain Alteration*

Wetlands and floodplains are often considered more appropriate downstream in catchments (Figure 12.1), as opposed to the usual upstream application of the two previous approaches (Hess *et al.*, 2010). In terms of designing a management train, wetlands and floodplain alteration measures are often located in association with large-scale regional retention areas that are much larger in size than other NFRM measures (Avery, 2012). Johnstonova (2007) acknowledged the significance of these areas for contributing to flood alleviation at the catchment scale, noting the multiple benefits of the Insh Marshes, RSPB nature reserve, Strathspey, Scotland, the largest and most naturally functioning floodplain

Figure 12.3   Off-line storage pond, Honeydale Farm.

in the UK. This extends to 8 km in length, accommodating out of bank flow from the River Spey as well as contributing surface flow and tributaries. Re-naturalising of this floodplain increased habitat provision for migrating birds, including varieties of breeding waders (over 1000 pairs), wildfowl (over 50% of the UK goldeneye population), spotted crakes, populations of wintering whooper swans and hen harriers, as well as a rich diversity of plants and invertebrates (Davies, 2004). Taken as an NFRM measure, during times of snowmelt and heavy precipitation in winter and spring, it can regularly be covered in floodwater up to 10 km^2 (Doak, 2008). However, even though there is potential for these measures to provide flood alleviation, as well as other benefits, there are barriers to their uptake when it comes to practical application, as discussed in the following section.

12.3.2 Practical Application

In terms of seeing results of the implementation of NFRM measures, a rather limited literature (Fitton *et al.*, 2015; Holstead *et al.*, 2015) has begun to investigate the significant role that society, stakeholders and policy intervention can play in its uptake. It is seen that

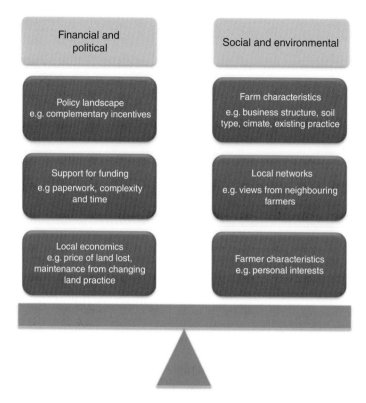

Figure 12.4 Factors that influence farmer's decisions on the implementation of NFRM features (adapted from Holstead *et al.*, 2015). This illustrates the balance of all factors for farmer engagement and support.

NFRM can support interconnectivity of ecosystem services, playing a role in 'reconnecting people with the landscape' (Nicholson *et al.*, 2012).

Holstead *et al.* (2015) found that the uptake of NFRM in the Scottish Borders was a sensitive issue across a multitude of scales (Figure 12.4). This study primarily focused on measures that impacted farmers, who are often the key decision-makers when it comes to land alteration practice. Larger, catchment-scale implementation would thus require more engagement with relevant partners, including landowners and farmers, to support delivery mechanisms including Countryside Stewardship and Catchment Sensitive Farming grants, as advised by Environment Agency guidance (Avery, 2012).

12.3.3 The Importance of the Study Approach

As Table 12.3 shows, methods and approaches that have either identified or analysed the role of NFRM have used one or more of four elements: mapping, modelling and to a lesser extent monitoring and engagement. All four approaches recognise the importance of carefully considering NFRM locations in order to provide the greatest benefit. For a completely 'sustainable' approach to flood-risk management, measures must be considered as a 'no-regrets' investment with a great deal of evidence (Werritty, 2006). The following sections consider each of these approaches in the study of NFRM and their contribution to a better understanding of their benefits.

12.3.3.1 Mapping

Mapping studies have utilised GIS techniques that consider 'opportunities' for implementing measures. An example of this is the Forestry Research (The Research Agency of the Forestry Commission) national opportunity mapping of areas for woodland planting and attenuation features. Broadmeadow *et al.* (2013) initially focused on removing areas unsuitable for NFRM measures, calling them areas of 'constraints'. These included:

- Ministry of Defence (MoD) Land
- wash lands
- floodplain buffer around urban centres (often 500 m around towns and villages and 300 m along highways)
- Ramsar sites
- areas of natural beauty (AONBs), sights of special scientific interest (SSSIs), special protection areas (SPAs), special areas of conservation (SACs) and biodiversity action plan (BAP) areas
- national parks and battlefields
- Grade 1 agricultural land

Removing these constraint areas prevents any potential measure having detrimental impacts on the existing landscape, therefore placing limitations on where such measures are applicable. Other GIS techniques have been used to inform of suitable areas to install NFRMs, such as the use of satellite imagery in the form of LiDAR, which enables surface runoff pathways to be mapped (SNIFFER, 2011). These pathways are widely recognised to significantly increase hydraulic conductivity across a catchment, causing increased magnitude and frequency of peaks over thresholds (POTs) (Wilkinson *et al.*, 2013). This is indicated in Boorman *et al.*'s (1995) hydrology of soil type (HOST) data classification, which reflects the interaction between soil and water. A significant issue with this data is that, while it is freely available, resolution can be poor, leading to substantial levels of uncertainty when attempting to determine water quantity as well as specific catchment characteristics.

Public bodies, including the UK Environment Agency, Natural England and local authorities, hold a wealth of data useful for understanding base flow. These include publications such as updated flood maps for surface water (uFMfSW) and current land use practice, and agricultural land classification (ALC) data. The Forestry Research group of the Forestry Commission (Broadmeadow *et al.*, 2013) has also published comprehensive data for potential new woodland (PNW), either riparian, floodplain or wider catchment.

12.3.3.2 Modelling

Modelling contributes to both a visual and, to some extent, statistical evidence base that can be used to support the preliminary mapping process. Previous studies have generally focused on floodplain/in channel interventions, where modelling methodologies are more fully understood (Acreman, 1985). However, as shown by LiDAR data, surface runoff of the wider catchment also contributes and therefore needs to be considered and accurately modelled when naturally managing flood risk across a catchment (McIntyre and Thorne, 2013). This therefore requires consideration of both in-channel conveyance mechanisms (1D) and floodplain flows across detailed elevation grids (2D), namely 1D-2D linked hydraulic models. The 2D elevation grids consider features existing in the floodplain itself, and Rose (2011) assigned Mannings 'n' roughness values based on land use classifications across 2D grids, which could be altered dependent on the features present. For example, the presence of woodland planting increases frictional drag on flowing water and requires a greater 'n' value than pasture land (Nisbet and Thomas, 2008). The majority of more recent studies (Table 12.2), apply this type of assessment, but a significant level of data is needed of both in-channel dynamics and ground surveys, in order to accurately calculate flood flows (Shaw *et al.*, 2011). As a result, most study sites have only undertaken this level of investigation at the small sub-catchment scale (<50 km²).

The Pickering study, in North Yorkshire, adopted a coupled hydrological–hydraulic model called OVERFLOW, developed by Durham University (Odoni *et al.*, 2010, and Odoni and Lane 2010). This process used flow accumulation and routing algorithms to calculate how rain falling on the catchment would flow through the system, converting flows to depths based on a 'Mannings roughness map' that was altered, as in other 2D models, to replicate changes in vegetation type. A further benefit of effective modelling is the determination of 'backwater' effects or impacts upstream (Odoni *et al.*, 2010). While this is considered in the GIS mapping buffer, based on the quality of existing data, modelling can determine the extent to which interception will impact upstream. This is important because measures, such as large woody debris dams, can extend backwater for up to 0.2– 0.5 km (Nisbet and Thomas, 2008). This quantitative evidence base can also allow studies to note the relationship between storm hydrographs within sub-catchments to reflect the impact NFRM measures can have on hydraulic conductivity and the possibility of synchronising peaks in the lower reaches of the catchment.

12.3.3.3 *Monitoring*

The reference to monitoring in this chapter just considers hydraulic conductivity in terms of catchment-systems engineering. Addy *et al.* (n.d.) made use of empirical data to show the impact of NFRM on hydrology, water quality and ecology in Scotland. Monitoring was based on 42 different measures across the Bowmont catchment, of varying catchment sizes, using time-lapse cameras to record geomorphological changes with time after implementation of NFRM. Young *et al.* (2015) note that time-lapse cameras offer a unique opportunity to remotely monitor flow discharge at a high resolution in areas downstream of NFRM implementation.

Existing monitoring studies (Table 12.2) are currently inconclusive as to the effectiveness of NFRM measures at reducing flood peaks and desynchronising catchment flows larger than the QMED return interval, or the median annual maximum flow series – the flow that has an annual exceedance probability of 50%, or a return period of two years (Shaw *et al.*, 2011). Therefore, these unique approaches considered by Addy *et al.* (n.d.) reflect a need to understand the influence that NFRM measures have on flow dynamics, as well as potential for wider ecosystem-services derived from multiple-benefits.

12.3.3.4 Engagement

Community engagement is paramount for successful FRM and is recommended as a key area of improvement (Cornell, 2006; Pitt, 2008; POST, 2011), since successful engagement with stakeholders can assist in the decision-making process. For example, Evans *et al.* (2014) worked closely with landowners and farmers to implement NFRM measures in the catchment of the River Darent, Kent, UK, which allowed for a more informed process, in spite of it being rather a lengthy procedure overall. SEPA (2011) engaged with farmers and landowners in the Allan Water catchment, Scotland, using citizen science to raise awareness. Wide engagement across the catchment was also a key influence in NFRM uptake in the Scottish borders (Howgate and Kenyon, 2009).

12.4   Significance of NFRM in Meeting Policy Agendas

FRM and the associated policies that drive it are recognised as key elements in supporting stakeholders to adhere to them (Johnson and Priest, 2008). This is just as applicable to NFRM, with key agendas at various scales and a focus on supporting the implementation of such measures. In recent years, the emphasis has shifted in the UK towards whole catchment flood risk management planning, as proposed in the Flood and Water Management Act (2010), which requires delivery of FRM plans under EU Directive 2007/60/EC on the assessment and management of floods (Ball 2008).

The Water Framework Directive (2000) targets 'good ecological status' by 2027, and recognises the importance of considering a 'no-regrets' FRM that also improves water quality. This is supported through financial incentives (e.g. Countryside Stewardship and Catchment Sensitive Farming grants). Failure to achieve these targets could result in large fines from the EU, with the most contributing pollutant to watercourses considered to be diffuse agricultural sources (RGS, 2012; POST, 2014). Therefore, while this chapter has primarily discussed flood alleviation, NFRM has a role in other policy agendas in the rural environment.

On a UK national scale, and after the 2007 floods, Pitt (2008, recommendation 27) suggested that Defra, the Environment Agency and Natural England should work with partners to establish a programme through catchment flood management plans and shoreline management plans to achieve greater working with natural processes. This has been promoted through river basin management plans (RBMPs) and via the Flood and Water Management Act (2010) to maintain or restore natural processes, 'where possible', as a method of reducing flood risk, and permits the implementation of natural features that control this risk (POST, 2011). This is similarly the case in Scotland with The Flood Risk Management (Scotland) Act (2009), which sets out the Scottish Government's long-term outcomes for flood management, which is to promote 'rural and urban landscapes with space to store water and slow down the progress of floods.'

Aside from policy, there has also been strong support for NFRM through 'Making Space for Water' (Defra, 2004), which encourages natural catchment-based runoff control and flood generation control at the source, and that go 'beyond traditional engineering solutions'. Combined, these government and intergovernmental agendas recognise the significance that NFRM can play in developing flood resilience but, as recommended by Section 12.3, they need development with a greater understanding of the key effects that NFRM can play in flood alleviation now and with the potential impacts predicted for climate change. These impacts are recognised in the Climate Change Act (2008) with the requirement for 'adaptation' to be considered in the National Adaptation Programme (Section 58). The governmental calls for adaptation were particularly recommended when it came to flood risk. The UK Climate Projections 2009 (UKCP09) (Murphy *et al.*, 2010) and the UK Climate Change Risk Assessment (UKCCRA, HM Government, 2012) both suggest increasing concerns for greater frequency and magnitude of high return interval events, likely to exceed defences designed to deal with the 1 in 100 year storm (Wilby *et al.*, 2007; Crooks *et al.*, 2009). Therefore, NFRM has the potential to adapt existing practices in regions that are likely to face worsening AEPs as a result of climate change.

12.5   Conclusions

NFRM is a fairly new approach to FRM, although its processes and benefits are recognised in other similar concepts, including rural SuDS. However, the ability to work with natural processes has multiple benefits, in its abilities to provide flood resilience, wide-ranging ecosystem services, meeting the challenges of a changing climate and meeting national and international policies, such as WFD targets, Floods Directives and Policies. Thus far, much of the focus for research has been on the theoretical benefits gained by implementing NFRM; there is therefore a requirement for grounded evidence to further its support and wider uptake, progressing the existing evidence included in this chapter. This is also reflected in support from delivery mechanisms, such as funding initiatives, including Countryside Stewardship that could have the potential to support such an evidence base.

References

Acreman, M. (1985) The effects of afforestation on the flood hydrology of the upper Ettrick valley. *Scottish Forestry.* 39, 89–99.

Addy, S., Wilkinson, M., Watson, H. and Stutter, M. (n.d.) *Implementation and Monitoring of Natural Flood Management in Scottish Upland Catchments.* The James Hutton Institute. Available at: http://tinyurl.com/oh36faq.

Armbruster, J., Muley-Fritze, A., Pfarr, U., Rhodius, R., Siepmann-Schinker, D., Sittler, B., Spath, V., Tremolieres, M., Rennenberg, H. and Kreuzwieser, J. (2006) *FOWARA: Forested Water Retention Areas, guideline for decision-makers, forest managers and land owners.* The FOWARA-project.

Avery, D. (2012) *Rural Sustainable Drainage Systems.* Environment Agency, Bristol.

Ball, T. (2008) Management approaches to floodplain restoration and stakeholder engagement in the UK: a survey *Ecohydrology and Hydrobiology*, 2-4, Elsevier BV, UK, pp 273–280.

Ballard, C., McIntyre, N. and Wheater, H. (2010) Peatland drain blocking: Can it reduce peak flood flows? *Proc. BHS Third Int. Symp., Managing Consequences of a Changing Global Environment, Newcastle.* 698–702.

Blanc, J., Wright, G. and Arthur, S. (2012) *Natural Flood Management knowledge system: Part 2 – The effect of NFM features on the desynchronising of flood peaks at a catchment scale.* CREW report. Available at: http://www.crew.ac.uk/projects/naturalflood-management.

Boorman, D.B., Hollis, J.M. and Lilly, A. (1995) *Hydrology of Soil Types: A Hydrologically-based Classification of the Soils of the United Kingdom.* IH Report No 126. Institute of Hydrology, Wallingford.

Broadmeadow, S., Thomas, H. and Nisbet, T. (2013) *Midlands Woodland for Water Project.* The Research Agency of the Forestry Commission.

Carter, V. (2014) *Catching the flood. Chartered Forester/Flood Planning.* Spring 2014, pp. 14–16.

Centre for Ecology and Hydrology (CEH) (2009) *The Flood Estimation Handbook*. HR Wallingford.

Chisholm, A. (2014) *A mature oak will drink 50 gallons of water a day*. Sylva. Spring/Summer.

Collingwood Environmental Planning (CEP) (2010) *Farming and water for the future in the Trent catchment – Innovative solution for local flood risk management*. Department for Environment, Food and Rural Affairs.

Cornell, S. (2006) *Improving Stakeholder Engagement in Flood Risk Management Decision Making and Delivery*. Environment Agency and Defra R&D Technical Report SC040033/SR2.

Cotswolds Honeydale Farm (2015) *Natural Flood Management*. Available at: http://cotswoldhoneydale. blogspot.co.uk/2015/07/natural-flood-management-other-news.html.

Couldrick, L.B., Granger, S., Blake, W., Collins, A. and Browning, S. (2014) WFD and the catchment based approach – going from data to evidence, in the Proc. of River Restoration Centre 5th Annual Network Conference 7th and 8th May 2014.

Crooks, S.M., Kay, A.L. and Reynard, N.S. (2009) *Regionalised Impacts of Climate Change on Flood Flows: Hydrological Models, Catchments and Calibrations*. Joint Defra/Environment Agency Flood and Coastal Erosion Risk Management R&D Programme (FD2020).

Davies, C. (2004) *Go with the Flow: The Natural Approach to Sustainable Flood Management in Scotland*. RSPB Scotland.

Defra (2004) *Making Space for Water: Developing a new Government strategy for flood and coastal erosion risk management in England*. Defra, London.

Defra (2009) *The Government's Response to Sir Michael Pitt's Review of the Summer 2007 Floods*. Defra, London.

Defra, RSPB and Hedgecott, S. (2010) *Working with natural processes to manage flood and coastal erosion risk*. Environment Agency, Peterborough. 21–22.

Doak, G. (2008) *The Way Forward for Natural Flood Management in Scotland*. Scottish Environment LINK.

Environment Agency (2014) Working with natural processes to reduce flood risk: a research and development framework. Available at: https://www.gov.uk/government/publications/working-with-natural-processes-to-reduce-flood-risk-a-research-and-development-framework

Evans, L., Davies, B., Brown, D. and Smith, L. (2014) *Using local community accounts and topographic analyses to identify optimal locations for natural flood management techniques in the Upper River Darent Catchment, Kent*. At: http://www.therrc.co.uk/sites/default/files/files/Conference/2015/Outputs/posters/bella_davies.pdf

Faustini, J.M. and Jones, J.A. (2003) Influence of large woody debris on channel morphology and dynamics in steep, boulder-rich mountain streams, Western Cascades, Oregon. *Journal of Geomorphology* 51, 187–205.

Feyen, L., Dankers, R., Bódis, K., Salamon, P. and Barredo, J.I. (2012) Fluvial flood risk in Europe in present and future climates. *Climatic Change*, 1, 47–62.

Fitton, S., Moncaster, A. and Guthrie, P. (2015) Investigating the social value of the Ripon rivers flood alleviation scheme. *Journal of Flood Risk Management*. DOI: 10.1111/jfr3.12176. At: http://on-linelibrary.wiley.com/doi/10.1111/jfr3.12176/pdf.

Francés, F., García-Bartual, R., Ortiz, E., Salazar, S., Miralles, J., Blöschl, G., Komma, J., Habereder, C., Bronstert, A. and Blume, T. (2008) *Efficiency of non-structural flood mitigation measures: 'room for the river' and 'retaining water in the landscape'*. CRUE Research Report No I-6 London. 172–213.

Freitag, B., Bolton, S., Westerland, F. and Clark, J. (2009) *Floodplain Management: A new Approach for a New Era*. The University of Chicago Press, 137.

Hess, T.M., Holman, I.P., Rose, S.C., Rosolova, Z. and Parrott, A. (2010) Estimating the impact of rural land management changes on catchment runoff generations in England and Wales. *Journal of Hydrological Processes*, 24 (10), 1357–1368.

HM Government (2012) *UK Climate Change Risk Assessment (CCRA): Government Report*. London: The Stationery Office.

Holstead, K.L., Waylen, K.A., Hopkins, J. and Colley, K. (2015) *The Challenges of Doing Something New: Barriers to Natural Flood Management*. Presentation at XVth IWRA World Water Congress, Edinburgh, 25–29 May 2015.

Howgate, O.R. and Kenyon, W. (2009) Community cooperation with natural flood management: a case study in the Scottish Borders Area, *Royal Geographical Society* 41(3), 329–340.

Iacob, O., Rowan, J., Brown, I. and Ellis, C. (2012) *Natural flood management as a climate change adaptation option assessed using an ecosystem services approach*. PhD thesis, Centre for Environmental Change and Human Resilience, University of Dundee, UK.

JBA (2007) *Ripon land management project – Final report*. Department for Environment, Food and Rural Affairs.

Johnson, C.L. and Priest, S.J. (2008) Flood risk management in England: a changing landscape of risk responsibility, *Water Resources Development* (24) 513–525.

Johnstonova, A. (2009) *Meeting the challenges of implementing the Flood Risk Management (Scotland) Act 2009*. A report by RSPB Scotland.

Lambe, R., Faulkner, D.S. and Zaidman, M.D. (2009) *Fluvial Design Guide* Chapter 2. Environment Agency. Available at: http://tinyurl.com/z5ed7pj.

Lane, S.N., Morris, J., O'Connell, P.E. and Quinn, P.E. (2006) Managing the rural landscape. In: Thorne, C (2006) *Future Flood and Coastal Erosion Risk*, Publisher: Thomas Telford Ltd.

Leeks, G. and Roberts, G. (1987) The effects of forestry on upland streams – with special reference to water quality and sediment transport. In: *Environmental Aspects of Plantation Forestry in Wales*, Good, J. (ed.). Institute of Terrestrial Ecology Symposium, 9–24.

Linstead, C. and Gurnell, A.M. (1998) *Large woody debris in British headwater rivers: Physical habitat and management guidelines*. R&D Technical Report W185, School of Geography and Environmental Sciences, University of Birmingham, UK. Management 2010 Conference, the International Centre, Telford, 29 June – 1 July 2010, 10 pp.

Mauch, C. and Zeller, T. (2009) *Rivers in History: Perspectives on Waterways in Europe and North America*. University of Pittsburgh Press.

McBain, W., Wilkes, D. and Retter, M. (2010) Flood resilience and resistance for critical infrastructure. *CIRIA SuDS training*.

McIntyre, N. and Thorne, C. (2013) *Land Use Management Effects on Flood Flows and Sediments-Guidance on Prediction*. CIRIA, London.

Morris, J. and Wheater, H. (2006) Catchment Land Use, in Thorne, C. (2006) *Future Flood and Coastal Erosion Risk*, Thomas Telford.

Murphy, J., Sexton, D., Jenkins, G., Boorman, J., Booth, B., Brown, K., Clark, R., Collins, M., Harris, G. and Kendon, L. (2010) *UK Climate Projections science report: Climate change projections*. Meteorological Office Hadley Centre.

National River Flow Archive (NRFA) (2015) *Catchment Spatial Data: FEH Catchment Descriptors*. Available at: http://nrfa.ceh.ac.uk/feh-catchment-descriptors.

Nicholson, A.R., Wilkinson, M.E., O'Donnell, G.M. and Quinn, P.F. (2012) Runoff attenuation features: a sustainable flood mitigation strategy in the Belford catchment, UK. *Area*. 463–469.

Nisbet, T.R. and Thomas, H. (2008) Project SLD2316: *Restoring Floodplain Woodland for Flood Alleviation*. Department for Environment, Food and Rural Affairs.

Nisbet, T.R., Marrington, S., Thomas, H., Broadmeadow, S. and Valatin, G. (2011) *Project RMP5455: Slowing the Flow at Pickering*. Department for Environment, Food and Rural Affairs.

O'Connell, P.E., Beven, K.J., Carney, J.N., Clements, R.O., Ewen, J., Hollis, J., Morris, J., O'Donnell, G.M., Packman, J.C., Parkin, A., Quinn, P.F., Rose, S.F. and Shepherd, M. (2004) *Review of Impacts of Rural Land Use and Management on Flood Generation. Part B: Research Plan*. R&D Technical Report FD2114/TR, Defra, London, UK.

Odoni, N. (2014) Can we plant our way out of flooding? *Sylva*. Spring/Summer 2014, 19–20.

Odoni, N.A. and Lane, S.N. (2010) *Assessment of the impact of upstream land management measures on flood flows in Pickering beck using OVERFLOW*. Durham University, UK.

Odoni, N.A., Nisbet, T.R., Broadmeadow, S.B., Lane, S.N., Huckson, L.V., Pacey, J. and Marrington, S. (2010) *Evaluating the effects of riparian woodland and large woody debris dams on peak flows in Pickering Beck, North Yorkshire*. In: Proceedings of the Flood and Coastal.

Park, J., Cluckie, I. and King, P. (2006) *The Parrett Catchment Project (PCP): Technical Report on the Whole Catchment Modelling Project*. The University of Nottingham, Nottingham, 69–84.

Pitt, M. (2008) *The Pitt review: learning lessons from the 2007 floods*.

POST (2011) *Natural Flood Management*, Postnote 396, Parliamentary Offices of Science and Technology, London.

POST (2014) *Diffuse Pollution of Water by Agriculture*. Postnote 478, Parliamentary Offices of Science and Technology, London.

Potter, K.M. (2006) *Where's the Space for Water? – How Floodplain Restoration Projects Succeed*. Unpublished MSc Thesis, Liverpool University, UK.

Quinn, P., O'Donnell, G., Nicholson, A., Wilkinson, M., Owen, G., Jonczyk, J., Barber, N., Hardwick, M. and Davies, G. (2013) *Potential use of runoff attenuation features in small rural catchments for flood mitigation: evidence from Belford, Powburn and Hepscott*. Joint Newcastle University, Royal Haskoning and Environment Agency Report, UK.

Robertson, R.A., Nicholson, I.A. and Hughes, R. (1968) *Runoff studies on a peat catchment*. Proc. 2nd International Peat Congress, HMSO, London. 161–166.

Rose, S.F. (2011) Natural Flood Management: Measures and Multiple Benefits. *The River Restoration Centre*.

Royal Geographical Society (with IBG) (2012) *Water policy in the UK: The challenges*. RGS-IBG Policy Briefing.

Scottish Environment Protection Agency (SEPA) (2011) The Eddleston Water Project, *The Tweed Forum*. Available at: http://www.tweedforum.org/projects/current-projects/eddleston.

Scottish Environment Protection Agency (SEPA) (2011) *Allan Water Natural Flood Management Techniques and Scoping Study*. Halcrow Group Ltd.

Scottish Environment Protection Agency (SEPA) (2016) The Natural Flood Management Handbook. Available at: http://www.sepa.org.uk/media/163560/sepa-natural-flood-management-handbook1.pdf

Sharp, R. (2014) *Investigating hillslope afforestation as a potential natural flood management strategy in the Eddleston Water catchment, Scottish Borders*. Unpublished MSc Thesis, University of Edinburgh, UK.

Shaw, E.M., Beven, K.J., Chappell, N.A. and Lamb, R. (2011) *Hydrology in Practice*. CRC Press.

SNIFFER (2011) *Understanding the opportunities and constraints for implementation of natural flood management features by farmers*. Project FRM21, Edinburgh, Scotland.

Spence, C. and Sisson, J. (2015) River Elwy Catchment: emulating nature for flood risk management – methods for analysis and monitoring the River Elwy catchment, North Wales. *The UK Water Projects*.

Thieken, A.H., Mariana, S., Longfield, S. and Vanneuville, W. (2014) Flood resilient communities – managing the consequences of flooding. *Natural Hazards and Earth System Sciences*. 14, 33–39.

Thomas, H. and Nisbet, T. (2012) Modelling the hydraulic impact of reintroducing large woody debris into watercourses *Journal of Flood Risk Management*, 5, 164–174.

Thorne, C.R., Evans, E.P. and Penning-Rowsell, E.C. (2007) *Future flooding and coastal erosion risks*. Thomas Telford.

Wainwright, J. and Mulligan, M. (2004) *Environmental Modelling: Finding Simplicity in Complexity*, John Wiley and Sons.

Werritty, A. (2006) 'Sustainable flood management: oxymoron or new paradigm' *Royal Geographical Society*. 38 (1), 16–23.

Westcountry Rivers Trust (2015) *Upstream Thinking*. available online: http://wrt.org.uk/project/upstream-thinking/

Wheater, H., Reynolds, B., McIntyre, N., Marshall, M., Jackson, B., Forgbrook, Z., Solloway, I., Francis, O. and Chell, J. (2008) *Impacts of upland management on flood risk: Multi-scale modelling methodology and results from the Pont Bren experiment*. Flood Risk Management Research Consortium.

Wilby, R.L., Beven, K.J. and Reynard, N.S. (2007) Climate change and fluvial flood risk in the UK: more of the same? *Journal of Hydrological Processes*, 22(14), 2511–2523.

Wilkinson, M.E., Quinn, P.F. and Hewett, C.J.M. (2013) The Floods and Agriculture Risk Matrix (FARM): a decision support tool for effectively communicating flood risk from farmed landscapes. *The International Journal of River Basin Management*, 11, 237–252.

Wilkinson, M.E., Quinn, P.F., Barber, N.J. and Jonczyk, J. (2014) A framework for managing runoff and pollution in the rural landscape using a catchment systems engineering approach. *Science of the Total Environment*, 468–469, 1245–1254.

World Wildlife Fund (WWF) (2007) *Flood planner: A manual for the natural management of River floods*. WWF Scotland.

Young, D.S., Hart, J.K. and Martinez, K. (2015) Image analysis techniques to estimate river discharge using time-lapse cameras in remote locations. *Computers & Geosciences*, 76, 1–10.

Statuses

European Union Water Framework Directive, 2000
European Union Floods Directive, 2007 (c. 22)
Flood and Water Management Act 2010 (c. 29)
Flood Risk Management (Scotland) Act 2009 (c. 6)
The EU Green Infrastructure Strategy, 2013
The Climate Change Act, 2008 (c. 58)

Sustainable Drainage Systems and Energy: Generation and Reduction

Amal Faraj-Lloyd, Susanne M. Charlesworth and Stephen J. Coupe

13.1 Introduction

The world's energy generation depends predominantly on burning fossil fuels (coal, oil and gas), and global demand for energy is only likely to rise. However, they are non-renewable resources, which will dwindle, as they are finite, and while they are still being used, they release greenhouse gases (GHG) such as carbon dioxide (CO_2), which is estimated to have increased between 1970 and 2004 by 80%; De Boeck *et al.* (2015) predicted that they would further rise by 52% between 2005 and 2050. Domestic energy consumption accounts for 47% of UK CO_2 emissions and of this 75% is due to energy consumed to provide heating and cooling (POST, 2010). Tackling domestic CO_2 emissions can be achieved by applying an efficient renewable energy (RE) system that can provide heating and cooling to buildings, leading to a reduction in GHGs and contributing to the reduction of the effects of climate change (HM Government, 2009; Song *et al.*, 2015). The use of RE is being encouraged by many governments; in fact, the European Parliament increased the proportion of Member States' sourcing their energy from RE to 27% by 2020. Shafiei and Salim (2014) suggest that investing in renewable sources of energy, in general, has the potential to reduce GHG emissions overall, and CO_2 in particular.

Evidence of the efficiency of SuDS in reducing the volume of runoff and improving water quality has come through a number of studies, such as Yu *et al.*, (2001), Deletic and Fletcher (2006), Abdulla and Al-Shareef (2009), Charlesworth *et al.*, (2012), Kazemi *et al.*, (2011), Bressy *et al.*, (2014). Similarly, studies of RE as natural resources that have successfully generated and provided comfortable indoor temperatures were published by Lund *et al.* (2004), Curtis *et al.* (2005), Hwang *et al.* (2009), Saner *et al.* (2010), Wang *et al.* (2015) and Reboredo (2015). Charlesworth (2010) provided a review of the multiple benefits of the SuDS approach in its ability in mitigating and adapting to global climate change.

Sustainable Surface Water Management: A Handbook for SuDS, First Edition.
Edited by Susanne M. Charlesworth and Colin A. Booth.
© 2017 John Wiley & Sons, Ltd. Published 2017 by John Wiley & Sons, Ltd.

This chapter focuses on two areas associated with energy generation and use reduction: first it outlines the feasibility of integrating SuDS and energy in one combined infrastructure that can reduce runoff but also generate energy to provide heating and cooling to buildings sustainably, and second, it reviews the use of blue and green infrastructure in reducing energy demand for both heating and cooling of buildings.

The SuDS approach, the processes involved and its efficiency and efficacy are covered in other chapters of this volume and will therefore not be covered in detail here. The following sections discuss ground source heat extraction and its specific integration into a SuDS pervious paving system to provide a combined system, which can address both flood resilience and provide a source of RE at the building scale.

13.2   Ground Source Heat Extraction

Ground source heat (GSH) is an abundant and constantly renewable source of energy, which is relatively easy to harvest (Self *et al.*, 2013). The extraction and concentration of this heat is by using GSH pumps (GSHP), which are a 'highly efficient renewable energy technology' (Omer, 2008) and which can be used in both heating and cooling modes. The temperature of this heat initially is relatively low, but once it is concentrated (Omer, 2008; Self *et al.*, 2013) it provides heat that is 'environmentally and economically advantageous' (Self *et al.*, 2013). Particularly suited to under-floor heating, GSHPs are suitable for many types of building worldwide (Omer, 2008). Furthermore, specifically extracting GSH has the potential to reduce CO_2 emissions and hence mitigate the impacts of climate change (Bayer *et al.*, 2002). While using this technology has been predicted by Bayer *et al.* (2002) to save up to 30% of GHG emissions in comparison with conventional heating methods across Europe, this is dependent on the efficiency of the pump, the electrical mix and the substituted heat. These potential savings are country-specific and depend on a saturated market for the technology and the use of renewables (e.g. solar or wind) to provide power for the pump. A problem with their use in dense urban settlements may be a lack of space; thus the ability to integrate it with other technologies to provide multiple benefits and flexibility in application needs to be explored.

Pervious paving can provide a means of integrating GSHE into the kind of built environment, whether domestic, retail or industrial, by providing hardstanding for parking, access roads for lightly trafficked areas and pedestrian walkways. The following section outlines types of pervious hardstanding, their role in reducing runoff quantity and the integration of GSHE during their construction.

13.3   Pervious Paving Systems

Pervious paving systems (PPS) can be grouped into two categories, according to the method of infiltration: water either infiltrates through the entire surface of the material, in which case it is termed 'porous', or through gaps between the impermeable block pavers, which is then called 'permeable' (Charlesworth *et al.*, 2014).

In the UK, homeowners are not supposed to lay more than $50\,m^2$ of impermeable surfacing on the driveways in front of their homes due to these 'sealed' driveways exacerbating flooding problems in urban areas (Wright, 2010). Using a permeable surface such as PPS has the potential to have a significant impact on rainwater runoff quantity and quality.

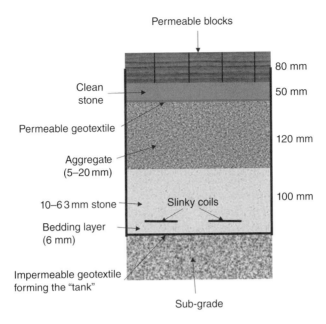

Permeable blocks

Clean stone

Permeable geotextile

Aggregate (5–20 mm)

10–6 3 mm stone

Bedding layer (6 mm)

Impermeable geotextile forming the "tank"

Slinky coils

Sub-grade

80 mm

50 mm

120 mm

100 mm

Figure 13.1   Vertical structure of a pervious paving system showing the position of the slinky coils that can be used to harvest GSH.

While excavating the PPS, GSHE could be laid in the bottom of the trench at the same time, using horizontal 'slinky' coils as the means of harvesting heat. Figure 13.1 shows the typical structure of a PPS with a surface layer of concrete block pavers with vertical gaps between, representing 8–20% of the total surface area. The gaps are filled with 2–4 mm pea gravel, to allow water to seep down through an open-grade base. The ASTM (2001) C936 specification states that the pavers should be at least 60 mm thick with a compressive strength of 55 MPa or greater, depending on the purpose of use. The function of the geotextile beneath the bedding layer is covered in detail in Chapter 11.

Under the geotextile is the compacted sub-base, of clean crushed stone, gravel or concrete, which has spaces in between to store water, and this is sometimes referred to as the water saturation zone or reservoir course. The depth of the sub-base can vary according to specific site conditions, and may have two layers in order to control infiltration rate. The overall storage capacity depends on the depth of the sub-base, the size of the aggregate and the ratio of voids. If designed and implemented correctly, PPS can allow a large proportion of stormwater to infiltrate, thus reducing peak runoff volumes and flows (Andersen et al., 1999; Sansalone and Teng, 2005; Sansalone et al., 2008). As described, with a single geotextile the PPS is an infiltrating pavement, but if the whole structure is enclosed in an impermeable membrane, it becomes a storage tank, and is thus called a tanked or attenuation system. The water infiltrating into the PPS can thus be retained or harvested by the tank, where it can be used for other purposes, such as garden watering, car washing or toilet flushing.

Water harvested by a tanked PPS can also be used as a means of accessing heat since, during winter, the temperature of the ground is higher than the overlying air and therefore stored water can be utilised as a heat source; conversely, in summer, the ground temperature is lower than overlying air and therefore the harvested water can act as a heat sink (Healy and Ugursal, 1997; Hepbasli, 2005; Ozgener and Hepbasli, 2007; Singh *et al.*, 2010).

PPS and GSHE, taken separately, are not new, but the multiple benefits of PPS and RE in providing a use for surface water which would otherwise be wasted, particularly in the face of the predicted impacts of global climate change, would appear to have potential. Tota-Maharaj and Paul (2015) call the combination of PPS and GSHE the 'next generation' of PPS. Laboratory-based studies of test rigs confirmed the potential for such a combined system in that they found that the inclusion of heat exchangers at depth in PPS did not compromise their water quality improvement capabilities (Tota-Maharaj *et al.*, 2009, 2010) or encourage the growth of potentially toxic microorganisms (Coupe *et al.*, 2009; Scholz *et al.*, 2012). Other studies have monitored the distribution of heat in PPS in the field (Novo *et al.*, 2010, 2013; del Castillo-Garcìa, 2013), finding that the process of evaporation in the sub-base, and the surface course thermal properties, were the most important factors when a combined PPS/GSHP was designed. Slinky coil size, energy efficiency and tank volume were determined by Tota-Maharaj *et al.* (2011), by modelling the temperature and energy balances in the combined systems and thus optimised their design. Many of these studies were undertaken in the laboratory, but the following sections detail two case studies where combined PPS and GSHE were installed in individual buildings and monitored to assess their ability to provide heat in domestic and office settings.

13.3.1 Hanson EcoHouse, Building Research Establishment, Watford, UK

An EcoHouse was constructed using prefabricated components including precast concrete flooring systems and prefabricated masonry cavity walls (2.4×9 m), together with traditional building materials (clay blockwork and concrete brickwork) to form an average-sized family home (Figure 13.2). The wall panels included openings in which the high-performance doors and windows with three-layered, argon gas-filled glazing were fitted. The key properties of the finished walls include higher flexural strength for both brick and block, increased vertical strength – the walls were about twice as strong as that of traditional masonry – and increased resistance to rain penetration due to continuous mortar joints. The last factor contributes to the air-tightness of the building, which is superior to that achieved with traditional masonry. The rate of heat loss through a material is known as its U-value: the lower the U-value, the better the insulation provided by the material. The walls of the house achieved a U-value of $0.18\,W/m^2K$, with a U-value of $0.15–0.27\,W/m^2K$ for external walls, sufficient to meet Energy Service Directive No 2006/32/EC. This states that EU countries must achieve a 9% annual energy saving (2008–2016) by using new energy services and other energy-efficiency measures (INFORSE, 2010). The U-value for the insulated steel-framed pyramidal roof was $0.15–0.18\,W/m^2K$, with the triple-glazed windows achieving $0.8\,W/m^2K$. The total heat loss for the EcoHouse was $6.6\,kWh/m^2/year$ (fabric heat loss was $77.9\,W/K$ and the ventilation heat loss was $62.12\,W/K$) (Hanson customer services, pers. comm., 2013). Based on its construction and the inclusion of the PPS/GSHE in an assessment under the Code of Sustainable Homes, the EcoHouse achieved a Level 4. With a total internal floor area of $143\,m^2$, the EcoHouse was constructed 'upside down', with three

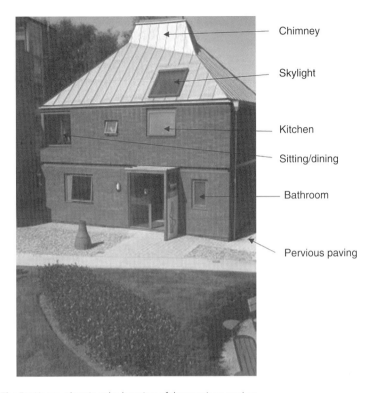

Figure 13.2   The EcoHouse, showing the location of the pervious paving.

bedrooms and two bathrooms downstairs, and a large open plan space on the upper floor where the kitchen, dining and living areas were located; this is shown in Figure 13.2.

13.3.2 PPS/GSHP

Rainwater falling on the pavement surface, and runoff from rooftops of the surrounding buildings, was collected in a 350 mm deep, 65 m² PPS underground tank. It is usually recommended that the horizontal GSHE is buried in trenches of around 1 m depth (Energy Saving Trust, 2004), but ground conditions at the EcoHouse dictated that the trench could not be that deep. The design of the PPS/GSHP is similar to that explained above, with the addition of the slinky coil inside the PPS tank (Figure 13.1).

Fifteen sensors were installed on site, in order to measure the temperature of the EcoHouse, both inside the building at the four cardinal points, and also inside and above the PPS/GSHP tank. Nine constantan (copper/nickel alloy) thermistor sensors were embedded in the external and internal panels, on the four sides of the EcoHouse, north, south, east and west, as well as inside on the partition wall located on the ground floor, to monitor the temperatures inside and outside the house. Four sensors were also installed inside the PPS/GSHP tank at 60, 130, 220, and 350 mm from the surface, in order to measure the temperature within the PPS tank. A sensor was installed in a bollard at 1300 mm above the paving to measure ambient air temperature.

Data was sent to a data-logger via a fibre connection to a community digital management centre based in the visitors' centre on-site. The data was then downloaded to a PC in DOS format. The data-logger in this study was set to collect readings every 10 minutes for three years (2008–2010). This resulted in more than 1.3 million observations in total.

13.4   Results of Monitoring the EcoHouse

13.4.1   The Habitated Space

The variation between indoor and outdoor temperatures during the heating period is presented in Figure 13.3, which shows how the temperatures varied on a daily basis inside the EcoHouse, in comparison with the ambient air at 1300 mm above the PPS surface. The data in Figure 13.3 represents when the house was being heated by the combined system, the gaps in the data being either when it was being heated from the electricity supply or on the occasions when there was a breakdown of some kind.

During the heating period, the outdoor average temperature was 12.0°C with minimum and maximum temperatures of 0.9 and 17.9°C, while the indoor temperature was an average of 21.5°C (min and max 18.0 and 28.2°C). This average indoor value could be considered as being 'comfortable', but Figure 13.3 also shows that the temperature was not stable and there were times when the indoor space was either too hot or too cold, in comparison with CIBSE's (2006) classification of 'comfortable' (19.5 ± 0.5°C in winter and 21 ± 1°C in summer).

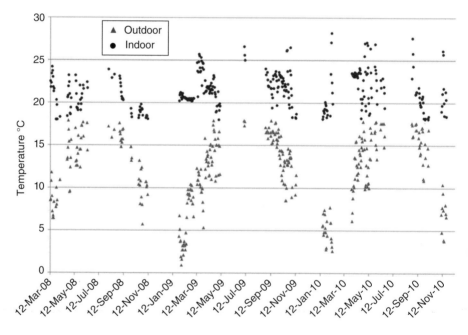

Figure 13.3   Differences between daily indoor and outdoor temperatures for days on which heating was provided to the EcoHouse ($n = 702$).

13.4.2 The PPS/GSHP

Ambient air and the four ground temperatures were collected, while heat was provided to the domestic setting, with the results presented in Figure 13.4 and summarised in Table 13.1. These show minimum temperatures for all depths were below zero (i.e. stored rainwater was in the frozen state), and that ground temperature nearest the surface (60 mm deep) recorded the highest maximum and the lowest minimum temperatures, and these characteristics reduced with increasing reservoir depth. Figure 13.4 also shows that there was minimal difference between the averages given for the ground temperature at the various depths. The daily average temperature for the four different depths were not significantly different from the ambient air daily average temperature ($t = 3.931$; 4.718; 8.074; 10.541; $p < 0.001$, for the four depths).

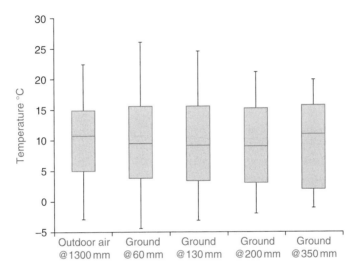

Figure 13.4   The main temperature features of the ambient air and the ground throughout the monitoring period ($n = 2449$).

Table 13.1   Statistical analysis of the ambient air and the ground temperatures (°C) throughout the monitoring period ($n = 2,449$).

	Temperatures throughout the monitoring period				
	Outdoor air @ 1300 mm	Ground @ 60 mm	Ground @ 130 mm	Ground @ 200 mm	Ground @ 350 mm
Minimum (°C)	−3	−4.4	−3.1	−1.9	−1.1
Maximum (°C)	22.5	26.2	24.7	21.2	20.0
Average (°C)	10.0	9.7	9.5	8.8	9.6
Median (°C)	10.8	9.5	9.2	9.1	11.0
Standard Deviation (°C)	6.0	6.9	6.9	6.6	6.7

13.4.3 The coefficient of performance

The efficiency of extracting heat and providing it to the building space is expressed as the coefficient of performance (CoP). A CoP of three, therefore, indicates that the output of the heat pump is three times that of the input, an efficiency of 300%. CoPs for geothermal systems typically vary between three and five with the occasional value of six being reported (Lund et al., 2003; O'Connell and Cassidy, 2003). Temperature data from the bottom of the PPS/GSHP tank recorded on days for which heating was being provided to the EcoHouse by the combined system was used to determine the CoP of the heat pump, which was found, on average, to have a value of 2.3 (ranging between 4.8 and 1.0), so the system cannot be considered a satisfactory renewable source of energy under the 2009 EU Renewable Energy Directive, since a CoP of 2.875 is required. For the days on which the ground temperature was less than 1°C (varying between −1.1 and 0.9), the CoP value was 1.0 or less. This indicates that heating provided to the EcoHouse on such days was derived completely from the electricity mains without any heat energy being derived from the ground. On the days on which heating was provided by the combined system, and the temperature of the ground was greater than 1.9°C, CoP varied between 1.1 and 3.8, with an exceptional day at 4.8. The latter was achieved when the indoor temperature was 19.4°C and the temperature of the ground was 15.3°C, whereby the lower heat load and the higher water temperature produced the high CoP. The lowest CoP value (1.1) (this is the next lowest value of the CoP, after the days on which CoP value of one were extracted) occurred when the daily average for the ground temperature was 1.9°C, and the indoor temperature was 21.8°C.

13.5   The Hanson Stewartby Office, Bedford, UK

Following on from applying the combined PPS/GSHP at the EcoHouse, an integrated system was selected as the energy solution aspect of a three-storey office block located at Stewartby, Bedford, UK (Figure 13.5).

A heating and cooling combined PPS/GSHP system with five 130 kW units was installed in a 285 bay (6500 m^2) car park, serving a 7000 m^2 office block. This included 8.4 km of slinky pipes set 1 m apart in a PPS trench embedded into 200 mm of permanently saturated stone, the base of which had to be flat to ensure that all the coils were covered with water. Any potential overflow was directed into a nearby lake. With lessons learnt from the EcoHouse at Watford, the coiled pipes for the GSHP were laid at a depth of 700 mm, to ensure that they did not freeze, in contrast to a typical car park excavation depth of about 300 mm. The five GSHPs were able to provide energy to meet the demands of the entire building via under-floor heating at the optimum efficiency of 45°C. Special geothermal radiators with a large surface area were used on the upper floors to efficiently distribute the harvested heat. Working from a thermostat inside the building, the five GHSPs work in series, with each one working in turn until the required temperature is reached, then they shut off individually. It was possible to provide both heating and cooling using a sliding header valve, which was achieved at the same time if necessary in different areas of the building, whereby offices may require heating, or a gym may need to be cooled. The building is 30% more efficient than is found in the usual new-build, and has a rating of 35 kg CO$_2$/m^2/year; this compares favourably to a typical new-build figure of 50 kg CO$_2$/m^2/year. It is estimated that pay-back on the system should be achievable within 5–6 years.

Figure 13.5 Three-storey office block at Stewartby, Bedfordshire, UK, showing the construction of the combined PPS/GSHP system.

The site also included SuDS attenuation systems, such as wetlands and ponds, amenity provision and biodiversity enhancement. All of these devices, together with the integrated PPS/GSHP system, maximised the efficiency of the build and contributed to it achieving an 'Excellent' BREEAM rating (Building Research Establishment Environmental Assessment Method) and an Energy Performance Certificate rating of B (http//www.laysells.co.uk). As was shown by Tota-Maharaj *et al*. (2012), these additional devices, such as ponds and wetlands, can be used to house surface water heat pumps. This technology was tested at laboratory scale by Tota-Maharaj *et al*. (2012), using a vertical flow constructed wetland test rig including *Phragmites australis* (common reed). Municipal wastewater was added to the system, which resulted in the removal of >75% suspended solids, chemical oxygen demand was reduced by 50%, ammonia-nitrogen and nitrate-nitrogen reduced by 50–60% and orthophosphate-phosphorus reduced by 40%. It should therefore be possible to incorporate GSHPs in suitable SuDS devices, and harvest the energy in a clean and sustainable manner.

13.6   Reducing Energy Use: The Use of Green and Blue Infrastructure on Buildings

The previous sections illustrated the flexibility of SuDS devices by the incorporation of RE in the provision of energy in a sustainable way. This section shows the multiple benefits of SuDS devices in reducing the need for energy in heating and cooling buildings. If these approaches are used together, there is substantial potential for SuDS to be able to be integrated into the built environment and provide not only flood resilience, but also energy resilience. According to Refahi and Talkhabi (2015), 40% of the world's energy expenditure is on the heating and cooling of residential and commercial buildings; so addressing this fact is becoming an urgent issue. As a result, many studies are being carried out worldwide, under different climatic conditions, with different SuDS infrastructures and seasonalities. For example, there have been studies of the benefits of green roofs in reducing energy demands in temperate (Virk et al., 2015), Mediterranean (Fioretti et al., 2010) (Figure 13.6), semi-arid (Issa et al., 2015) and sub-tropical (Yang et al., 2015) climates. All of these studies found reduction of energy

Figure 13.6   Green roof in Valencia, Spain.

use in buildings with an associated green roof, although some studies were less positive, in regions with particularly cold winters, even when having a green roof had distinct advantages during the summer (Coma *et al.*, 2016). For example, in a comparative study of six types of green roof in a humid sub-tropical region, Simmons *et al.* (2008) found in all cases that internal temperature was significantly reduced on warmer days, but that there was no discernible difference between the green roofs and controls on the cold ones.

In a review of the performance of green roofs, Hashemi *et al.* (2015) stated that one of the key roles of both intensive and extensive green roofs was in the reduction of energy requirements for both heating and cooling buildings. Green roofs insulate the building against escape of heat or they can cool the air inside the building, making the energy expended on heating and air conditioning more efficient. Virk *et al.* (2015) state that they do this via two main mechanisms and one associated impact:

1. There are direct heat flow changes through the roof due to changes in surface temperature as well as insulating effects.
2. There are indirect changes to the temperature of the air brought into the building as a means of refreshing breathable air qulity for the occupants.
3. Dependent on the efficiency of the above two mechanisms, boundary conditions associated with heat transfer throughout the fabric of the building are affected.

This is green infrastructure, because green roofs cool the air above the building, air conditioning requirements are less and the urban heat island (UHI) effect could also be reduced (Charlesworth, 2010; Costanzo *et al.*, 2015). In fact, if half of the flat roofs in New York, USA were greened, the UHI could be reduced by as much as 0.8°C (Rosenzweig *et al.*, 2006). A study of three different climates in Iran, Refahi and Talkhabi (2015) found that energy consumption was reduced by 6.6–9.2%; taking this fact alone, and not taking account of any other benefits that would reduce financial considerations, the pay-back was calculated at 25–57 years.

While not seen as often as green roofs, green walls have similar benefits in retaining heat inside buildings during cooler periods, and cooling the building during warmer ones. Ip *et al.* (2010) reported on a study of a 'vertical deciduous climbing plant canopy' in the UK, and found that there were distinct seasonal benefits because of shading during the summer, which reduced the internal temperature of the building by 4–6°C. Once the leaves had fallen, during the autumn, incident solar radiation was able to enter the building through the windows, providing natural heat to the room inside.

Urban trees can assist in attenuating the storm peak by intercepting rain falling on their leaves, stems and trunks. They also absorb water into their tissues, retaining as much as 380 litres or more. It is estimated that an urban forest can reduce annual runoff by up to 2–7%. At the individual building scale, energy savings due to shade trees can be considerable. It has also been known for some time that trees can shade buildings and therefore directly reduce energy use by reducing incident solar radiation (Simpson, 2002). For example, Akbari *et al.* (1997) monitored the affects of installing 16 shade trees at two domestic dwellings, finding that they provided 30% cooling energy savings, with daily average savings of 3.6 and 4.8 kWh/d. Peak demand savings were 0.6 and 0.8 kW, which equated to 27 and 42% savings for each house individually. Figure 13.7 illustrates how using street trees can cool outdoor temperatures and indirectly reduce air conditioning use by reducing the effect of the UHI. Akbari *et al.* (2001) found that the peak in urban electric demand rose by 2–4% for every 1°C rise in temperature above a threshold of 15–20 °C. Thus, a UHI of 3–4°C would represent an increase in electric demand of 6–16% due to increased use of air-conditioning to reduce indoor temperatures. This does not take account of the underlying effects of global climate change.

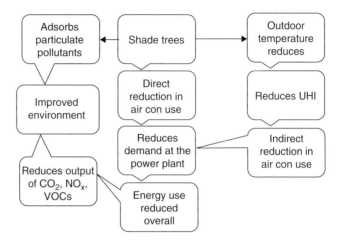

Figure 13.7 Relationship between properties of street or urban trees, improvements to the environment and reductions in overall energy use (after Akbari (2002)). UHI = Urban Heat Island.

Urban trees improve air quality by adsorbing polluted particulates and thus have multiple benefits by reducing contamination, the storm peak and energy demand. Furthermore, in addressing these issues, the value of these ecosystem service provisions can be substantial (Donovan and Butry, 2009). Pandit and Laband (2010), for instance, calculated that an individual homeowner could save 9.3% on their electricity bills if their house was almost 20% shaded, and up to 14.4% savings with 50% shading in summer. These estimates depend of course on a great number of factors, including the type of tree, house construction and orientation, climate, space available, land prices, possibility of retrofit, etc. (Donovan and Butry, 2009).

These devices cool buildings not only passively by means of insulation, but also actively by the process of evaporative cooling (Robitu *et al.*, 2006), which occurs when moisture is evaporated from a wet surface into the overlying air. This releases latent heat, which has a substantial cooling effect, not only associated biologically with vegetation, but also physically with the surface of ponds and porous paving (Asaeda and Ca, 2000). While PPS will cool the surroundings of a building in this way, ponds installed onto building roofs have been used as a means of cooling the insides of buildings, hence reducing the need for air-conditioning. In fact, Robitu *et al.* (2006) have recommended the use of ponds in order to improve the internal thermal comfort of buildings, and have quoted the difference between a pond and a road surface in full sun in the mid-afternoon as 29 K (Robitu *et al.*, 2004). In a study of the temperatures within a building with a roof-pond, Givoni (1998) found the difference in temperature between the ceilings underneath the pond and that indoors was between 2 and 3°C, showing the potential of this approach.

13.7   Conclusions

To provide resilience in a changing world, and for that resilience to be sustainable, any interventions have to be multiple benefit and flexible in application. SuDS has proven to be advantageous on both counts, combining the flood attenuation and water quality improvement attributes of the PPS with the RE harvesting of GSH extraction in a single device, which can be used successfully at the small domestic or the large office-block scale. The case study of the domestic building enabled lessons to be learnt, which were fully implemented in the case of the office block, with the result that energy use was reduced, carbon similarly so, and a renewable source of heat was harvested in a sustainable manner.

It has also been shown that SuDS devices can reduce energy use by using blue and green infrastructure in urban environments such as green roofs and walls, street trees and roof ponds to cool the urban environment, insulate individual buildings and improve the environment overall, again due to the multiple benefits of the SuDS approach.

If these devices were carefully designed together into cities, their flexibility would enable substantial reductions in energy use, and in the utilisation of RE, leading to substantial monetary savings based on their wide-ranging ecosystem services provision.

References

Abdulla, F.A. and Al-Shareef, A.W. (2009) Roof rainwater harvesting systems for household water supply in Jordan. *Desalination* 243(1-3), pp.195–207.

Akbari, D.M., Kurn, H., Bretz, S.E. and Hanford, J.W. (1997) Peak power and cooling energy savings of shade trees. *Energy and Buildings* 25, 139–148.

Akbari, H., Pomerantz, M. and Taha, H. (2001) Cool surfaces and shade trees to reduce energy use and improve air quality in urban areas. *Solar Energy.* 70, 3, 295–310.

Andersen, C.T., Foster, I.D.L. and Pratt, C.J. (1999) The role of urban surfaces (permeable pavements) in regulating drainage and evaporation: Development of a laboratory simulation experiment. *Hydrological Processes* 13(4), 597–609.

Asaeda, T. and Ca, V.T. (2000) Characteristics of permeable pavement during hot summer weather and impact on the thermal environment. *Build. Environ.* 35, 363–375.

ASTM, 2001. Standard specification for solid concrete interlocking paving units. ASTM C936, West Conshohocken, Pa. American Society for Testing and Materials (ASTM).

Bayer, P., Saner, D., Bolay, S., Rybach, L. and Blum, P. (2002) Greenhouse gas emission savings of ground source heat pump systems in Europe: A review. Renewable and Sustainable Energy Reviews. 16(2): 1256–1267.

Bressy, A., Gromaire, M., Lorgeoux, C., Saad, M., Leroy, F. and Chebbo, G. (2014) Efficiency of source control systems for reducing runoff pollutant loads: Feedback on experimental catchments within Paris conurbation. *Water Research* 57, 234–246.

Charlesworth, S. (2010) A review of the adaptation and mitigation of global climate change using sustainable drainage in cities. Journal of Water Climate Change 1(3), 165–180.

Charlesworth, S.M., Nnadi, E., Oyelola, O., Bennett, J., Warwick, F., Jackson, R. and Lawson, D. (2012) Laboratory based experiments to assess the use of green and food based compost to improve water quality in a sustainable drainage (SUDS) device such as a swale. *Science of The Total Environment* 424, 337–343.

Charlesworth, S.M., Lashford, C. and Mbanaso, F. (2014) Hard SUDS Infrastructure. Review of Current Knowledge, Foundation for Water Research.

CIBSE (2006) Guide A: environmental design, heating, air conditioning and refrigeration, 7th edition. Chartered Institute of Building Services Engineers. Available at: http://tinyurl.com/oll2w8a accessed 1 November 2015.

Coma, J., Perez, G., Sole, C., Castell, A. and Cabeza, L.F. (2016) Thermal assessment of extensive green roofs as passive tool for energy savings in buildings. *Renewable Energy.* 85, 1106–1115.

Coupe, S., Tota-Maharaj, K., Scholz, M. and Grabiowiecki, P. (2009) Water stored within permeable paving and the effect of ground source heat pump applications on water quality. Proceedings 9th International Conference on Concrete Block Paving. Buenos Aires, Argentina, 18–21 Oct 2009. Available at: http://tinyurl.com/pbpt2oh Accessed 30/11/15.

Curtis, R., Lund, J., Sanner, B., Rybach, L. and Hellström, G. (2005) Ground source heat pumps – geothermal energy for anyone, anywhere: Current worldwide activity. World Geothermal Congress Antalya, Turkey, 24–29 April 2005.

De Boeck, L., Verbeke, S., Audenaert, A. and De Mesmaeker, L. (2015) Improving the energy performance of residential buildings: A literature review. Renewable and Sustainable Energy Reviews. 52: 960–975 (in progress).

del Castillo-Garcìa G., Borinaga-Treviño R., Sañudo-Fontaneda L.A. and Pascual-Muñoz P. (2013) Influence of pervious pavement systems on heat dissipation from a horizontal geothermal system. European Journal of Environmental and Civil Engineering. 17(10): 956–967.

Deletic, A. and Fletcher, T. (2006) Performance of grass filters used for stormwater treatment – a field and modelling study. Journal of Hydrology 317, 261–275.

Donovan, G.H. and Butry, D.T. (2009) The value of shade: Estimating the effect of urban trees on summertime electricity use. Energy and Buildings. 41, 662–668.

Energy Saving Trust (2004). Energy efficiency best practice in housing: Domestic ground source heat pumps: Design and installation of closed-loop systems. Available from: http://www.gshp.org.uk/documents/CE82-DomesticGroundSourceHeatPumps.pdf

Fioretti, R., Palla, A., Lanza, L.G. and Principi, P. (2010) Green roof energy and water related performance in the Mediterranean climate. Building and Environment 45, 1890–1904.

Givoni, B. (1998) Climate Considerations in Building and Urban Design. Wiley, New York.

Government, H.M. (2009) The UK Renewable Energy Strategy, Cm 7686, Stationery Office, July 2009 [on-line]. Available at: http://tinyurl.com/nlvmbst Accessed 30/11/15

Hashemi, S.S.G., Bin Mahmud, H. and Ashraf, M.A. (2015) Performance of green roofs with respect to water quality and reduction of energy consumption in tropics: a review. Renewable and Sustainable Energy Reviews 52 669–679.

Healy, P.F. and Ugursal, V.I. (1997) Performance and economic feasibility of ground source heat pumps in cold climate. International Journal of Energy Research. 21, 10, 857–870.

Hepbasli, A. (2005) Thermodynamic analysis of a ground-source heat pump system for district heating. International Journal of Energy Research 29, 671–687.

Hwang, Y., Lee, J., Jeong, Y., Koo, K., Lee, D., Kim, I., Jin, S. and Kim, S.H. (2009) Cooling performance of a vertical ground-coupled heat pump system installed in a school building. Renewable Energy 34, 578–582.

INFORSE, 2010. EU Directive on Energy End-use Efficiency and Energy Services. International Network for Sustainable Energy (INFORSE). [on-line] Available at: http://tinyurl.com/pmht3v6 Accessed 30/11/15

Ip, K., Lam, M. and Miller, A. (2010) Shading performance of a vertical deciduous climbing plant canopy. Build. Environ. 45 (1), 81–88.

Issa, R.J., Leitch, K. and Chang, B. (2015) Experimental heat transfer study on green roofs in a semi-arid climate during summer. Journal of Construction Engineering. Article ID 960538, 15 pages. 10.1155/2015/960538.

Kazemi, F., Beecham, S. and Gibbs, J. (2011) Streetscape biodiversity and ther role of bioretention swales in an Australian urban environment. Landscape and Urban Planning 101 (2), 139–148.

Lund, J., Sanner, B., Rybach, L., Curtis, R. and Hellströmm, G. (2004) Geothermal (ground-source) heat pumps, a world overview. GHC Bulletin 1–10.

Lund, J., Sanner, B., Rybach, L., Curtis, R. and Hellströmm, G. (2003) Geothermal (ground-source) heat pumps a world overview. Renewable Energy World. 6(4): 218–227.

Novo, A.V., Bayón, J.B., Castro-Fresno, D. and Rodríguez-Hernández, J. (2013) Temperature performance of different pervious pavements: rainwater harvesting for energy recovery purposes. Water Resour. Manag., 27: 5003–5016.

Novo, A., Gomez-Ullate, E., Bayón, J.B., Castro-Fresno, D. and Rodríguez-Hernández, J. (2010) Monitoring and evaluation of thermal behaviour of permeable pavement under northern Spain climate. Proceedings: Sustainable Techniques and Strategies in Urban Water Management. 7th International Conference Novatech, Lyons, France. Available at: http://tinyurl.com/q68azxz Accessed 1 November 2015.

O'Connell, S. and Cassidy, S.F. (2003) Recent large scale ground source heat pump installations in Ireland. Proceedings: International Geothermal Conference, Reykjavik, Iceland. Available at: http://tinyurl.com/o6yxkj3 accessed 1 November 2015.

Omer, A.M. (2008) Ground-source heat pumps systems and applications. Renewable and Sustainable Energy Reviews. 12(2): 344–371.

Ozgener, O. and Hepbasli, A. (2007) Modelling and performance evaluation of ground source (geothermal) heat pump systems. *Energy and Buildings* 39(1), 66–75.

Pandit, R. and Laband, D.N. (2010) Energy savings from tree shade. *Ecological Economics*. 69, 1324–1329.

POST (Parliamentary Office of Science and Technology) (2010). Renewable Heating. London

Reboredo, J.C. (2015) Renewable energy contribution to the energy supply: Is there convergence across countries? *Renewable and Sustainable Energy Reviews* 45, 290–295.

Refahi, A.H. and Talkhabi, H. (2015) Investigating the effective factors on the reduction of energy consumption in residential buildings with green roofs. *Renewable Energy* 80, 595–603.

Robitu, M., Inard, C., Groleau, D. and Musy, M. (2004) Energy balance study of water ponds and its influence on building energy consumption. *Build. Services Eng. Res. Technol.* 25, 171–182.

Robitu, M., Musy, M., Inard, C. and Groleau, D. (2006) Modeling the influence of vegetation and water pond on urban microclimate. *Sol. Energy* 80, 435–447.

Rosenzweig, C., Gaffin, S. and Parshall, L. (2006) Green Roofs in the New York Metropolitan Region. Executive Summary. Colombia University Centre for Climate Systems Research. NASA Goddard Institute for Space Studies, New York.

Saner, D., Juraske, R., Kübert, M., Blum, P., Hellweg, S. and Bayer, P. (2010) Is it only CO_2 that matters? A life cycle perspective on shallow geothermal systems. *Renewable and Sustainable Energy Reviews* 14,7, 1798–1813.

Sansalone, J., Kuang, X. and Ranieri, V. (2008) Permeable pavement as a hydraulic and filtration interface for urban drainage. *Urban Storm-Water Management* 134, 5, 666–674.

Sansalone, J. and Teng, Z. (2005) Transient rainfall-runoff loadings to a partial exfiltration system: implications for urban water quantity and quality. *Environmental Engineering-ASCE* 131, 8, 1155–1167.

Scholz, M., Tota-Maharaj, K. and Grabiowiecki, P. (2012) Modelling of retrofitted combined permeable pavement and ground source heat pump systems. Retrofit Conference, University of Salford, Manchester. Available at: http://tinyurl.com/nmohxwj Accessed 30/11/15.

Self, S.J., Reddy, B.V. and Rosen, M.A. (2013) Geothermal heat pump systems: Status review and comparison with other heating options. *Applied Energy*, 101: 341–348.

Shafiei, S. and Salim, R.A. (2014) Non-renewable and renewable energy consumption and CO_2 emissions in OECD countries: A comparative analysis. *Energy Policy*. 66: 547–556.

Simpson, J.R. (2002) Improved estimates of tree-shade reductions on residential energy use. *Energy and Buildings*. 34, 1067–1076.

Singh, H., Muetze, A. and Eames, P.C. (2010) Factors influencing the uptake of heat pump technology by the UK domestic sector. *Renewable Energy* 35(4), 873–878.

Song, J., Yang, W., Higano, Y. and Wang, X. (2015) Introducing renewable energy and industrial restructuring to reduce GHG emission: Application of a dynamic simulation model. Energy Conversion and Management 96, 625–636.

Tota-Maharaj, K. and Paul, P. (2015) Sustainable approaches for stormwater quality improvements with experimental geothermal paving systems. *Sustainability* 7(2) 1388–1410.

Tota-Maharaj, K., Grabiowiecki, P., Babatunde, A. and Devi Tumula P. (2012) Constructed wetlands incorporating surface water heat pumps (SWHPS) for concentrated urban stormwater runoff treatment and reuse. 16th International Water Technology Conference, IWTC 16, Istanbul, Turkey. Available at: http://tinyurl.com/qb8jhod Accessed 30/11/015.

Tota-Maharaj, K., Scholz, M. and Coupe, S.J. (2011) Modelling temperature and energy balances within geothermal paving systems. *Road Materials and Pavements Design.* 12(2) 315–344.

Tota-Maharaj, K., Scholz, M. and Coupe, S.J. (2010) Utilisation of geothermal heat pumps within permeable pavements for sustainable energy and water practices. Zero Emission Buildings –

Proceedings of Renewable Energy Conference, Trondheim, Norway. Available at: http://tinyurl.com/p2tooxy accessed 1 November 2015.

Tota-Maharaj, K., Grabiowiecki, P. and Scholz, M. (2009) Energy and temperature performance analysis of geothermal (ground source) heat pumps integrated with permeable pavement systems for urban run-off reuse. *International Journal of Sustainable Engineering.* 2(3): 201–213.

Virk, G., Jansz, A., Mavrogianni, A., Mylona, A., Stocker, J. and Davies, M. (2015) Microclimatic effects of green and cool roofs in London and their impacts on energy use for a typical office building. *Energy and Buildings* 88, 214–228.

Wang, S., Liu, X. and Gates, S. (2015) Comparative study of control strategies for hybrid GSHP system in the cooling dominated climate. *Energy and Buildings* 89, 222–230.

Wright, G. (2010) Extent and cost of designing and constructing small areas of hard-standing around new and existing, domestic and non-domestic buildings. The Scottish Government. Directorate for the Built Environment, Building Standards Division

www.laysells.co.uk. Hanson Formpave helps reduce annual fuel costs by a minimum of 42%. Available at: http://tinyurl.com/owkvjw7 Accessed 30/11/015.

Yang, W., Wang, Z., Cui, J., Zhu, Z. and Zhao, X. (2015) Comparative study of the thermal performance of the novel green(planting) roofs against other existing roofs. *Sustainable Cities and Society.* 16, 1–12.

Yu, S.L., Kuo, J.T., Fassman, E.A. and Pan, H. (2001) Field test of grassed-swale performance in removing runoff pollution. Journal of Water Resources *Planning and Management* 127(3), 168–171. Available at: http://tinyurl.com/ncby3b8 Accessed 30/11/15

Carbon Sequestration and Storage: The Case for Green Roofs in Urban Areas

Brad Rowe

14.1 Introduction

A green roof, the practice of growing plants on rooftops, is a proven method that reduces stormwater runoff and is employed by many municipalities, especially in Germany. In addition to their stormwater management benefits, they also sequester CO_2, one of the greenhouse gases that are believed to contribute to climate change. They can provide many ecosystem services and long-term economic benefits to those municipalities that implement them widely.

14.2 The Importance of Carbon Sequestration

The earth is warming. There could be many reasons for this, but there is little doubt that these higher global temperatures have coincided with the industrial revolution and the burning of fossil fuels, and most scientists believe that human activity is the main culprit (IPCC, 2014). Concentrations of carbon dioxide (CO_2) in the atmosphere have increased by 32% since 1750 and this increase in CO_2, along with other greenhouse gases, are believed to be the main cause (IPCC, 2007). As fossil fuels are burned, CO_2 is released as a by-product of combustion. As it builds up in the atmosphere the greenhouse effect is evident as the CO_2 keeps terrestrial energy from escaping into space, thus resulting in increased temperatures.

Not only has atmospheric CO_2 increased dramatically since the dawn of the industrial revolution, but it appears to be accelerating. The burning of fossil fuels and biomass, cultivation of soil for agriculture, deforestation and drainage of wetlands, as well as other changes in land use, increased CO_2 emissions by approximately 80% from 1970 to 2004 (IPCC, 2007). Unless this issue is addressed, emissions of CO_2 will likely continue to increase. For example, in the USA the Department of Energy (2011) has proposed that over 100 new coal-fired power plants be constructed by 2017 to meet expected demands for energy.

Sustainable Surface Water Management: A Handbook for SuDS, First Edition.
Edited by Susanne M. Charlesworth and Colin A. Booth.
© 2017 John Wiley & Sons, Ltd. Published 2017 by John Wiley & Sons, Ltd.

These anthropogenic-driven greenhouse gas emissions and the resultant heating of the planet could lead to serious problems. Warmer global temperatures will inevitably lead to the melting of polar ice caps that could result in coastal flooding and the disruption of marine and freshwater systems, impact biological systems, alter precipitation patterns and distribution, increase the rate of heat-related illnesses and increase the spread of infectious disease vectors, insect pests and invasive weed species (IPCC, 2014).

Some have tried to address this issue by attempting to sequester and store more carbon, reducing carbon emissions and through public policy such as carbon trading programmes. However, in order to discuss options for reducing the amount of CO_2 in the atmosphere, a definition of carbon sequestration and storage must be given. Carbon is sequestered in plants through the process of photosynthesis, by soils as organic compounds, and by the oceans in the form of dissolved carbon as part of the natural carbon cycle. Carbon dioxide is removed from the atmosphere during photosynthesis and is stored as plant biomass, a process commonly referred to as terrestrial carbon sequestration. Carbon is also sequestered and stored in the soil as plant tissues die and produce plant litter as well as through root exudates. If net primary production of organic matter exceeds decomposition, then the ecosystem is a net carbon sink, at least in the short term. Carbon is constantly being sequestered and then released. The rate that this occurs depends on many environmental factors, and the presence of microorganisms. It could be argued that lumber used to build a house is storing carbon for a longer period of time than a tree with a shorter lifespan than the house. Although the house is not sequestering any further carbon, it is not being released through decomposition.

Much of the research regarding carbon sequestration has been conducted on natural and agricultural ecosystems, and to a lesser degree on urban forests or landscapes. When agricultural fields are cultivated under no-till practices the mineralisation of organic matter can be reduced by half, compared to conventional tillage systems (Balesdent et al., 2000, West and Marland, 2003). Cultivation can reduce microbial populations, root biomass, and the overall amount of organic matter in the soil by up to 55% (Balesdent et al., 2000; Rhoades et al., 2000; Matamala et al., 2008). Furthermore, agricultural lands are often originally prepared by cutting down forests that contain large quantities of carbon in favour of herbaceous annual crops or pasture (Rhoades et al., 2000). Urban landscapes can store a significant amount of carbon, and it is estimated that urban forests in the USA store 712 million tonnes of carbon (Rowntree and Nowak, 1991; Nowak, 1993).

The ability of a particular landscape to sequester carbon depends on many variables, including species composition and diversity (Kaye et al., 2000; Tilman et al., 2006), ecosystem age (Matamala et al., 2008), plant morphology (Rhoades et al., 2000; Fang et al., 2007), plant density (Fang et al., 2007; Matamala et al., 2008), climate (Matamala et al., 2008) and management practices (Wu, et al., 2008). Tilman et al. (2006) reported that 160% more carbon was sequestered in the root systems in their research plots when 16 species were planted together compared to when they were grown as individual species in monoculture. Likewise, in an example from the practice of agroforestry, *Eucalyptus saligna* stored twice as much carbon when interplanted with the nitrogen-fixing legume, *Albizia falcataria*, than when grown as a monoculture (Kaye et al., 2000). Ecosystem age is also important, especially in young landscapes. Leaf litter, root biomass and microbial activity all increase with time until equilibrium is reached (Matamala et al., 2008). This can be more pronounced when woody plants are part of the plant community because they add biomass (Fang et al., 2007). In addition, management practices such as spacing and harvesting schedules for trees and supplemental irrigation can be a factor (Fang et al., 2007; Wu et al., 2008). When arid cropland was irrigated it increased soil organic carbon to a level 133% above that of the native unirrigated soil over a period of 55 years (Wu *et al.*, 2008).

Much less is known about the potential for carbon sequestration in urban landscapes relative to natural or agricultural landscapes; this is especially true for ornamental areas (Marble *et al.*, 2011). Trees in urban areas have been shown to provide a significant contribution to the reduction of air pollutants such as CO_2 (Scott *et al.*, 1998; Akbari *et al.*, 2001; Nowak, 2006). However, taking advantage of the services that plants provide is a challenge, since urban sites tend to be covered with impervious surfaces such as roads, car parks and rooftops. Thus there is limited space at ground level to plant trees or other landscaping. For example, impervious surfaces cover 94% of the land in the mid-Manhattan west section of New York (Rosenzweig *et al.*, 2006). One option to remedy this problem is to use rooftops – which often account for 40–50% of the impermeable surface in urban sites – to grow vegetation (Dunnett and Kingsbury, 2004). These typically wasted spaces provide a unique opportunity to sequester carbon.

14.3   Coupling the Stormwater Management Benefits of Green Roofs with Carbon Sequestration

The concept of green roofs involves growing plants on rooftops, which partially replaces the vegetation that was destroyed when the building was constructed, thus allowing for the sequestration of carbon. In addition to carbon storage, they can provide numerous benefits such as energy conservation (Sailor, 2008; Castleton *et al.*, 2010;), mitigation of the urban heat island (Susca *et al.*, 2011), a reduction in air and noise pollution (Van Renterghem and Botteldooren, 2008; Rowe, 2011), increased longevity of roofing membranes (Kosareo and Ries, 2007), increased urban biodiversity (Brenneisen, 2006; Eakin *et al.*, 2015), providing a place to grow local vegetables (Whittinghill and Rowe, 2012; Whittinghill *et al.*, 2013), providing a more aesthetically pleasing environment in which to work and live (Getter and Rowe, 2006), and improved return on investment compared to traditional roofs (Kosareo and Ries, 2007; Clark *et al.*, 2008; Peri *et al.*, 2012; Chenani *et al.*, 2015). However, most would agree that the greatest service they provide is in stormwater management by reducing runoff and in improving water quality (VanWoert *et al.*, 2005; Getter *et al.*, 2007; Oberndorfer *et al.*, 2007; Czerniel Berndtsson, 2010; Rowe, 2011).

The reduction in stormwater runoff is due primarily to the substrate layer. Water is absorbed by the soil particles and is held within the pore spaces until the substrate reaches field capacity. During light rain events, 100% of the precipitation may be held in the substrate and is eventually removed by surface evaporation or through plant transpiration which replenishes its water-holding capacity. During heavier rain events, water will run off, once the substrate reaches field capacity, but the runoff is delayed, thus reducing the peak runoff, which can exceed the capacity of a municipal stormwater system. The reduction in runoff generally ranges from 50–100%, depending on the type of green roof system, substrate composition and depth, roof slope, plant species, pre-existing substrate moisture and the intensity and duration of the rainfall (Rowe, 2011).

In communities that do not have separate stormwater and sewage systems, one of the major problems resulting from stormwater runoff is the occurrence of a combined sewage overflow (CSO). When this occurs, raw untreated sewage flows directly into our waterways because the volume of runoff exceeds the capacity of the stormwater system. This is quite common in the USA, as there are 772 communities that do not have separate sewer and stormwater systems (USEPA, 2014). For example, approximately 40 billion gallons of untreated wastewater are dumped every year into New York's waterways due to CSO events. In fact, half of all rainfall events in New York result in a CSO event (Cheney, 2005). In urban areas where stormwater and sewer systems are separated, pollutants are still washed into waterways because of the preponderance of impervious surfaces. Green roofs work because both the total amount and the peak runoff are reduced, so that municipal stormwater systems do not have to be as large and costly.

In regards to the water quality that runs off a green roof, quality depends on numerous factors such as roof age, plant community, substrate depth and composition, management practices such as fertilisation and maintenance, the intensity and duration of the rainfall, local pollution sources and the physical and chemical properties of those pollutants (Rowe, 2011). The main pollutants tend to be nitrogen and phosphorus, which come from decomposing organic matter that was incorporated into the original substrate mix. After the first year or two this problem tends to decrease to where green roofs have a positive influence on water quality (Rowe, 2011).

Of course not all green roofs are the same. They are generally categorised as either 'intensive' or 'extensive'. Intensive green roofs may include shrubs and trees, and appear similar to landscapes found at ground level (Figure 14.1). In order to sustain these plants, substrate depths greater than 15 cm are usually required (Snodgrass and McIntyre, 2010). In contrast, extensive green roofs are generally built with substrate depths <15 cm, and because of the shallower depth, plant species are limited to grasses, herbaceous perennials,

Figure 14.1   Intensive green roofs, such as the roof of the Church of Jesus Christ of Latter-day Saints Conference Center in Salt Lake City, Utah, include shrubs and trees, and appear similar to landscapes found at ground level.

Figure 14.2   Extensive green roofs such as this one on the Plant and Soil Sciences Building at Michigan State University are generally built with substrate depths less than 15 cm. Because of the shallower depth, plant species are limited to grasses, herbaceous perennials, annuals and drought-tolerant succulents such as sedum. They have less potential to sequester carbon.

annuals and drought-tolerant succulents such as sedum (Figure 14.2). The type of roof installed can have a significant impact on the amount of carbon that can be sequestered. In addition to normal carbon sequestration into plant biomass as described above, green roofs can also reduce atmospheric CO_2 by reducing energy consumption in individual buildings and by mitigating the urban heat island. A green roof will eventually reach a carbon equilibrium (plant growth = plant decomposition), but initially this artificial ecosystem will serve as a carbon sink.

14.4   Carbon Sequestration on Green Roofs

Getter *et al.* (2009) conducted two studies to quantify the carbon storage potential of extensive green roofs and the effect of species selection on carbon accumulation. In the first study, above-ground biomass was measured on 12 sedum-based green roofs ranging from one to six years in age and 2.5–12.7 cm in substrate depth. The amount of carbon sequestered ranged from 73–276 g C m^{-2} with an average of 162 g C m^{-2} in above-ground biomass. Both substrate depth and the age of a green roof have been shown to influence plant growth on a green roof (Durhman *et al.*, 2007; Getter and Rowe, 2008; Rowe *et al.*, 2012).

In the second study, carbon was determined for above- and below-ground biomass, as well as carbon present in the soil substrate (Getter *et al.*, 2009). Twenty replicated plots were utilised, with four replications each of four plant species and a substrate-only control. All plots were at a depth of 6.0 cm, and plant material and substrate were harvested seven times over the course of two growing seasons. Results at the end of the second year showed that above-ground plant material storage varied by species, ranging from $64\,g\,C\,m^{-2}$ (*S. acre*) to $239\,g\,C\,m^{-2}$ (*S. album*), with an average of $168\,g\,C\,m^{-2}$. Below-ground biomass ranged from $37\,g\,C\,m^{-2}$ (*S. acre*) to $185\,g\,C\,m^{-2}$ (*S. kamtschaticum*) and averaged $107\,g\,C\,m^{-2}$. Substrate carbon content averaged $913\,g\,C\,m^{-2}$. In total, this entire extensive green roof system held $1188\,g\,C\,m^{-2}$ in combined plant material and substrate. However, after subtraction of the $810\,g\,C\,m^{-2}$ that existed in the original substrate, net carbon sequestration totalled $375\,g\,C\,m^{-2}$.

The Getter *et al.* (2009) study quantified the carbon sequestered in a shallow sedum-based extensive green roof, but this is the lower limit of a green roof's potential (Rowe, 2011). A third study quantified the carbon sequestration potential of nine in-ground and four green roof landscape systems with increasing levels of complexity, ranging from sedum to woody shrubs (Whittinghill *et al.*, 2014). The landscape systems examined at ground level included (1) succulent rock garden consisting of sedum, (2) prairie consisting of native perennials and grasses, (3) a mulched ornamental bed of herbaceous perennials and grasses, (4) vegetable and herb garden, (5) Kentucky bluegrass lawn, (6) woody ground covers, (7) deciduous shrubs, (8) broad-leaf evergreen shrubs and (9) narrow-leaf evergreen shrubs. The first four landscape systems were also replicated on green roof platforms. The objectives of this study were to quantify the amount of carbon sequestered by ornamental and green roof landscapes of varying complexity and then determine if there were differences in carbon sequestration between green roof landscapes and similar landscape systems at ground level.

There were differences in carbon content in above-ground biomass, below-ground biomass and substrate contents for all systems, but the three shrub landscape systems and herbaceous perennial and grasses contained the greatest amount of carbon. This makes sense because wood contains more carbon (4.7–16.7% more) than other plant structures (Fang *et al.*, 2007). However, these woody systems did have the lowest below-ground carbon contents, contrasting with their high above-ground biomass, soil/substrate and total carbon contents. In addition, in most cases, the green roof landscape systems contained less carbon than their corresponding in-ground landscape systems. The shallower substrate may have inhibited root growth, which would have reduced the size of plant that the plots could support, limiting plant above-ground biomass volume.

14.5   Embodied Energy

There is little doubt that the plants and soil on a green roof will sequester carbon. The 6.0 cm deep extensive green roof in the Getter *et al.* (2009) study sequestered $375\,g\,C\,m^{-2}$, ($168\,g\,C\,m^{-2}$ in above-ground plant biomass, $107\,g\,C\,m^{-2}$ in below-ground plant biomass, and $100\,g\,C\,m^{-2}$ in substrate carbon) beyond what was stored in the initial substrate. However, we must also consider the embodied energy required to initially construct it. Embodied energy is defined as the total energy consumed, or carbon released, by a product over its life cycle. The components necessary to construct a typical green roof (root barrier, drainage layer, growing substrate and even the plants) all require energy during the manufacturing and shipping process. This carbon cost is in addition to those required for a conventional roof (Kosareo and Ries, 2007). Based on an embodied energy analysis of building materials by Hammond and Jones (2008), the total embodied energy cost of the green roof components in the Getter *et al.* (2009) study was 23.6 kg CO_2 per square meter of green roof. This equates to approximately 6.5 kg C m^{-2} which is considerably greater than the 375 g C m^{-2} that was sequestered. Also, the 375 g C m^{-2} represents an equilibrium point where carbon assimilation equals carbon decomposition, so no further net carbon sequestration will occur on this roof.

Even so, the carbon that is sequestered by the plants and the growing substrate is only part of the equation. Because green roofs acts as insulators, they reduce the heat transfer into and out of buildings, so they reduce energy consumption. The lower demand for heating and air conditioning results in less CO_2 released into the atmosphere from power plants that produce this electricity. Therefore, the carbon emissions avoided due to energy savings should eventually pay for those costs and swing the energy balance equation in the positive direction. How long this takes depends on many factors, such as type of green roof, building specifications and local climate.

Based on Energy Plus, a building energy balance model supported by the US Department of Energy, green roofs can reduce electricity and natural gas consumption by 2% and 9–11%, respectively (Sailor, 2008). If green roofs were implemented throughout an urban area then an additional 25% reduction in electricity consumption is also possible due to a lessening of the urban heat island (Akbari and Konopacki, 2005). Using the Energy Plus model, a typical building with a 2000 m^2 green roof would save a minimum of 27.2 GJ of electricity and 9.5 GJ of natural gas annually. Accounting for the greenhouse gas potential of generating electricity and burning natural gas (USEPA, 2007, 2008), these figures translate to savings of 702 g C m^{-2} of green roof per year. Remembering that the embodied energy cost of 6.5 kg C m^{-2} for the Getter *et al.* (2009) green roof, nine years would be needed to offset the carbon debt of the green roof materials used in this installation. After this time, the emissions avoided would simply add on to the sequestration potential of the roof. The carbon sequestered by growing biomass (375 g C m^{-2} in the Getter study) shortens the carbon payback period in this scenario by two years.

The influence of plant biomass on carbon sequestration is demonstrated when one compares the results of the Getter *et al.* (2009) with the Whittinghill *et al.* (2014) studies. The 6.0 cm deep sedum roof of the Getter study had an embodied energy value of 6.5 kg C m^{-2} with a payback period due to energy savings alone of nine years. When the carbon sequestered by the green roof vegetation was included it reduced that payback period to seven years (Getter *et al.*, 2009). Assuming a similar embodied carbon value for the substrate used in the Whittinghill study, the 10.2 cm depth would contain 10.5 kg C m^{-2} with a payback period of 15 years. However, because the deeper substrate allowed for greater biomass production, the carbon payback period for these roofs decreased to 2.2, 1.9, 1.2 and 0.2 years for the sedum, prairie, vegetable garden and mulched ornamental bed of herbaceous perennials and grasses, respectively.

14.6   Improving Carbon Sequestration Potential

Net carbon sequestration can be improved immensely by altering plant selection, as discussed above, and through changes in substrate depth, substrate composition and management practices such as supplemental irrigation, fertilisation and the use of power equipment. This holds true whether a landscape is on a roof or at ground level.

14.6.1 Substrate Depth

All of the green roof landscape systems in the Whittinghill *et al.* (2014) study exhibited greater carbon sequestration than that reported by Getter *et al.* (2009). One reason is that the plants in the Whittinghill study were grown in a substrate that was almost twice as deep as the sedum in the Getter study (10.5 cm vs 6.0 cm). This not only provides a larger substrate volume to store carbon, but it increased the potential plant species to include those with greater biomass such as herbaceous perennials. In addition, altering the composition of the growing substrate has been shown to affect plant growth on an extensive green roof (Rowe *et al.*, 2006), which in turn affects growth and carbon sequestration.

Succulents such as sedum are common green roof plants because of their tolerance to drought. Because they exhibit crassulacean acid metabolism (CAM), a form of plant metabolism that allows them to conserve water by opening their stomates during the night to take up CO_2 and closing them during the day to reduce transpiration, CAM plants are ideal for survival on shallow green roofs (Cushman, 2001; Getter and Rowe, 2006). However, this can limit growth and reduces their likelihood of sequestering large amounts of carbon. When operating under CAM mode, rates for daily carbon assimilation are half to one-third that of non-CAM species (Hopkins and Hüner, 2004).

14.6.2 Substrate Composition

Traditionally, the green roof industry has utilised lightweight expanded aggregates made from heat-expanded slate, shale, and clay for growing substrates (Rowe *et al.*, 2006). However, as noted in the Getter *et al.* (2009) study, the heat expanded slate used for that roof accounted for 80% of the embodied energy used to construct it. Similarly, Peri *et al.* (2012) concluded that the production of materials, in particular the extraction and kilning of the inert substrate components, accounted for the majority of the crude oil consumed during the life of a green roof located in Bagheria, Italy. Likewise, Chenani *et al.* (2015) conducted a life cycle analysis on a modelled extensive green roof located in Chicago, IL, and showed that the inclusion of expanded clay in the substrate was the major negative contributor to the environmental impact of the roof. In addition, Bianchini and Hewage (2012) concluded that the use of recycled materials resulted in a more than two-fold reduction on the time required to offset the environmental cost of producing those components.

If alternative materials were used, the embodied energy could be reduced substantially. Natural or recycled materials that are locally available are likely candidates. For example, in the Pacific Northwest, volcanic pumice is readily available and is often used as a component in substrates. The pumice has been heat expanded by nature and thus its embodied energy is vastly reduced (Rowe, 2011). Other potential materials that may be more sustainable include crushed brick from demolished buildings (Molineux *et al.*, 2009; Graceson *et al.*, 2014; Young *et al.*, 2014; Bates *et al.*, 2015), crushed shells and coco coir (Steinfeld and Del Porto, 2008), crushed tile (Graceson *et al.*, 2014), bottom ash from incinerators (Molineux *et al.*, 2009; Graceson *et al.*, 2014), recycled rubber (Steinfeld and Del Porto, 2008; Pérez *et al.*, 2012), and recycled aerated concrete (Bisceglie *et al.*, 2014). Results in these studies are difficult to compare, but they do serve as strong examples that replacement of conventional green roof substrates with sustainable alternatives is an achievable goal. Economics dictate that these materials must be suitable for the intended plant selection, climatic zone and anticipated level of maintenance of the roof.

14.6.3 Management Practices

All urban landscapes can sequester carbon, but management practices will affect their net carbon sequestration and the permanence of the carbon sequestered. Practices such as supplemental irrigation and fertilisation influence plant growth as soil moisture and nutrients are often a limitation in many plant ecosystems (Vitousek and Howarth, 1991; Marble *et al.*, 2011; Rowe *et al.*, 2014). For example, plants in the Whittinghill *et al.* (2014) study were irrigated, whereas those in the Getter *et al.* (2009) study were not.

Urban landscapes are more complicated than natural areas because of human inputs. Ornamentals such as trees, shrubs, herbaceous perennials and turf grass all have relatively short lifespans in an urban environment and require periodic replacement. For example, should an urban street tree die and need to be replaced, Nowak *et al.* (2002) suggest that if the wood of the old tree is not used in some more permanent capacity such as lumber, the new tree just offsets the carbon being released by the old tree as it decomposes. Also, the typical potting medium used to produce ornamental plants consists of mostly organic matter and has much higher carbon content than most field soils (Marble *et al.*, 2011). What happens to this carbon when the container is transplanted for the nursery to the landscape is not well understood. In addition, the use of power equipment for such practices as mowing the lawn burns fuel thus releasing carbon. Overall, species selection and management practices influence carbon sequestration and storage as species vary in their water use efficiency, nutrient needs, growth and biomass allocation and decomposition rates (Naeem *et al.*, 1996; Rowe *et al.*, 2006).

14.7   Conclusions

Increasing levels of CO_2 and other greenhouse gases are believed to be one of the main causes of climate change. Since plants naturally sequester carbon, the widespread implementation of green roofs would aid in the removal of some CO_2 from the atmosphere. Of course, not all green roofs are created equal. Larger plants with greater biomass equate to greater carbon sequestration. However, the structural weight capacity of many buildings often limits the depth of the growing substrate and in turn the plant species that can be grown. One must also consider the embodied energy that is required to construct the roof and the carbon that will be saved from future energy savings. Although it is inherently easier to install vegetation at ground level, green roofs are especially applicable to urban areas where there is little space at ground level to grow trees or other vegetation. Quantifying the carbon sequestration potential of green roofs could make carbon more prominent in certification programmes such as LEED (Leadership in Energy and Environmental Design), the Sustainable Sites Initiative and in any potential future carbon cap and trade programmes. Green roofs are one tool that can help mitigate the negative effects of urbanisation.

References

Akbari, H. and Konopacki, S. (2005) Calculating energy-saving potentials of heat island reduction strategies. *Energy Policy*, 33(6): 721–56.

Akbari, H., Pomerantz, M. and Taha, H. (2001) Cool surfaces and shade trees to reduce energy use and improve air quality in urban areas. *Sol Energy*, 70(3): 295–310.

Balesdent, J., Chenu, C. and Balabane, M. (2000) Relationship of soil organic matter dynamics to physical protection and tillage. *Soil and Tillage Research*, 53: 215–230.

Bates, A., Sadler, J., Greswell, R. and Mackay, R. (2015) Effects of recycled aggregate growth substrate on green roof vegetation development: A six year experiment. *Landscape and Urban Planning*, 135: 22–31.

Bianchini, F. and Hewage, K. (2012) How 'green' are the green roofs? Life cycle analysis of green roof materials. *Building and Environment*, 48: 57–65.

Bisceglie, F., Gigante, E. and Bergonzoni, M. (2014) Utilization of waste autoclaved aerated concrete as lighting material in the structure of a green roof. *Construction and Building Materials*, 69: 351.

Brenneisen, S. (2006) Space for Urban Wildlife: Designing Green Roofs as Habitats in Switzerland. *Urban Habitats*, 4: 27–36.

Castleton, H.F., Stovin, V., Beck, S.B.M. and Davison, J.B. (2010) Green roofs; building energy savings and the potential for retrofit. *Energy and Buildings*, 42: 1582–1591.

Chenani, S.B., Levävirta, S. and Häkkinen, T. (2015) Life cycle assessment of layers of green roofs. *Journal of Cleaner Production*. doi: 10.1016/j.jclepro.2014.11.070.

Cheney, C. (2005) New York City: Greening Gotham's Rooftops, pp. 130–133. In: EarthPledge. *Green Roofs: Ecological Design and Construction*. Schiffer Books, Atglen, PA.

Clark, C., Adriaens, P. and Talbot, F.B. (2008) Green Roof Valuation: A Probabilistic Economic Analysis of Environmental Benefits. *Environ. Sci. Technol.*, 42: 2155–2161.

Cushman, J.C. (2001) Crassulacean acid metabolism: A plastic photosynthetic adaptation to arid environments. *Plant Physiol.*, 127: 1439–1448.

Czerniel Berndtsson, J. (2010) Green roof performance towards management of runoff water quantity and quality: a review. *Ecological Engineering*, 36: 351–360.

Dunnett, N. and Kingsbury, N. (2004) *Planting Green Roofs and Living Walls*. Timber Press, Inc., Portland, OR.

Durhman, A.K., Rowe, D.B. and Rugh, C.L. (2007) Effect of substrate depth on initial coverage, and survival of 25 succulent green roof plant taxa. *HortScience*, 42: 588–595.

Eakin, C., Campa III, H., Linden, D., Roloff, G., Rowe, D.B. and Westphal, J. (2015) *Avian Response to Green Roofs in Urban Landscapes in the Midwestern US*. Wildlife Society Bulletin, 39(3): 574–587.

Fang, S., Xue, J. and Tang, L. (2007) Biomass production and carbon sequestration potential in poplar plantations with different management patterns. *Journal of Environmental Management*, 85: 672–679.

Getter, K.L. and Rowe, D.B. (2006) The role of green roofs in sustainable development. *HortScience*, 41(5): 1276–1285.

Getter, K.L. and Rowe, D.B. (2008) Media depth influences sedum green roof establishment. *Urban Ecosystems*, 11: 361–372.

Getter, K.L., Rowe, D.B. and Andresen, J.A. (2007) Quantifying the effect of slope on extensive green roof stormwater retention. *Ecological Engineering*, 31: 225–231.

Getter, K.L., Rowe, D.B., Robertson, G.P., Cregg, B.M. and Andresen, J.A. (2009) Carbon sequestration potential of extensive green roofs. *Environ. Sci. Technol.*, 43(19): 7564–7570.

Graceson, A., Monaghan, J., Hall, N. and Hare, M. (2014) Plant growth responses to different growing media for green roofs. *Ecological Engineering*, 69: 196–200.

Hammond, G.P. and Jones, C.I. (2008) Embodied energy and carbon in construction materials. *Energy*, 161(2): 87–98.

Hopkins, W.G. and Hüner, N.P.A. (2004) *Introduction to Plant Physiology*, 3rd edn, John Wiley & Sons, New York.

Intergovernmental Panel on Climate Change (2007) *Climate Change 2007: The Physical Science Basis*. Cambridge University Press. Cambridge, UK.

Intergovernmental Panel on Climate Change (2014) *Climate Change 2014: Impacts, Adaptation, and Vulnerability*. IPCC Working Group II contribution to AR5. Cambridge University Press. Cambridge, UK, and New York, NY.

Kaye, J.P., Resh, S.C., Kaye, M.W. and Chimner, R.A. (2000) Nutrient and carbon dynamics in a replacement series of *Eucalyptus* and *Albizia* trees. *Ecology*, 81(12): 3267–3273.

Kosareo, L. and Ries, R. (2007) Comparative environmental life cycle assessment of green roofs. *Building and Environment*, 42: 2606–2613.

Marble, S.C., Prior, S.A., Runion, G.B., Torbert, H.A., Gilliam, C.H. and Fain, G.B. (2011) The importance of determining carbon sequestration and greenhouse gas mitigation potential in ornamental horticulture. *HortScience*, 46(2): 240–244.

Matamala, R., Jastrow, J.D., Miller, R.M. and Garten, C.T. (2008) Temporal changes in C and N stocks of restored prairie: implications for C sequestration strategies. *Ecological Applications*, 18(6): 1470–1488.

Molineux, C., Fentiman, C. and Gange, A. (2009) Characterising alternative recycled waste materials for use as green roof growing media in the U.K. *Ecological Engineering*, 35: 1507–1513.

Naeem, S., Håkansson, K., Lawton, J.H., Crawley, M.J. and Thompson, L.J. (1996) Biodiversity and plant productivity in a model assemblage of plant species. *Oikos*, 76(2): 259–264.

Nowak, D.J. (1993) Atmospheric carbon reduction by urban trees. *J. Environ. Mgt.*, 37: 207–217.

Nowak, D.J. (2006) Air pollution removal by urban trees and shrubs in the USA. Urban *Forestry and Urban Greening*, 4: 115–123.

Nowak, D.J., Stevens, J.C., Sisinni, S.M. and Luley, C.J. (2002) Effects of urban tree management and species selection on atmospheric carbon dioxide. *Journal of Arboriculture*, 28(3): 113–122.

Oberndorfer, F., Lundholm, J., Bass, B., Connelly, M., Coffman, R., Doshi, H., Dunnett, N., Gaffin, S., Köhler, M., Lui, K. and Rowe, B. (2007) Green roofs as urban ecosystems: ecological structures, functions, and services. *BioScience*, 57(10): 823–833.

Pérez, G., Vila, A., Rincón, L., Solé, C. and Cabeza, L. (2012) Use of rubber crumbs as drainage layer in green roofs as potential energy improvement material. *Applied Energy*, 97: 347–354.

Peri, G., Traverso, M., Finkbeiner, M. and Rizzo, G. (2012) Embedding 'substrate' in environmental assessment of green roofs life cycle: evidences from an application to the whole chain in a Mediterranean site. *Journal of Cleaner Production*, 35: 274–287.

Rhoades, C.C., Eckert, G.E. and Coleman, D.C. (2000) Soil carbon differences among forest, agriculture and secondary vegetation in lower montane Ecuador. *Ecological Applications*, 10(2): 497–505.

Rosenzweig, C., Solecki, W., Parshall, L., Gaffin, S., Lynn, B., Goldberg, R., Cox, J. and Hodges, S. (2006) Mitigating New York City's heat island with urban forestry, living roofs, and light surfaces. In: Proceedings of Sixth Symposium on the Urban Environment, Jan 30 – Feb 2, Atlanta, GA. (http://www.giss.nasa.gov/research/news/20060130/103341.pdf).

Rowe, D.B. (2011) Green roofs as a means of pollution abatement. *Environmental Pollution*, 159(8–9): 2100–2110.

Rowe, D.B., Monterusso, M.A. and Rugh, C.L. (2006) Assessment of heat-expanded slate and fertility requirements in green roof substrates. *HortTechnology*, 16(3): 471–477.

Rowe, D.B., Getter, K.L. and Durhman, A.K. (2012) Effect of green roof media depth on Crassulacean plant succession over seven years. *Landscape and Urban Planning*, 104(3–4): 310–319.

Rowe, D.B., Kolp, M.R., Greer, S.E. and Getter, K.L. (2014) Comparison of irrigation efficiency and plant health of overhead, drip, and sub-irrigation for extensive green roofs. *Ecological Engineering*, 64: 306–313.

Rowntree, R.A. and Nowak, D.J. (1991) Quantifying the role of urban forests in removing atmospheric carbon dioxide. *J. Arbor.*, 17: 269–275.

Sailor, D.J. (2008) A green roof model for building energy simulation programs. *Energy and Buildings*, 40: 1466–1478.

Scott, K.I., McPherson, E.G. and Simpson, J.R. (1998) Air pollution uptake by Sacramento's urban forest. *J. Arboriculture*, 24: 224–234.

Snodgrass, E.C. and McIntyre, L. (2010) *The Green Roof Manual*. Timber Press, Portland, OR.

Steinfield, C. and Del Porto, D. (2008) Green roof alternative substrate pilot study. *In* Preliminary Report to the Leading by Example Program, Executive office of Energy and Environmental Affairs, Commonwealth of Massachusetts. At: http://www.fishisland.net/LBE-Green%20Roof%20report.pdf

Susca, T., Gaffin, S.R. and Dell'Osso, G.R. (2011) Positive effects of vegetation: Urban heat island and green roofs. *Environmental Pollution* doi: 10.1016/j.envpol.2011.03.007.

Tilman, D., Hill, J. and Lehman, C. (2006) Carbon-negative biofuels from low-input high diversity grassland biomass. *Science*, 314: 1598–1600.

US Department of Energy, National Energy Technology Laboratory (2011) *Tracking new coal-fired power plants*. http://www.netl.doe.gov/coal/refshelf/ncp.pdf. Accessed 11 May 2015.

US Environmental Protection Agency (2007) *Inventory of U.S. Greenhouse Gas Emissions and Sinks: Fast Facts 1990–2005*. Conversion Factors to Energy Units (Heat Equivalents) Heat Contents and Carbon Content Coefficients of Various Fuel Types. EPA-430-R-07-002. Washington, DC.

US Environmental Protection Agency (2008) *Climate Leaders Greenhouse Gas Inventory Protocol Core Module Guidance: Indirect Emissions from Purchases/Sales of Electricity and Steam*. EPA-430-K-03-006. Washington, DC.

US Environmental Protection Agency (2014) *Water: Combined Sewer Overflows (CSO) Home*. http://water.epa.gov/polwaste/npdes/cso. Accessed 11 May 2015.

Van Renterghem, T. and Botteldooren, D. (2008) Numerical evaluation of sound propagating over green roofs. *Journal of Sound and Vibration*, 317: 781–799.

VanWoert, N.D., Rowe, D.B., Andresen, J.A., Rugh, C.L., Fernández, R.T. and Xiao, L. (2005) Green roof stormwater retention: Effects of roof surface, slope, and media depth. *J. Environ. Quality*, 34(3): 1036–1044.

Vitousek, P.M. and Howarth, R.W. (1991) Nitrogen limitation on land and in the sea: How can it occur? *Biogeochemistry*, 13: 87–115.

Whittinghill, L.J. and Rowe, D.B. (2012) The role of green roof technology in urban agriculture. *Renewable Agriculture and Food Systems*, 27(4): 314–322.

Whittinghill, L.J., Rowe, D.B. and Cregg, B.M. (2013) Evaluation of vegetable production on extensive green roofs. *Agroecology and Sustainable Food Systems*, 37(4): 465–484.

Whittinghill, L.J., Rowe, D.B., Cregg, B.M. and Schutzki, R. (2014) Quantifying carbon sequestration of various green roof and ornamental landscape systems. *Landscape and Urban Planning*, 123: 41–48.

West, T.O. and Marland, G. (2003) Net carbon flux from agriculture: Carbon emissions, carbon sequestration, crop yield, and land-use change. *Biogeochemistry*, 63: 73–83.

Young, T., Cameron, D.D., Sorrill, J., Edwards, T. and Phoenix, G.K. (2014) Importance of different components of green roof substrate on plant growth and physiological performance. *Urban Forestry and Urban Greening*, 13: 507–516.

Wu, L., Wood, Y., Jiang, P., Li, L., Pan, G., Lu, J., Chang, A.C. and Enloe, H.A. (2008) Carbon sequestration and dynamics of two irrigated agricultural soils in California. *SSSAJ*, 72(3): 808–814.

Dual-Purpose Rainwater Harvesting System Design
Peter Melville-Shreeve, Sarah Ward and David Butler

15.1   Introduction

The implementation of rainwater harvesting (RwH) in England and Wales has historically been driven by water efficiency considerations, such as those imposed under building regulations or suggested by guidance schemes such as the Code for Sustainable Homes. Even then, water demand management measures such as dual flush toilets, low flow taps and waterless urinals are often used in preference (Grant, 2006), with RwH rejected on financial grounds when a whole life cost assessment is undertaken (Roebuck *et al.*, 2011). However, researchers and practitioners have suggested that further investigation of the stormwater source control benefits of RwH is warranted, for example, their role within sustainable drainage systems (SuDS) (Hurley *et al.*, 2008; Gerolin *et al.*, 2010; Kellagher, 2011; Melville-Shreeve *et al.*, 2014). When considered together, these dual benefits could enhance the uptake of residential systems, particularly if technological innovation enables them to be realised within a single proprietary system (Debusk *et al.*, 2013). Similarly, Debusk and Hunt's (2014) comprehensive review of international RwH literature concluded that further research is required into RwH's benefits as a stormwater management tool.

The basic configuration required to achieve these dual benefits is shown in Figure 15.1. The retention and throttle concept effectively integrates water demand management and stormwater management objectives into a single RwH installation. This includes dedicated storage for retaining runoff and limiting outflow, while protecting the volume required for non-potable water supply.

This chapter begins with a brief review of RwH and SuDS, as well as existing approaches to integrate RwH and SuDS in England and Wales. A new design method is proposed for the design of dual-purpose RwH systems, and the method is subsequently used to assess the benefits of such systems for a case study development in Exeter, England. Benefits and limitations are discussed and conclusions drawn.

Sustainable Surface Water Management: A Handbook for SuDS, First Edition.
Edited by Susanne M. Charlesworth and Colin A. Booth.
© 2017 John Wiley & Sons, Ltd. Published 2017 by John Wiley & Sons, Ltd.

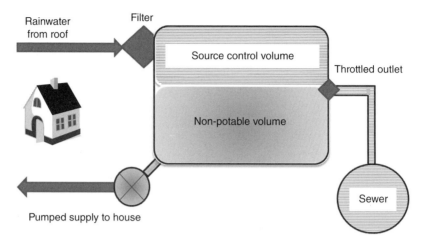

Figure 15.1   A dual-purpose rainwater harvesting system (adapted from Herrmann and Schmida, 1999).

15.2   RWH and SuDS in England and Wales

Stormwater management in England and Wales is strongly regulated by planning controls (DCLG, 2006) and associated guidance (Kellagher, 2012). To satisfy these regulations, new developments incorporate SuDS to manage stormwater runoff, with an emphasis on the need to attenuate flows to match those of the undeveloped site during the 1 in 100 year storm event. Typically, a 20–30% 'allowance for climate change' is included to enable drainage systems to cope during rainfall events in excess of those historically experienced. Where RwH systems are installed, they are designed solely to meet the non-potable water demand and do not usually provide significant source control or SuDS benefits (Kellagher, 2012). With the exception of the studies noted in this chapter, and following a comprehensive investigation over a number of years, the authors have not identified evidence of any RwH systems in England and Wales that have been installed with an emphasis on stormwater management.

Woods Ballard *et al.* (2007) defined the SuDS hierarchy in an effort to minimise stormwater runoff and pollution; (1) prevention, (2) source control, (3) site control, (4) regional control. Solutions such as green roofs, infiltration chambers, water butts and RwH can contribute to a source control strategy. Practitioners designing SuDS are encouraged to maximise source control opportunities before considering site-wide or regional control strategies such as attenuation tanks. Despite this, there remains a prevalence of end-of-pipe solutions that are frequently deemed to offer the 'easiest' way of complying with the legislation (Bastien *et al.*, 2009). Detention basins such as those illustrated in Figure 15.2 represent best practice, although perhaps 'most frequent practice' is a more appropriate epithet. Such basins are often inaccessible or even fenced off from the wider development and can be unattractive or not available for use, even as a green space.

In terms of the previously mentioned SuDS hierarchy, the benefits of a given source control technique need to be maximised, to minimise additional downstream storage volumes within a site-wide drainage design to achieve a best practice SuDS. RwH can reduce stormwater runoff volumes and rates (Leggett *et al.*, 2001; Debusk and Hunt, 2014; Campisano *et al.*, 2013), although the magnitude of such benefits cannot be generalised because a wide

Figure 15.2   A typical SuDS solution used in England and Wales – a detention basin at Newcourt, Exeter, Devon, England.

range of site-specific parameters need to be taken into account. Key design criteria include antecedent rainfall, yield, non-potable water demand and the RwH system configuration. However, the need for site specific design is not a barrier to implementation, because other SuDS techniques (e.g. detention ponds) are not universally appropriate for use at all new developments. Currently, the Environment Agency (EA) ensures regulatory compliance of new drainage systems through review of drainage strategy documents as part of planning applications (DCLG, 2012).

15.3   Approaches to Stormwater Source Control Using RwH in England and Wales

Early studies which sought to appraise the ability of RwH to control stormwater discharges were undertaken during the WaND project (Butler *et al.*, 2010). As part of this study, Kellagher and Maneiro Franco (2007) used hydraulic models to assess the overall reduction in stormwater runoff volumes and peak flow rates from a development where a large communal RwH storage tank was proposed. A stochastic rainfall series was used to model the source control benefits under a range of scenarios. The design assumed that non-potable water demands reduced the tank level and thus provided capacity for capturing rainfall runoff. It was concluded that tanks should be 1.5–2.5 times larger than standard RwH tanks to achieve 'considerable (sic) stormwater benefits'. In relation to managing extreme rainfall events, the study showed a notable reduction in runoff volumes for the 100 year return period rainfall event (23–55%). Similar research conducted by Memon *et al.* (2009) modelled a development of 200 properties and also concluded that RwH could reduce peak flows in downstream sewers.

In addition to the recent work conducted by Debusk *et al.* (2013), evidence of international interest in RwH as a source control tool is identified in a modelling study carried out by Huang *et al.* (2009). This research appraised the functionality of a RwH system designed with a 5 m³ tank capacity and a 50 mm diameter outlet throttle. A series of scenarios was considered, assuming that such a system was installed at each property in a development of 242 houses in Kuala Lumpur. It was demonstrated that the integration of the retention and throttle approach successfully limited peak rainfall discharges for the 30-minute duration rainfall event by 22% for the 100 year return period event (Huang *et al.*, 2009).

The first edition of the British Standard for RwH, *BS 8515:2009; Rainwater harvesting systems – Code of practice*, (BSI, 2009) focused on provision of an alternative water resource to meet water demand management drivers. The implementation of RwH as a stormwater source control technique is covered by suggesting that designers specify (intentionally) oversized RwH storage tanks to increase the likelihood that storage (empty space) is available at the beginning of a storm event. However, this is only viable when water demand (D) is greater than runoff yield (Y). Without significant water demand, the tanks are likely to remain full or close to full at all times. Even where D > Y, it cannot be guaranteed that the desired storage will be available because a number of other factors affect the demand. One major limitation of this approach is that it relied upon water user behaviour to be consistent with the core assumptions.

Building on the initial approaches documented in BS 8515:2009 (BSI, 2009) a further study into the benefit of RwH systems for source control in England and Wales was described in Gerolin *et al.* (2010). This method focused on demand for water from a RwH system freeing storage capacity for the next storm. The work has since been extended by Kellagher (2011), whereby a number of RwH systems in a residential housing development were installed and monitored. In the study, each RwH system was designed to comply with the Kellagher/Gerolin methodology. This methodology relies on the designer appraising the predicted non-potable demand for each property at the proposed development. The yield is calculated from the contributing roof area and average rainfall for the development. In summary, where the Y/D ratio is identified as less than 0.95 there is a high likelihood that storage will be available in the RwH tank at the commencement of an extreme storm event. Where Y/D is < 0.7 'there is usually considerable storage available' (Kellagher, 2011). It follows that this available storage volume can reduce the total volume of runoff during the next storm, and thus provides source control. A later study also concluded that stormwater can successfully be managed through implementation of this approach (Kellagher and Gutierrez-Andres, 2015).

To incorporate the growing evidence base, an updated British Standard was released in 2013 (BSI, 2013). A review of this update illustrates that few amendments have been made to the main body of the document, which continues to focus on water provision. However, Kellagher's (2011) technique for allowing source control benefits to be estimated has now been incorporated in the annex relating to source control (BSI, 2013). In addition, the annex suggests that active RwH systems can be implemented to maintain spare storage at all times. However, unlike the retention and throttle configuration, which is controlled by the mechanics of the water entering and leaving the tank, the updated British Standard suggests that actively managed RwH systems should include level sensors and some form of intelligent control system (BSI, 2013).

A year on from the publication of the updated British Standard a comprehensive review of the RwH market in the UK was undertaken by the authors. The research did not identify a product or case study site that complies with the active RwH concept or the retention and throttle specification. In contrast, physical trials were identified in the USA as reported in Debusk *et al.* (2013), which conclude that modifications in design can substantially improve the efficiency and stormwater management potential of RwH systems. This chapter seeks to address the first step towards improved control of stormwater using RwH systems by proposing a retention and throttle RwH design method for application at UK development sites.

15.4 Integrating Stormwater Source Control into RwH System Design

15.4.1 Defining the Design Process for a Case Study Development

Two drainage design options were proposed for a small residential development of seven houses in Exeter, south-west England, to evaluate the viability of using retention and throttle RwH tanks (Figure 15.1) as part of a wider drainage system. Both design options assume that all houses include a RwH system for non-potable reuse in the property's toilets. The non-potable reuse volume of the RwH tanks was conservatively assumed to be full at the start of all drainage simulations. Building on the modelling assessment carried out by Huang *et al.* (2009), the method enables the user to minimise the (stormwater) retention volume based on site characteristics such as the specified maximum discharge rate.

Step 1: Identify volume of RwH storage for non-potable reuse

RwH tank volumes were calculated using the 'intermediate approach' set out in BS8515:2013 (BSI, 2013) for each of the seven houses. This method defines the tank volume required for each RwH system as the lesser of two volumes (Y_R or DN) calculated using Equations 1 and 2;

$$Y_R = A \times e \times h \times f \times 0.05 \qquad (1)$$

where Y_R is 5% of the annual rainwater yield (l); A is the collecting area (m²); e is the yield coefficient (%); h is the annual depth of rainfall (mm); f is the hydraulic filter efficiency.

$$DN = Pd \times n \times 365 \times 0.05 \qquad (2)$$

where DN is 5% of the annual non-potable water demand (l); Pd is the daily requirement per person (l); and n is the number of persons.

Daily non-potable water demand was estimated assuming five flushes/person/day and an average flush volume of 4.5l (MTP, 2011; Waterwise, 2014). Occupancy was taken from the site's design drawings (Pell Frischmann, 2013) as either four or five people per house. No allowance was made for the use of irrigation or laundry water. For this stage, roof areas of 42–50m² were used to establish the non-potable reuse volume (V_{NP}) required for the RwH system at each house.

Step 2: Identify options for a compliant stormwater attenuation system

Option 1 (traditional SuDS) and option 2 (RwH as source control) were developed with all the drainage simulations carried out using the MicroDrainage software (XPSolutions, 2015) based on the input parameters set out in Table 15.1.

Table 15.1   Site characteristics, parameters and global design criteria.

Parameter	Input data
Location	Exeter, Devon, SW England
Total site area, m²	1230
Existing site runoff rate (1 in 100 year event), l/s	7.2
Maximum future site discharge rate (during 1 in 100 year rainfall event), l/s	7.2
Allowance for climate change, rainfall intensity %	30
Proposed roof areas, m²	334
Proposed parking and roadway areas, m²	340
Total impermeable area, m²	674
Design rainfall event	1 in 100 year, critical duration event
Design criteria	No above ground flood during Design Event
Runoff coefficient (all impermeable surfaces)	0.84
Runoff coefficient (all permeable surfaces)	0
Range of rainfall events tested, mins (mm/hour)	15 (155) to 168 (1.26)
Rainfall model	Flood Estimation Handbook (IoH, 1999)

Option 1: Traditional SuDS approach – site-wide attenuation tank

An outline drainage design was devised to route all stormwater from roofs and paved surfaces into a single on-line attenuation tank. The design was modelled as a geocellular storage tank located beneath the parking area with a proprietary vortex flow regulator controlling discharges. An iterative approach was implemented to reduce the volume of the required storage tank from an initial estimate of 30 m³ until a minimum size tank was identified that met the design criteria (Table 15.1). Each property was modelled with a RwH system intercepting all roof water, but these tanks were set to a status of 'full' before the simulations were run and thus had no capacity for stormwater source control.

Option 2: RwH as source control – decentralised retention and throttle RwH

An alternative drainage design was developed for Option 2, where the RwH tanks at each house were oversized to include an additional storage volume that drains down via an orifice following each storm event. The aim was to provide a RwH tank that can passively attenuate stormwater runoff from all roof areas, and thus provide 100% of the SuDS attenuation required to comply with the design characteristics (Table 15.1).

With V_{NP} established, the source control volume (V_{SC}) was identified and added to V_{NP} to obtain the total RwH tank volume (V_{RWH}). Limitations in the simulation model meant that a single calculation for a roof of 50 m² was used to represent a typical house. Firstly, a range of head–discharge relationships was developed for orifice outlet diameters 0–50 mm, at 5 mm increments, calculated using the standard orifice equation with a fixed coefficient of discharge of 0.6 (Butler and Davies, 2011). A series of tank volumes was assessed starting with an outlet orifice = 0 mm (i.e. zero tank discharge during a storm event). This identified that a maximum storage volume of 3.7 m³ was required to capture the critical rainfall event. All modelling for Option 2 was, therefore, carried out for a house with a roof area of 50 m² discharging roof runoff to a tank with a footprint of 4 m² and a depth of 1 m. Maximum water volumes were recorded for each rainfall event and the full range of orifice diameters was tested. This method allows for a maximum discharge rate to be plotted against the maximum tank storage volume. The critical rainfall event was identified as the event that generates the highest storage volume for a given orifice diameter. With the roof areas attenuated using the retention and throttle RwH, the remaining impermeable areas were addressed. Roadways and parking areas were modelled as draining into a geocellular storage tank, as it was assumed that the water quality from these areas may not be suitable for reuse without further treatment and thus could not be routed into the RwH tanks. A similar iterative approach to that used in Option 1 was used to minimise the tank size in light of the reduced input areas. The overall RwH tank volume identified in Option 2 is given in Equation 3.

$$V_{RWH} = V_{NP} + V_{SC} \tag{3}$$

The process of model simulations, orifice selection and ultimately sizing of the RwH tank is summarised as a flow chart in Figure 15.3.

15.4.2 Findings of Applying the Design

The design method generates a large number of output files, one for each scenario tested (e.g. orifice and design storm). The critical design storm for each scenario was identified as the storm that generated the largest storage volume. Table 15.2 illustrates a summary example of some of the simulation results, conducted to identify the V_{SC} for a RwH tank with a 20 mm orifice outlet.

Option 1: RwH tanks for non-potable use were designed to comply with the intermediate approach at each property. Six of the properties required a V_{NP} of 2.05 m³ with the seventh's lower occupancy indicating that 1.65 m³ would suffice. The drainage design approach modelled in Option 1 identified that a geocellular storage tank of 18 m³ was the minimum tank size required to accommodate stormwater runoff from the entire development to comply with the design criteria.

Option 2: Roof runoff was routed into individual RwH tanks at each house. Flows from the upper part of each tank were controlled using an orifice. Approximately half of the impermeable areas were roofs, and therefore half of the development's maximum permitted discharge rate (7.2 l/s) was allocated to the RwH outlets, equating to 0.5 l/s/RwH tank. Figure 15.4 illustrates the range of source control volumes required to accommodate the critical rainfall events for a range of discharging orifice sizes. At a peak discharge rate of 0.5 l/s, a 1.5 m³ V_{SC} is required with a 20 mm orifice outlet. The remaining 3.5 l/s of permissible discharge was allocated to the parking and roadways so that a separate storage tank could control runoff from these areas. A total of 10 m³ of storage was the minimum volume identified for this tank in combination with a suitable vortex flow controller.

15.4.3 Discussion of the Findings

In this chapter, a method has been developed and tested to allow passive stormwater source control capacity to be designed in RwH tanks using the retention and throttle configuration. The method seeks to support the initial source control concepts set out in BS 8515:2013 (BSI, 2013) by maintaining sufficient attenuation capacity in the RwH tanks

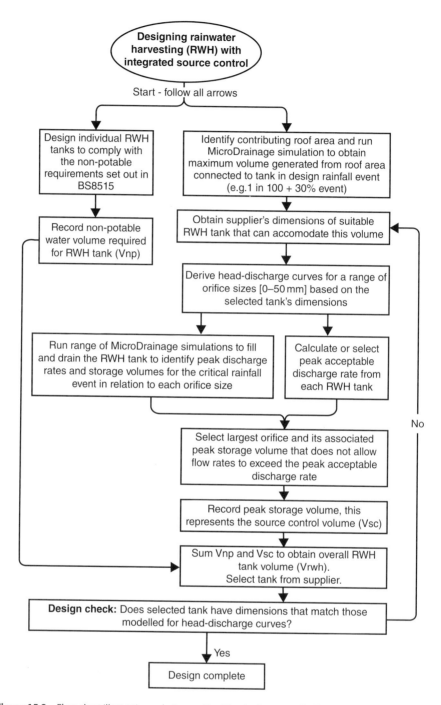

Figure 15.3   Flow chart illustrating a design method for dual-purpose RwH systems.

Table 15.2 Sample simulation results of 1 in 100 year storm events controlled by a RwH tank with a 20 mm orifice. (* = critical rainfall event).

Storm event	Modelled rain (mm/hr)	Time to vol peak (mins)	Max water level (m)	Max water depth (m)	Max discharge rate (l/s)	Total discharge volume (m³)	Overflow volume (m³)	Max volume required (m³)
15	155	16	0.337	0.337	0.5	1.6	0.0	1.3
30	95	28	0.363	0.363	0.5	2.0	0.0	1.5
60*	58	46	0.367	0.367	0.5	2.4	0.0	1.5
120	35	82	0.333	0.333	0.5	3.0	0.0	1.3
180	27	116	0.293	0.293	0.5	3.3	0.0	1.2
240	22	150	0.256	0.256	0.4	3.6	0.0	1.0
360	16	212	0.199	0.199	0.4	4.1	0.0	0.8
480	13	274	0.160	0.160	0.3	4.4	0.0	0.6
600	11	334	0.130	0.130	0.3	4.7	0.0	0.5
720	10	390	0.109	0.109	0.3	5.0	0.0	0.4
960	8	502	0.087	0.087	0.2	5.4	0.0	0.3
1440	6	748	0.065	0.065	0.2	5.9	0.0	0.3
2160	4	1100	0.048	0.048	0.1	6.5	0.0	0.2
2880	3	1468	0.039	0.039	0.1	7.0	0.0	0.2
4320	3	2180	0.028	0.028	0.1	7.6	0.0	0.1
5760	2	2912	0.022	0.022	0.1	8.0	0.0	0.1
7200	2	3568	0.019	0.019	0.0	8.4	0.0	0.1
8640	2	4400	0.016	0.016	0.0	8.7	0.0	0.1
10080	1	5024	0.014	0.014	0.0	8.9	0.0	0.1

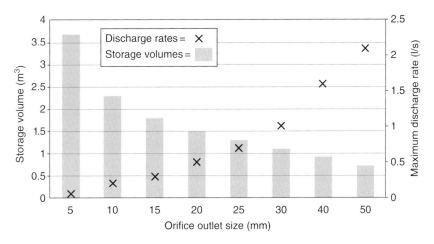

Figure 15.4 V_{SC} and maximum discharge rates for the range of orifices tested. For the development assessed, the final size of each RwH tank is illustrated in Figure 15.5 and a summary of the results described above is set out in Table 15.3.

without relying on user behaviour. If capacity can be maintained at all times, the method has potential to offer a more robust solution to achieving RwH tanks in a dual-purpose configuration. Empirical data is now warranted to appraise the concept, to establish the validity of its application as a dual-purpose RwH system.

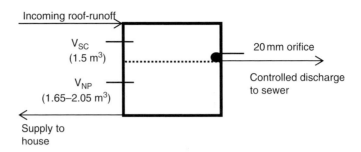

Figure 15.5 Illustration of the proposed retention and throttle RwH tanks for the case study development.

Table 15.3 Summary of results: Drainage designs for Option 1 and 2.

	Option 1	Option 2
	Large attenuation tank with throttled outlet discharging all drainage (RwH systems in houses assumed full)	Small attenuation tank with retention and throttle RwH tanks controlling roof runoff from each house
Maximum Discharge Rates (l/s)		
Roof areas	—	3.5
Paved areas	—	3.5
Total impermeable Area	<7.2	<7.2
Attenuation tank size (m³)	18	10
Minimum RwH tank sizes (m³)	6no. × 2.05 and 1no. × 1.65	6no. × 3.55 and 1no. × 3.15
Commercially available RwH tank sizes (m³)	7no. × 2.7	7no. × 3.8 and 20 mm orifice
Total V_{RWH} (m³)	18.9	26.6

The designs tested for the case study development showed that the concept is feasible and the method can be implemented using currently available software. Results from Option 2 demonstrated that a development's attenuation tank can be reduced in volume as a result of including retention and throttle RwH systems. By demonstrating a reduction in the overall volume of the attenuation tank, capital savings could be generated. For example, for Option 1, the smallest appropriate RwH tank identified from a supplier was 2.7 m³, costing £3440 installed. In Option 2, a 3.8 m³ tank was commercially available costing £3600 each. The integration of retention and throttle RwH would therefore cost an additional £1120 for the case study development. This cost would be offset by the reduced attenuation tank size, which is 8 m³ smaller than the design implemented in Option 1. Based on typical costs of geocellular storage crates of around £200/m³ (Graf Rain Bloc, 2015), it is conservatively estimated that the reduced size geocellular storage tank would generate a saving in excess of £2500.

With regard to water quality concerns, these may arise through risks of contamination from the sewer surcharging to the level of the tank's orifice. However, a non-return valve is frequently deployed to guard against this risk in existing RwH overflow systems. Secondly, the small orifice outlet is potentially at risk of blockage. However, installing appropriate filters for runoff entering the tank, a mechanical clearance mechanism or a blockage alarm could all help minimise this risk. By implementing retention and throttle RwH, runoff from roof areas can be fully attenuated without other SuDS being deployed. Furthermore, with a reduced attenuation tank proposed for the remainder of the development, it may be feasible to address the remaining runoff from hardstanding areas using swales, infiltration trenches or above-ground features integrated into the development layout, thus allowing the attenuation tank to be completely removed from the development arrangement.

Application of the design method to larger-scale developments would also be feasible. At such developments it is possible that the reduction in space needed for attenuation ponds/basins could permit additional properties to be constructed, generating financial benefit. The retention and throttle concept, alongside the proposed design methodology could potentially allow RwH to become more economically viable at new developments. Further work is now underway to support a supplier to develop and install systems in this novel configuration (RwH Ltd., 2015). The performance of these pilot systems will be empirically monitored to verify the modelling method and assumptions against physical observations.

15.5   Conclusions

Rainwater harvesting (RwH) at the residential property scale remains underexploited in England and Wales, as the benefits of RwH for both water demand management and source control applications have yet to be fully realised. Despite being encouraged to maximise source control opportunities before considering site-scale attenuation systems, practitioners designing SuDS often resort to end-of-pipe solutions that do not consider water reuse options. A recent update to the British Standard for Rainwater Harvesting (BSI, 2013) included a tank sizing method to enable RwH systems to contribute as a stormwater attenuation tool as well as providing an alternative water supply. However, to date there has been limited practical assessment of the retention and throttle configuration and this is not included in the British Standard. This chapter has contributed to this knowledge gap and seeks to further support the use of dual purpose RwH systems in England and Wales through the development of a new design method.

It was demonstrated that improved, passive stormwater source control could potentially be incorporated into RwH systems with a relatively small adjustment to the design configuration. The method outlined has also been used to develop a proprietary RwH system incorporating the retention and throttle configuration, which is now undergoing field trials. Research is ongoing to empirically test this system. In summary, the following conclusions can be drawn:

1. The retention and throttle concept can be used for source control by installing oversized RwH tanks that incorporate an additional mid-level outlet throttle.
2. Optimal sizing of the RwH volumes and outlet orifices is needed on a site-specific basis.
3. Retention and throttle RwH at residential properties in England and Wales may be achieved at a relatively low cost.
4. Technical barriers to implementation, such as site-specific constraints (e.g. ensuring that the throttle outlet can gravitate into existing drainage infrastructure) will prevent this method from providing an integrated solution that will suit all development plots.
5. A design tool could readily be developed to allow practitioners to select appropriate RwH attenuation volumes and orifice sizes for specific locations.

6. Passive RwH can contribute to source control in a SuDS system when either of the following factors is applicable: (a) the yield/demand ratio is < 0.95 or (b) passive source control is integrated into the tank design using the retention and throttle concept.

Acknowledgements

This work has been funded by sponsorship from Severn Trent Water plc and the UK Engineering and Physical Sciences Research Council through a STREAM Engineering Doctorate.

References

Bastien, N., Arthur, S., Scholz, M. and Wallace, S. (2009) Towards the best management of SuDS treatment trains. *13th International Diffuse Pollution Conference (IWA DIPCON 2009)*, 12–15 October 2009, Seoul, Korea.

BSI (2009) *BS 8515:2009 – Rainwater harvesting systems – Code of practice*. BSI, London.

BSI (2013) *BS 8515:2009 + A1:2013 Rainwater harvesting systems – Code of practice*. BSI, London.

Butler, D. and Davies, J.W. (2011) *Urban Drainage*, 3rd edition, Spon Press, London.

Butler, D., Memon, F.A., Makropoulos, C., Southall, A. and Clarke, L. (2010) *WaND. Guidance on Water Cycle Management for New Developments*, CIRIA Report C690, London.

Campisano, A., Cutore, P., Modica, C. and Nie, L. (2013) Reducing inflow to stormwater sewers by the use of domestic rainwater harvesting tanks. *Novatech*, June 23–27, 2013, Lyon. Available at: http://www.novatech.graie.org

DCLG (2006) *PPS25 Development and Flood Risk*. Department for Communities and Local Government, TSO Publications, London.

DCLG (2012) *National Planning Policy Framework*. Department for Communities and Local Government Crown Copyright. London.

Debusk, K.M., Hunt, W.F. and Wright, J.D. (2013) Characterizing rainwater harvesting performance and demonstrating stormwater management benefits in the humid southeast USA. *JAWRA Journal of the American Water Resources Association*, 49, 1398–1411.

Debusk, K.M. and Hunt, W.F. (2014) *Rainwater Harvesting: A Comprehensive Review of Literature*. Report 425 Water Resources Research Institute of the University of North Carolina. Available at: *http://www.lib.ncsu.edu/resolver/1840.4/8170*

Gerolin, A., Kellagher, R. and Faram, M.G. (2010) Rainwater harvesting systems for stormwater management: Feasibility and sizing considerations for the UK. *7th International conference on sustainable techniques and strategies in urban water management Novatech*, 27 June – 1 July, 2010, Lyon, France.

Graf Rain Bloc (2015) *Flood Attenuation Block Graf Rain Bloc 300 HGV traffic*. Available at: http://tinyurl.com/o8wmpmx

Grant, N. (2006) Water conservation products. In: Butler D., Memon F.A., editors. *Water Demand Management*. IWA Publishing, 82–106.

Huang, Y.F., Hashim, Z. and Shaaban, A.J. (2009) Potential and effectiveness of rainwater harvesting systems in flash floods reduction in Kuala Lumpur City, Malaysia. *Proceeding of the 2nd International Conference on Rainwater Harvesting and Management*, 7–12 September 2009, Tokyo, Japan.

Hurley, L., Mounce, S.R., Ashley, R.M. and Blanksby, J.R. (2008) No Rainwater in Sewers (NORIS): assessing the relative sustainability of different retrofit solutions. *Proceedings of the 11th International Conference on Urban Drainage*, Edinburgh, Scotland, UK.

IoH (1999) *Flood Estimation Handbook. Volume 3 – Statistical procedures for flood frequency estimation*, Wallingford, UK.

Kellagher, R. and Maneiro Franco, E. (2007) *Rainfall collection and use in developments; benefits for yield and stormwater control*, WaND Briefing Note 19 Release 3.0, WaND Portal CD-ROM, Centre for Water Systems, Exeter.

Kellagher, R. (2011) *Stormwater management using rainwater harvesting: testing the Kellagher/ Gerolin methodology on a pilot study*. Report SR 736, HR Wallingford Limited, UK.

Kellagher, R. (2012) *Preliminary rainfall runoff management for development*. RandD Technical Report W5–074/A/TR/1 Revision E, HR Wallingford Limited, UK.

Kellagher, R. and Gutierrez-Andres, J. (2015) Rainwater harvesting for domestic water demand and stormwater management. In: Memon F.A. and Ward S. (ed.) *Alternative Water Supply Systems*, IWA Publishing, 62–83.

Leggett, D.J., Brown, R., Stanfield, G., Brewer, D. and Holliday, E. (2001) *Rainwater and Greywater use in buildings: decision-making for water conservation*, CIRIA Report PR80, London.

Melville-Shreeve, P., Ward, S. and Butler, D. (2014) A preliminary sustainability assessment of innovative rainwater harvesting for residential properties in the UK. *Journal of Southeast University*, 30, 2, 135–142.

Memon, F.A., Fidar, A., Lobban, A., Djordjević, S. and Butler, D. (2009) Effectiveness of rainwater harvesting as stormwater management option. *Water Engineering for a Sustainable Environment – Proceedings of 33rd IAHR Congress*, Vancouver, Canada, 9–14 August 2009.

MTP (2011) *BNWAT01 WCs: Market Projections and Product Details*. Market Transformation Programme, Defra, London.

Pell Frischmann (2013) *Flood Risk Assessment R63059Y001B*.

Rainwater Harvesting Ltd. (2015) *RainActiv™. Rainwater Harvesting and Active Attenuation Combined*. Available at: http://tinyurl.com/hh52cmq

Roebuck, R.M., Oltean-Dumbrava, C. and Tait, S. (2011) Whole-life cost performance of domestic rainwater harvesting systems in the United Kingdom. *Water and Environment Journal*, 25, 355–365.

Waterwise (2014) Water calculator. Available at: http://www.thewatercalculator.org.uk/

Woods Ballard, B., Kellagher, R., Martin, P., Jefferies, C., Bray, R. and Shaffer, P. (2007) *The SuDS Manual*. CIRIA Report C69, London.

XPSolutions (2015) *Microdrainage*. Available at: http://xpsolutions.com/Software/MICRODRAINAGE/

Progress with Integration of Ecosystem Services in SuDS

Mark Everard, Robert J. McInnes and Hazem Gouda

Society is slowly transitioning from environmental management and resource use addressing single or few outcomes, towards recognition that all interventions have systemic impact (de Groot *et al.*, 2010; Norgaard, 2010). Ecosystem services comprise the interconnected human benefits provided by the natural world, spanning interlinked value systems and societal needs (Millennium Ecosystem Assessment, 2005). International commitments encouraging governments to undertake an ecosystem approach include the Convention on Biological Diversity (2000, 2010), the EU Biodiversity 2020 strategy (European Union, 2011) and the Ramsar Convention (Resolution IX.1, 2005). Many countries have transposed these obligations into national-level strategy, for example the UK's Natural Environment White Paper (HM Government, 2011). However, societal transition to systemic decision-making remains challenging (Armitage *et al.*, 2008) due to knowledge gaps, narrow legacy assumptions, legislation, regulatory implementation, technical solutions, vested interests and decision-support models founded on reductive paradigms (Everard *et al.*, 2014). Tools to expose the wider ramifications of policies, designs and actions, also highlighting the benefits and opportunities of systemic practice, are needed to promote systemic, sustainable practice (Smith *et al.*, 2007). Failure to achieve this transition perpetuates risks from economic, social and environmental externalities (Robinson *et al.*, 2012).

Water management in urban environments presents particular challenges related to growing populations accommodated by finite land area, with trends suggesting increasingly dense urbanisation (United Nations, 2011). Drivers include adequate water supply sourced from substantially beyond the urban catchment area (Fitzhugh and Richter, 2004), management of flood risk (surface water and groundwater) compounded by climatic instability (Scholz, 2006) and processing and treating wastewater and water-vectored pollutants (Figure 16.1) (Niemczynowicz, 1999).

Sustainable Surface Water Management: A Handbook for SuDS, First Edition.
Edited by Susanne M. Charlesworth and Colin A. Booth.
© 2017 John Wiley & Sons, Ltd. Published 2017 by John Wiley & Sons, Ltd.

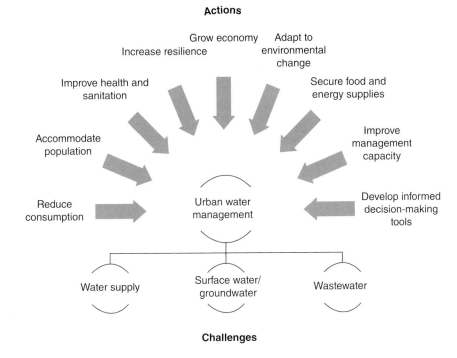

Figure 16.1   Challenges of urban water management.

These principal challenges operate within a wider operational landscape of urban land use planning and decision-making (Figure 16.1). Disconnected, single-solution outcomes still predominate (Everard, 2014) despite the need for the built environment to be planned to operate synergistically with functioning ecosystems (UN Habitat, 2012) accommodating water-mediated ecosystem services including maintenance of equable microclimate, food production and amenity (Bolund and Hunhammer, 1999) and reduced carbon and ecological footprints (Secretariat of the Convention on Biological Diversity, 2012).

Flood management policy and practice has morphed from localised 'defence' of assets towards an ecosystems-based, adaptive approach working with natural processes (Colls *et al.*, 2009), partly responding to severe flooding, for instance in the UK where established flood management norms were insufficient (Defra, 2005), but also resulting from longer-term recognition of the importance of working with natural processes rather than reliance on increased 'hard engineering' defences (Palmer *et al.*, 2009). Step-wise progress towards natural floodwater retention and dissipation mark an ongoing transition across the developed world at catchment scale and within urban environments (Wong, 2006; Everard *et al.*, 2009).

The evolving philosophy of SuDS and similar approaches such as WSUD (water sensitive urban design) underpin a significant transition in urban flood risk management (Wong, 2006). Published guidance (Woods Ballard *et al.*, 2007) highlights intent 'to manage the environmental risks resulting from urban runoff and to contribute wherever possible to environmental enhancement', working 'upstream' in the drainage chain and progressively taking account of wider outcomes for water quantity, water quality, amenity and biodiversity. 'Water regulation' (scale and timing of flows) is also an ecosystem service, interconnected with a broader spectrum of potential societal benefits from management interventions.

Studies contrasting life cycle assessment (LCA) outcomes between conventional and sustainable approaches to urban drainage, highlight the need for a transition to consideration of whole life cycle cost and performance. These factors depend on detailed scheme design, but particularly the importance of systematic assessment addressing frequently neglected dimensions of sustainability essential for meeting the challenges of growing urban populations and changing climate (Ellis *et al.*, 2003; Zhou, 2014). They conclude that SuDS principles contribute to sustainable development by averting unintended negative impacts, particularly life cycle material inputs, environmental emissions and energy use, also potentially optimising outcomes across ecosystem services (Everard and Street, 2001; Natural England, 2009; McInnes, 2013; Everard and McInnes, 2013).

Even in more ecologically aware cities, urban environmental management systems often overlook many ecosystem benefits (McInnes, 2013). Implicit in SuDS design is protection and improvement of the environment (Woods Ballard *et al.*, 2015) and implementation of green infrastructure can also promote human health and deliver multiple benefits (Tzoulas *et al.*, 2007). This chapter analyses the potential contribution of selected urban drainage solutions to delivery of ecosystem services.

16.2   Potential Contribution of SuDS Types to Ecosystem Services

The SuDS Manual (Woods Ballard *et al.*, 2015) specifies techniques ranging from simply increasing floodwater storage capacity in dense, constrained urban settings (for example underground gabions beneath hard infrastructure) through to incorporating multiple ecosystem service outcomes additional to drainage. Filter drains and pervious pipes, pervious surfaces, infiltration basins and constructed wetlands were selected as representative techniques. The following descriptions derive largely from Woods Ballard *et al.* (2015):

- *Filter drains and pervious pipes* comprise trenches filled with permeable material receiving water falling on paved areas, filtering and conveying it elsewhere on site. This slows and provides some physical filtration of stormwater, though without significant chemical purification or habitat for amenity and biodiversity.
- *Pervious surfaces* allow water to infiltrate an underlying storage layer, detaining it before infiltration to the ground, reuse or release to surface waters. These systems offer no habitat for wildlife or amenity (beyond the paved surface which is built infrastructure rather than ecosystem service).
- *Infiltration basins* are depressions in landscapes, constructed to store runoff during intense precipitation, enabling it to infiltrate progressively into the ground. Infiltration basins may be landscaped, providing aesthetic and amenity value; however, due to necessary regular maintenance, only simple low-sward habitat tends to form.
- *Constructed wetlands* are diverse, typically comprising ponds with shallow vegetated areas which improve pollutant removal and provide wildlife habitat. They may accumulate organic matter, recycle nutrients and become attractive features in urban developments. Constructed wetlands range from simple stilling ponds and reed-filled hollows through to extensive semi-natural systems. Potential outcomes from 'best practice' constructed wetlands designed to achieve multi-functional benefits are used as reference points from Australia (Wong and Brown, 2009) and Ireland (Doody *et al.*, 2009).

All SuDS techniques vary in specific detail and potential service production depending on locational constraints and design, but each potentially achieves multiple ecosystem service outcomes. Assessment of modelled outcomes from traditional piped drainage solutions, featuring in the case study in Gouda *et al.* (unpublished data), are included as a comparative baseline. The potential contribution of SuDS types to each ecosystem service is scored using a 'traffic lights' approach (Table 16.1):

- *green*: has the potential to make a contribution to the service, with foresighted planning and implementation
- *amber*: has a limited potential to contribute to the service; and
- *red*: does not contribute to the service, or may undermine it.

Two broad areas of subjectivity are acknowledged in the assessments. (1) No SuDS method is uniform, varying in detail and outcome with location and design. (2) Assessment of potential contribution to ecosystem service outcomes is challenging due substantially to a paucity of indicators and data (Burkhard *et al.*, 2012). Burkhard *et al.* (2009) and Busch *et al.* (2012) propose using expert evaluations to garner an overview and identify trends, an approach successfully applied elsewhere, where assessments have been based on intensive literature searches, stakeholder interviews and partially on expert estimates (Vihervaara *et al.*, 2010). While detailed quantitative analysis of SuDS schemes would add rigour, each case study represents a 'snapshot' of the potential of each approach. Therefore, while acknowledging some subjectivity, we suggest that the lack of detailed scheme-level appraisal under each SuDS option does not undermine the inherently systemic nature of the comparative analysis. Indeed, the key challenge being addressed in this analysis is to represent a systemic perspective of the contributions of drainage techniques, not a detailed reductive analysis of each service outcome.

This simplification of assessment via three traffic lights, building from a simpler pass/fail scoring system of other water management techniques (Everard, 2014), presents likely outcomes in illustrative yet intuitive terms that may be useful in guiding non-technical development proponents towards more sustainable methods. This articulation therefore usefully represents potential outcomes for design options in decision-support models. The traffic lights approach has proved useful previously to represent the potential contribution of water management strategies to the 12 principles of the Ecosystem Approach without implying a greater degree of certainty than analyses can support, serving the important purpose of illustrating systemic coverage (Everard *et al.*, 2014).

16.3 Analysis of Ecosystem Service Outcomes from SuDS Schemes

Potential ecosystem service contributions of traditional piped drainage and the four selected SuDS design approaches is presented in Table 16.1. The simplified traffic lights colour-coding reveals a spectrum from a low range of services for piped drainage to potentially far broader service contributions from constructed wetlands. Traditional piped solutions perform some services well (local removal of stormwater including pollutant loads), but few other services are addressed, though some potential co-benefits arise (fire regulation by avoidance of combustible materials, and an educational resource) together with several negative externalities (displacement of stormwater and contaminant concentration). Conversely, constructed wetlands potentially produce multiple co-benefits, albeit with some risks from inappropriate context-specific design and/or management.

Table 16.1 'Traffic lights' signalling outcomes for ecosystem services from drainage options. (*See insert for colour representation of this table*).

Services and service categories	Traditional piped drainage solutions (included here for comparison based on life cycle study in Gouda et al., unpublished data)	Filter drains and pervious pipes	Pervious surfaces	Infiltration basins	Constructed wetlands (optimally designed for access and amenity, biodiversity, etc.)
Provisioning services					
Fresh water	Does not recharge groundwater, though may feed into surface water resources, albeit likely to aggregate pollutants associated with concentrated stormwater surges	Makes a partial contribution to recharging water resources if there is significant infiltration to groundwater or egress to surface water bodies	Makes a partial contribution to recharging water resources if there is significant infiltration to groundwater or egress to surface water bodies	Supports recharge of groundwater and can attenuate pollutants	Supports recharge of groundwater or surface resources and can attenuate pollutants
Food (e.g. crops, fruit, fish, etc.)	Heavy infrastructure reduces potential for peri-urban farming	Makes no provision for peri-urban farming	Pervious surfaces make no provision for peri-urban farming	Simple low-sward habitat provides little opportunity for food growing, including peri-urban farming, though grazing animals such as rabbits and geese may be harvested at low frequency	Constructed wetlands may allow for harvesting of limited amounts of and possibly for limited peri-urban farming, though both are rarely realised in practice
Fibre and fuel (e.g. timber, wool, etc.)	Heavy infrastructure reduces potential for peri-urban fibre and fuel production	Makes no provision for peri-urban fibre and fuel production	Pervious surfaces make no provision for peri-urban fibre and fuel production	Simple low-sward habitat provides little opportunity for peri-urban fibre and fuel production, beyond low potential for composting of green waste from management and its potential use in bioenergy production as noted below	Potential for peri-urban fibre and fuel production though rarely realised in practice
Genetic resources (used for crop/stock breeding and biotechnology)	Heavy infrastructure displaces ecosystems potentially hosting genetic resources	Drain and pipe infrastructure displaces ecosystems potentially hosting genetic resources	Pervious paving displaces ecosystems potentially hosting genetic resources	Simple low-sward habitat provides no opportunity for ecosystems potentially hosting genetic resources	Constructed wetlands may host genetic resources, particularly if this is an element of design considerations
Biochemicals, natural medicines, pharmaceuticals	Heavy infrastructure displaces ecosystems potentially hosting medicinal resources	Drain and pipe infrastructure displaces ecosystems potentially hosting medicinal resources	Pervious paving displaces ecosystems potentially hosting medicinal resources	Simple low-sward habitat provides no opportunity for ecosystems potentially hosting medicinal resources	Constructed wetlands may host medicinal resources, particularly if this is an

Ornamental resources (e.g. shells, flowers, etc.)	Heavy infrastructure displaces ecosystems potentially hosting ornamental resources	Drain and pipe infrastructure displaces ecosystems potentially hosting ornamental resources	Pervious paving displaces ecosystems potentially hosting ornamental resources	Simple low-sward habitat provides no opportunity for ecosystems potentially hosting ornamental resources	Constructed wetlands may host ornamental resources, particularly if this is an element of design considerations
Energy harvesting	While there may be opportunities for energy harvesting from water egressing at pace through pipework, this is not seen as a common or significant ecosystem services	Does not provide opportunities for energy harvesting	Does not provide opportunities for energy harvesting	Limited potential for biomass harvesting for energy production during routine management, though little evidence that this practice occurs operationally	Potential for biofuel production based on reuse of green waste from wetland management
Regulatory services					
Air quality regulation	Heavy drainage infrastructure does not regulate air quality	Artificial infrastructure does not support ecosystems regulating air quality	Artificial infrastructure does not support ecosystems regulating air quality	Simple low sward can influence air quality, though to a limited extent compared to complex habitat	Constructed wetlands can comprise complex vegetation efficient at 'scrubbing' and transforming air quality pollutants
Climate regulation (microclimate)	Heavy drainage infrastructure makes no contribution to modification of local climate	Artificial infrastructure does not support ecosystems modifying local climate	Removal of habitat reduces potential to modify local climate	Simple low sward can influence microclimate, though to a limited extent compared to complex habitat	Constructed wetlands can comprise complex vegetation efficient at modifying microclimate
Climate regulation (global climate)	Heavy drainage infrastructure does not sequester carbon and requires heavy energy consumption during construction and maintenance	Artificial infrastructure does not support ecosystems that sequester carbon and moderate energy consumption required during construction phase, with lower maintenance energy consumption than traditional piped drainage	Removal of habitat reduces potential to sequester carbon and moderate energy consumption during construction phase, with lower maintenance energy consumption than traditional piped drainage	Simple low sward can sequester carbon and transform other climate-active gases, though to a limited extent compared to complex habitat and lower energy consumption than piped drainage during construction phase, and moderate consumption during maintenance stage	Constructed wetlands can comprise complex vegetated systems efficient at sequestering carbon and transforming other climate-active gases, though with a limited risk of generation of some climate-active gases (methane, nitrous oxide) in anaerobic sections and lower energy consumption than piped drainage during construction phase, and higher consumption during maintenance stage

(Continued)

Table 16.1 (Continued)

Services and service categories	Traditional piped drainage solutions (included here for comparison based on life cycle study in Gouda et al., unpublished data)	Filter drains and pervious pipes	Pervious surfaces	Infiltration basins	Constructed wetlands (optimally designed for access and amenity, biodiversity, etc.)
Water regulation (timing and scale of runoff, flooding, etc.)	Piped drainage infrastructure addresses local flood relief but may relocate flooding problems elsewhere	Produces local flood relief, and may slow the pace of runoff partially averting the risk of relocating flooding problems elsewhere	Produces local flood relief, and may slow the pace of runoff partially averting the risk of relocating flooding problems elsewhere	Produces local flood relief, and slows the pace of infiltration and runoff averting the risk of relocating flooding problems elsewhere	Constructed wetlands provide mosaic semi-natural habitat that detains floodwater and slows its release to ground and surface flows
Natural hazard regulation (i.e. storm protection)	Makes no contribution to hazard protection, and can displace complex vegetation performing storm buffering services	Artificial infrastructure makes no contribution to hazard protection, and can displace complex vegetation performing storm buffering services	Artificial infrastructure makes no contribution to hazard protection, and can displace complex vegetation performing storm buffering services	Simple low sward makes no contribution to hazard protection, and displaces complex vegetation performing storm buffering services	The complex habitat structure of constructed wetlands contributes to hazard protection, and displaces complex vegetation performing storm buffering services
Pest regulation	Makes no positive contribution to pest regulation, but may eliminate habitat for natural pest predators	Artificial infrastructure makes no positive contribution to pest regulation, but may eliminate habitat for natural pest predators	Artificial infrastructure makes no positive contribution to pest regulation, but may eliminate habitat for natural pest predators	Simple low sward provides only limited habitat for the predators of pest organisms, and very low risk of supporting colonies of pest organisms	The complex habitat structure of constructed wetlands may support the predators of pest organisms, though there is a risk of poor management supporting populations of potential pest organisms such as rats and mosquitoes
Disease regulation	May help to rid contaminated water locally but may displace downstream risks concentrating pollutants	Helps rid contaminated water locally with limited attenuation of potential disease organisms, though may displace risks downstream	Helps rid contaminated water locally with limited attenuation of potential disease organisms, though may displace risks downstream	Simple low sward can eliminate some pathogenic organisms though mixed uses of infiltration basins can expose people to accumulated risks, and little opportunity for hosting potential disease vectors	The complex habitat structure of constructed wetlands attenuates pathogenic organisms, though poor design and management can support populations of potential disease vectors such as mosquitoes

Erosion regulation	More concentrated flows of piped water may exacerbate erosion both upstream and downstream of piped solution	Buffering of flows reduces the risks of erosion on site and downstream, albeit that this solution applies where land surfaces are sealed for development	Buffering of flows reduces the risks of erosion on site and downstream, albeit that this solution applies where land surfaces are sealed for development	Infiltration basins still water allowing for progressive infiltration, so avert local erosion of soil except where overtopping in extreme precipitation	Constructed wetlands still water allowing for progressive runoff and infiltration, averting local erosion of soil except where overtopping in extreme precipitation
Water purification and waste treatment	Habitat performing purification processes is lost, with no substantial within-infrastructure purification	Limited physical filtration but artificial infrastructure performs purification processes suboptimally	Limited physical filtration but artificial infrastructure performs purification processes suboptimally	Simple low sward provides limited physico-chemical purification of accumulated water, albeit not as effectively as more complex habitat	The complex, mosaic habitat structure of constructed wetlands can be highly efficient at a range of physical and chemical water purification processes
Pollination	Habitat supporting pollinators is lost	Habitat supporting pollinators is lost, and artificial drainage infrastructure provides no compensatory habitat	Habitat supporting pollinators is lost, and artificial drainage infrastructure provides no compensatory habitat	Simple low sward provides only limited habitat for natural pollinating organisms	The complex habitat structure of constructed wetlands can support populations of pollinating organisms
Salinity regulation (mainly arid landscapes)	No impact	Emulation of natural hydrology may help avert risks of salinisation of land downstream through buffered infiltration and runoff rates	Emulation of natural hydrology may help avert risks of salinisation of land downstream through buffered infiltration and runoff rates	Emulation of natural hydrology may help avert risks of salinisation of land downstream through buffered infiltration rates	Emulation of natural hydrology may help avert risks of salinisation of land downstream through buffered runoff and infiltration rates
Fire hazard regulation (mainly arid landscapes)	Lack of potentially combustible vegetation reduces overall urban fire risk	Lack of potentially combustible vegetation reduces overall urban fire risk	Lack of potentially combustible vegetation reduces overall urban fire risk	The simple sward structure of managed basins provides only a low biomass of potentially combustible vegetation, reducing overall urban fire risk	Complex habitat may create fire risks in urban settings, albeit that wetland habitat generally tends to be low risk
Cultural services					
Cultural heritage	Heavy pipework has little or no cultural value, with some exceptions such as old drainage infrastructure of localised heritage interests, but is more generally perceived as a threat	Artificial infrastructure does not of itself provide cultural heritage, but may be a solution that works sympathetically with heritage structures and landscapes	Artificial infrastructure does not of itself provide cultural heritage, but may be a solution that works sympathetically with heritage structures and landscapes	Infiltration basins contribute minimal cultural heritage, but may work sympathetically with heritage structures and landscapes	Constructed wetland ecosystems can potentially be a source of cultural value in urban settings

(Continued)

Table 16.1 (Continued)

Services and service categories	Traditional piped drainage solutions (included here for comparison based on life cycle study in Gouda et al., unpublished data)	Filter drains and pervious pipes	Pervious surfaces	Infiltration basins	Constructed wetlands (optimally designed for access and amenity, biodiversity, etc.)
Recreation and tourism	Heavy engineering pipework lacks recreation and tourism value	Artificial infrastructure makes no contribution to recreation and tourism	Artificial infrastructure makes no contribution to recreation and tourism	Infiltration basins commonly serve as valuable recreational spaces except when holding water after heavy precipitation events	Constructed wetland ecosystems may be of recreational value, depending on wetland design and conflicts with other values (such as habitat for wildlife), particularly where semi-natural spaces are limited in urban settings
Aesthetic value	Heavy engineering pipework lacks aesthetic value	Artificial infrastructure is not itself of aesthetic value, but may be a solution that works sympathetically with mixed use landscapes	Artificial infrastructure is not itself of aesthetic value, but may be a solution that works sympathetically with mixed use landscapes	Infiltration basins may form part of the landscaping of public and private spaces	Constructed wetland ecosystems can potentially be of aesthetic value in otherwise impoverished urban environments, with the added benefit of complex vegetation providing visual and noise buffering services
Spiritual and religious value	Heavy engineering pipework lacks, and may positively erode, spiritual and religious value	Artificial infrastructure is not itself of aesthetic value, but may be a solution that works sympathetically with spiritually valued landscapes	Artificial infrastructure is not itself of aesthetic value, but may be a solution that works sympathetically with spiritually valued landscapes	Infiltration basins contribute minimally to spiritual values, but may work sympathetically with spiritually valued landscapes	Constructed wetland ecosystems can potentially be a limited source of spiritual value in urban settings
Inspiration of art, folklore, architecture, etc.	Heavy engineering pipework generally lacks artistic inspiration value, though there is some potential interest and inspiration of local heritage structures	Artificial infrastructure is not itself of inspirational value, but may be a solution that works sympathetically with landscapes	Artificial infrastructure is not itself of inspirational value, but may be a solution that works sympathetically with landscapes	Infiltration basins may fit sympathetically with landscapes valued for inspirational qualities	Constructed wetland ecosystems can potentially be a limited source of creative value in urban settings

Social relations (e.g. fishing, grazing or cropping communities)	Heavy engineering pipework lacks support for local communities other than those united by an interest in urban drainage	Minimal support for local communities other than those united by an interest in urban drainage	Minimal support for local communities other than those united by an interest in urban drainage	Infiltration basins may provide communal places in dry conditions for social gatherings, albeit of lesser value than natural spaces	Constructed wetland ecosystems can be a focus for common interests and community activities in urban areas
Education and research	All drainage solutions may support learning and research activities	All drainage solutions may support learning and research activities	All drainage solutions may support learning and research activities	All drainage solutions may support learning and research activities	All drainage solutions may support learning and research activities
Supporting services					
Soil formation	Heavy pipework makes no contribution to soil formation	Filter drains and pervious pipes make no contribution to soil formation	Pervious paving makes no contribution to soil formation	Infiltration basins support only simple low sward which contributes to soil formation, albeit suboptimally	Constructed wetlands can be efficient at soil formation
Primary production	Heavy pipework makes no contribution to primary production	Filter drains and pervious pipes make no contribution to primary production	Pervious paving makes no contribution to primary production	Infiltration basins support only simple low sward which contributes suboptimally to primary production	Constructed wetlands can perform significant primary productivity
Nutrient cycling	Heavy pipework makes no contribution to nutrient cycling	Filter drains and pervious pipes make no contribution to nutrient cycling	Pervious paving makes no contribution to nutrient cycling	Infiltration basins support only simple low sward which contributes suboptimally to nutrient cycling	Constructed wetlands can perform significant nutrient transformation and cycling services
Water recycling	Heavy pipework makes no contribution to local-scale water recycling	Filter drains and pervious pipes make no contribution to local-scale water recycling	Pervious paving makes no contribution to local-scale water recycling	Infiltration basins support only simple low sward which contributes suboptimally to local-scale water recycling	Constructed wetlands can perform significant local-scale water recycling and retention services
Photosynthesis (production of atmospheric oxygen)	Heavy pipework makes no contribution to photosynthesis	Filter drains and pervious pipes make no contribution to photosynthesis	Pervious paving makes no contribution to photosynthesis	Infiltration basins support only simple low sward which contributes suboptimally to photosynthesis	Constructed wetlands can perform significant photosynthetic processes
Habitat for wildlife	Heavy pipework provides no useful habitat for wildlife, other than for organisms such as rats that may generally be considered as pest rather than assets	Filter drains and pervious pipes provide no useful habitat for wildlife	Pervious paving provides no useful habitat for wildlife	Infiltration basins support only simple low sward which provides some habitat for wildlife, albeit not supporting diverse biodiversity	Constructed wetlands can provide a diversity of habitats for wildlife, which may be of particular value both directly and as a 'stepping stone' where habitat is scarce in urban settings

Identical narrow outcomes were identified for filter drains and pervious pipes and for pervious surfaces, reflecting their essentially volumetric role, albeit with some physical filtration. The low managed grass sward of infiltration basins performs some additional ecological and physico-chemical functions and fit into multi-functional landscapes of wider value.

Certain ecosystem services (including provision of genetic, biochemical or ornamental resources) only potentially arise from constructed wetlands. Conversely, a high level of regulation of water and erosion control was provided by all approaches except traditional piped drainage. Education and research benefits extended across all approaches. Different approaches lend themselves to different ranges of ecosystem services, though net benefits to society (breadth of ecosystem services) increases with performance or emulation of natural processes. Each solution has its place in a mix of approaches in urban environments, yet also has a different 'footprint' of environmental impact and net provision of value to society within and beyond the discipline of urban drainage.

16.4 Recognising the Multi-Functional Opportunities of SuDS

Notwithstanding the ethical, environmental and net societal value implications of considering multiple service outcomes in decision-making, consistent with international and national requirements to take an ecosystem approach, significant challenges remain in 'mainstreaming' ecosystem services into planning, policy and implementation (Apitz et al., 2006), with significant impediments to sustainable water management persisting particularly in urban contexts (Farrelly and Brown, 2011). Philosophical and practical progress, supported by a growing body of case studies (for example, as reviewed by Grant, 2012), is promoting incremental integration of SuDS and related techniques into urban drainage policy and practice. This can contribute to wider uptake of ecosystem service considerations in urban design and management (Grant, 2012) helping overcome ignorance surrounding the potential values of urban biodiversity (Rodríguez et al., 2006).

Recognising multiple service outcomes differentiates the net consequences of hard engineering (traditional piped drainage) versus ecosystem-based approaches to urban drainage. Individual scheme design has to be fit for purpose, though the definition of 'purpose' remains open to debate. Drainage design will be steered in a particular direction if the purpose is framed solely as dealing with flood events, assumed to become increasingly episodic due to climatic instability, and to deliver within specific urban context and policy requirements. Hard engineering approaches that are focused narrowly on drainage may be locally appropriate in dense built infrastructure, with constrained opportunities for multi-benefit solutions, though unintended consequences have to be addressed and mitigated wherever possible. However, in greenfield development, or where other design considerations permit, the benefits of a multi-benefit approach are compelling for delivering water management and wider societal benefits (Steiner, 2014). Nevertheless, narrow or otherwise ill-informed economic or perceived technical constraints still frequently shape decision-making (Barbosa et al., 2012). Yet more sustainable approaches to drainage are increasingly required by planning policies, whether because developers or planners see direct benefits from taking a more sustainable approach or because development proponents recognise wider benefits. Converting these aspirations into practical implementation requires clear communication of the advantages of taking a multi-benefit approach, requiring illustrative communications tools to engage diverse stakeholders influencing scheme design, operation and regulation.

Systemic intent, seeking optimal public value across services, can promote a multi-benefit vision. The systemic approach supersedes narrower paradigms that may generate unintended externalities, which nonetheless represent real costs and lost opportunities for multiple constituencies. Where a vision of multiple ecosystem service outcomes is successfully shared among stakeholders this may promote collaborative funding from multiple sources which, though formerly managed independently (e.g. estates management, public health, flood management, air quality and public amenity provision), may cost-effectively optimise societal benefits.

A shift to more sustainable practice will be driven not merely conceptually but by more integrated regulatory requirements, practical methods assuring developers and regulators that techniques are robust broader economic considerations, and also pragmatic decision-support tools. This situation was highlighted in Australia, where historical entrenchment, sectorial barriers, perceived risks and lack of experimentation in policy decisions present perceived barriers to implementing sustainable urban water management (Farrelly and Brown, 2011).

As most design and regulatory decisions are based on modelled outcomes, drainage models need to incorporate potential public benefits and externalities across ecosystem services to promote cross-sector and multi-disciplinary working (Ward *et al.*, 2012). Evolution of practical design guidance and evaluation models has an important role in accelerating transition to sustainable approaches to drainage.

As SuDS design progresses to encompass diverse ecosystem services, distinctions between SuDS and other urban water and environmental management solutions (green infrastructure, urban forests, urban river restoration, etc.) blur as narrow disciplinary interests coalesce into net contribution to sustainable urban design (Everard and Moggridge, 2012). Everard and McInnes (2013) identify a 'systemic solutions' approach, defined as 'low-input technologies using natural processes to optimise benefits across the spectrum of ecosystem services and their beneficiaries', that can contribute to sustainable development by recognising and averting unintended negative impacts and through optimisation of outcomes, increasing net economic value.

The principle of fitness for purpose of drainage scheme design expands beyond ensuring sufficient drainage capacity, to also accommodate implications for all ecosystem services. Failure of vision in design can constrain the value of ostensibly more sustainable approaches (McInnes, 2013). For example, constructed wetlands designed with steep sides substantially limits area available for establishment of functional habitats (such as for wildlife and regulation of air quality – Becerra Jurado *et al.*, 2010) also representing a potential hazard for people (a disbenefit rather than a service). Wetland design has also to balance outcomes across services (Harrington *et al.*, 2011), for example optimising outcomes for climate change by promoting the sequestration of carbon while averting methane and nitrous oxide generation in extensive anaerobic areas (Mander *et al.*, 2011).

Local setting has also to be considered to prevent unintended consequences outweighing potential benefits (Wong, 2006). For example, permanent open water in a tropical constructed wetland in an urban environment may provide efficient drainage and other services, but could present malarial risks and substantial water loss through evapotranspiration (Greenway *et al.*, 2003; Knight *et al.*, 2003), where SuDS techniques promoting groundwater recharge may deliver service value additional to drainage including water resource recharge, disease regulation and other cultural benefits (Yang *et al.*, 2008).

16.5   Conclusions and Recommendations

The diversity of potential outcomes for any sustainable drainage design set in its geographic, climatic and demographic context demands case-by-case consideration. However, every design presents an opportunity to optimise benefits rather than maximise single outcomes (Everard and McInnes, 2013) and to integrate across the multiple actions demonstrated in Figure 16.1. To make this workable, we recommend the following.

1. Decision-makers, planners and managers need to adopt systemic approaches to urban water management challenges. These should optimise societal value, including elimination of unintended disbenefits, across the full spectrum of ecosystem services.
2. Cross-disciplinary models that can optimise design, accounting for all ecosystem services and context-specific risk factors and interdependences, will be invaluable in navigating this complexity. Such models must provide options, warnings and guidance to help developers shape design, supporting optimal public value across ecosystem services.
3. The simplistic traffic lights illustration of likely outcomes of drainage options used in this chapter represents a useful and intuitive means for models to represent potential outcomes to non-specialist users and audiences.

References

Apitz, S.E., Elliott, M., Fountain, M. and Galloway, T.S. (2006) European environmental management: moving to an ecosystem approach. *Integrated environmental assessment and management*, 2(1) 80–85.

Armitage, D.R., Plummer, R., Berkes, F., Arthur, R.I., Charles, A.T., *et al.* (2008) Adaptive co-management for social-ecological complexity. *Frontiers in Ecology and the Environment*, 7(2) 95–102.

Barbosa, A.E., Fernandes, J.N. and David, L.M. (2012) Key issues for sustainable urban stormwater management. *Water research*, 46(20) 6787–6798.

Bolund, P. and Hunhammer, S. (1999) Ecosystem services in urban areas. *Ecological Economics*, 29, 293–301.

Burkhard, B., Kroll, F., Müller, F. and Windhorst, W. (2009) Landscapes' capacities to provide ecosystem services – a concept for land-cover based assessments. *Landscape On-line*, 15, 1–22.

Burkhard, B., Kroll, F., Nedkov, S. and Müller, F. (2012) Mapping ecosystem service supply, demand and budgets. *Ecological Indicators*, 21, 17–29.

Busch, M., La Notte, A., Laporte, V. and Erhard, M. (2012) Potentials of quantitative and qualitative approaches to assessing ecosystem services. *Ecological Indicators*, 21, 89–103.

Colls, A., Ash, N. and Ikkala, N. (2009) *Ecosystem-based Adaptation: a natural response to climate change*. Gland: IUCN.

Convention on Biological Diversity. (2000) *Ecosystem Approach*. UNEP/CBD COP5 Decision V/6 (http://www.cbd.int/decision/cop/?id=7148, accessed 30th December 2013).

Convention on Biological Diversity (2010) *Operational guidance for application of the Ecosystem Approach*. (http://www.cbd.int/ecosystem/operational.shtml, accessed 30th December 2013).

Defra (2005) *Making space for water: Taking forward a new Government strategy for flood and coastal erosion risk management in England*. First Government response to the autumn 2004 'Making space for water' consultation exercise, March 2005. Department for Environment, Food and Rural Affairs, London.

De Groot, R.S., Alkemade, R., Braat, L., Hein, L. and Willemen, L. (2010) Challenges in integrating the concept of ecosystem services and values in landscape planning, management and decision-making. *Ecological Complexity*, 7(3) 260–272.

Doody, D., Harrington, R., Johnston, M., Hofman, O. and McEntee, D. (2009) Sewerage treatment in an integrated constructed wetland. *Municipal Engineer*, 162(4) 199–205.

Ellis, J.B., Shutes, R.B.E. and Revitt, M.D. (2014) *Constructed Wetlands and Links with Sustainable Drainage Systems*. Environment Agency R&D Technical Report P2-159/TR1. Environment Agency, Bristol.

European Union. (2011) *The EU Biodiversity Strategy to 2020*. European Union: Belgium. 28pp.

Everard, M. (2014) Integrating integrated water management. *Water Management* (DOI: http://dx/doi.org/10.1680/wama.12.00125).

Everard, M., Bramley, M., Tatem, K., Appleby, T. and Watts, W. (2009) Flood management: from defence to sustainability. *Environmental Liability*, 2, 35–49.

Everard, M., Dick, J., Kendall, H., Smith, R.I., Slee, W., Couldrick, L., Scott, M. and MacDonald, C. (2014) Improving coherence of ecosystem service provision between scales. *Ecosystem Services*. DOI: 10.1016/j.ecoser.2014.04.006.

Everard, M. and McInnes, R.J. (2013) Systemic solutions for multi-benefit water and environmental management. *The Science of the Total Environment*, 461(62) 170–179.

Everard, M. and Moggridge, H.L. (2012) Rediscovering the value of urban rivers. *Urban Ecosystems*, 15(2) 293–314.

Farrelly, M. and Brown, R. (2011) Rethinking urban water management: Experimentation as a way forward? *Global Environmental Change*, 21(2) 721–732.

Everard, M. and Street, P. (2001) *Sustainable drainage systems (SuDS): An Evaluation Using The Natural Step Framework*. The Natural Step, Cheltenham.

Fitzhugh, T.W. and Richter, D.D. (2004) Quenching urban thirst: growing cities and their impacts on freshwater ecosystems. *BioScience*, 54, 741–754.

Government, H.M. (2011) *The Natural Choice: Securing the Value of Nature*. The Stationery Office, London.

Grant, G. (2012) *Ecosystem Services Come to Town: Greening Cities by Working with Nature*. Wiley-Blackwell, Chichester, UK.

Greenway, M., Dale, P. and Chapman, H. (2003) An assessment of mosquito breeding and control in four surface flow wetlands in tropical-sub-tropical Australia. *Water Science ad Technology*, 48(5) pp. 249–256.

Harrington, R., Carroll, P., Cook, S., Harrington, C., Scholz, M. and McInnes, R.J. (2011) Integrated constructed wetlands: water management as a land-use issue, implementing the 'Ecosystem Approach'. *Water Science and Technology*, 63(12).

Jurado, G.B., Callanan, M., Gioria, M., Baars, J.R., Harrington, R. and Kelly-Quinn, M. (2010) Comparison of macroinvertebrate community structure and driving environmental factors in natural and wastewater treatment ponds. In: *Pond Conservation in Europe* B. Oertli, R. Cereghino, A. Hull and R. Miracle (eds), Springer Netherlands. pp. 309–321.

Knight, R.L., Walton, W.E., O'Meara, G.F., Reisen, W.K. and Wass, R. (2003) Strategies for effective mosquito control in constructed treatment wetlands. *Ecological Engineering*, 21(4) 211–232.

Mander, Ü., Maddison, M., Soosaar, K. and Karabelnik, K. (2011) The impact of pulsing hydrology and fluctuating water table on greenhouse gas emissions from constructed wetlands. *Wetlands*, 31(6) 1023–1032.

McInnes, R.J. (2013) Recognising wetland ecosystem services within urban case studies. *Marine and Freshwater Research*, 64, 1–14.

Millennium Ecosystem Assessment. (2005) *Ecosystems and Human Well-being: General Synthesis*. Island Press, Washington DC.

Natural England. (2009) *Green Infrastructure Guidance*. Natural England, Peterborough.

Niemczynowicz, J. (1999) Urban hydrology and water management – present and future challenges. *Urban Water*, 1(1) 1–14.

Norgaard, R.B. (2010) Ecosystem services: From eye-opening metaphor to complexity blinder. *Ecological Economics*, 69(6) 1219–1227.

Palmer, M.A., Lettenmaier, D.P., Poff, N.L., Postel, S.L., Richter, B. and Warner, R. (2009) Climate change and river ecosystems: protection and adaptation options. *Environmental Management*, 44(6) 1053–1068.

Resolution, I.X.1. (2005) Additional scientific and technical guidance for implementing the Ramsar wise use concept. Resolutions of the 9th meeting of Conference of the Contracting Parties. Kampala, Uganda, 8–15 November, 2005.

Robinson, D.A., Hockley, N., Dominati, E., Lebron, I., Scow, K.M. *et al.* (2012) Natural capital, ecosystem services, and soil change: Why soil science must embrace an ecosystems approach. *Vadose Zone Journal*, 11(1) 6, pp. 10.2136/vzj2011.0051.

Rodríguez, J.P., Beard, T.D., Bennet, E.M., Cumming, G.S., Cork, S.J. *et al.* (2006) Trade-offs across space, time and ecosystem services. *Ecology and Society*, 11(1) 28. (http://wwwecologyand society. org/vol11/art28).

Scholz, M. (2006) *Wetland Systems to Control Urban Runoff*. Amsterdam: Elsevier.

Secretariat of the Convention on Biological Diversity (2012) *Cities and Biodiversity Outlook – Executive Summary*. Montreal, 16pp.

Smith, A.D.M., Fulton, E.J., Hobday, A.J., Smith, D.C. and Shoulder, P. (2007) Scientific tools to support the practical implementation of ecosystem-based fisheries management. *ICES Journal of Marine Science: Journal du Conseil*, 64(4) 633–639.

Steiner, F. (2014) Frontiers in urban ecological design and planning research. *Landscape and Urban Planning*, 125, 304–311.

Tzoulas, K., Korpela, K., Venn, S., Yli-Pelkonen, V., Kaźmierczak, A., Niemela, J. and James, P. (2007) Promoting ecosystem and human health in urban areas using green infrastructure: a literature review. *Landscape and Urban Planning*, 81(3) 167–178.

UN Habitat, (2012) *Urban Patterns for a Green Economy – Working With Nature*. UN Habitat, Kenya: UNON. 74pp.

United Nations (2011) *World Population Prospects: The 2010 Revision, Highlights and Advance Tables*. Department of Economics and Social Affairs, Population Division. New York: United Nations.

Vihervaara, P., Kumpula, T., Tanskanen, A. and Burkhard, B. (2010) Ecosystem services – a tool for sustainable management of human-environment systems. Case study Finnish Forest Lapland. *Ecological Complexity*, 7(3) 410–420.

Ward, S., Lundy, L., Shaffer, P., Wong, T., Ashley, R. *et al.* (2012) Water sensitive urban design in the city of the future. Proceedings of the 12 Porto Alegre, Brazil, 11–16 September 2011. 79–86.

Wong, T.H. (2006) Water sensitive urban design-the journey thus far. *Australian Journal of. Water Resources*, 10(3) 213.

Wong, T.H.F. and Brown, R.R. (2009) The water sensitive city: principles for practice. *Water Science and Technology*, 60(3).

Woods Ballard, B., Wilson, S., Udale-Clarke, H., Illman, S., Ashley, R. and Kellagher, R. (2015) *The SuDS Manual*. CIRIA, London.

Yang, W., Chang, J., Xu, B., Peng, C. and Ge, Y. (2008) Ecosystem service value assessment for constructed wetlands: A case study in Hangzhou, China. *Ecological Economics*, 68(1) 116–125.

Zhou, Q. (2014) A review of sustainable urban drainage systems considering the climate change and urbanization impacts. *Water*, 6, 976–992.

Section 5 Integrating Sustainable Surface Water Management into the Built Environment

Whole Life Costing and Multiple Benefits of Sustainable Drainage

Jessica E. Lamond

17.1   Introduction

When planning major engineering works, urban water management schemes, flood defence systems and drainage for new developments, the use of public or shareholders' money is subject to scrutiny, and so it warrants selecting the option that delivers best value for money. Traditional schemes to manage urban flooding and urban drainage have been subject to project appraisal methods that typically include some measure of economic costs and benefits of the schemes (Department of the Environment Food and Rural Affairs, 2009). The natural inclination, therefore, is to judge alternative approaches to flood and stormwater using the same appraisal techniques and measure their costs against the reduction in flood damage and loss or against the cost of other projects that could yield the same stormwater control.

It has, however, been recognised that the blue-green approach, with facilities at surface level and distributed across wider areas, may yield multiple benefits over and above stormwater management and equally that these facilities may require greater maintenance to function efficiently and deliver those benefits (HR Wallingford, 2004). If grey and blue-green approaches are being compared, then it is crucial to use a methodology that can compare like with like and is equally robust to assumptions for both types of scheme. It is more critical, when comparing across widely diverging approaches, to think about the whole life costs and the other benefits outside water management that might tip the balance in favour of one type of installation over another. Suitable approaches to cost–benefit analysis have therefore been the subject of much study and debate within the sustainable drainage literature, and parallel concepts such as ecosystem services and multi-criteria analysis have been proposed. This chapter discusses the most common and established, as well as emerging trends and summarises recent findings and literature around cost–benefit whole life costing and valuation of multiple benefits.

Sustainable Surface Water Management: A Handbook for SuDS, First Edition.
Edited by Susanne M. Charlesworth and Colin A. Booth.
© 2017 John Wiley & Sons, Ltd. Published 2017 by John Wiley & Sons, Ltd.

There is a great deal of uncertainty surrounding the future performance needs and other requirements for sustainability in an ever-evolving urban environment, therefore it is important to continue to develop plurality in evaluation for different purposes. First, better technical understanding of the installation, function and maintenance of systems will lead to improved cost estimation and thereby economic appraisal. Second, as the benefits from SuDS accrue at diverse spatial scales and to multiple beneficiaries including commercial entities, local statutory bodies and communities, models that can equitably apportion costs and benefits across different stakeholder groups are gaining more importance. Third it is important to recognise that, by their very nature, benefits such as amenity value are contextually specific and culturally weighted, so decisions are unlikely to be reduced to comparison of a single ratio without a great amount of local consultation and site specific observation. In many cases the prohibitive cost of such studies will result in the use of human judgement or proxy valuation of benefits drawn from other locations, as being the benefit transfer approach. Decision-makers will need better understanding of the evaluation process, and evaluators are developing alternative methods for communication to decision-makers that foster transparency and engagement in the selection of project goals.

17.2   Whole Life Costing

According to Constructing Excellence (2006) whole life costing (WLC) is 'the systematic consideration of all relevant costs and revenues associated with the ownership of an asset'. As such it is highly relevant to the issue of assessing the costs of SuDS in relation to urban drainage and it is further defined by the relevant British Standard BS ISO 15686-5:2008: Buildings and constructed assets. Service life planning. lifecycle costing:

> economic assessment considering all agreed projected significant and relevant cost flows over a period of analysis expressed in monetary value. The projected costs are those needed to achieve defined levels of performance, including reliability, safety and availability.

The WLC method is particularly relevant if there is a planned maintenance and replacement regime for assets. Indeed, the Environment Agency investment plans for flood defence investment (Environment Agency, 2009) include WLC estimates under different maintenance assumptions. For built assets designed to last for decades, it is clear that maintenance and operational costs may easily exceed construction costs. Early costings by the Environmental Protection Agency (EPA) of maintenance requirements for SuDS yielded annual estimates of up to 20% of construction costs (EPA, 1999) so capital costs would be exceeded after five years. It is generally the case that the most cost-effective opportunities for SuDS installation exist during new construction and development (for example, Bloomberg and Strickland, 2012) but Gordon-Walker et al. (2007) conclude that, despite higher capital costs of installation, the longer expected life of permeable paving rendered retrofit of permeable paving during planned replacement in the UK a cost-effective option.

If it is argued that SuDS have a smaller capital investment requirement but involve a greater investment in maintenance over their lifetime (HR Wallingford, 2004) then ignoring the WLC for a project may overstate the economic suitability of SuDS as replacement for traditional drainage systems. Some sources such as CIRIA (2009) and Everett *et al.* (2015) note that behaviour around SuDS can detrimentally impact their amenity and functionality through, for example, dumping of grass clippings or litter. However, others assert that SuDS may be better able to tolerate shortfalls in planned maintenance than some conventional systems (Stevens and Ogunyoye, 2012). Sources of cost data to aid calculation of costs of installation and/or maintenance of SuDS include HR Wallingford (2004), CIRIA (2009), Lampe *et al.* (2004) and Stevens and Ogunyoye (2012) in the UK. US sources include EPA sources (EPA, 1999) and Narayanan and Pitt (2005); also, a benchmarking report by Barr Engineering Company (2011) for the Minnesota Pollution Control Agency (USA) give both construction and maintenance costs for a range of best management practices (BMPs) in the US context. Taylor and Wong (2002) have provided information for the Australian market, based to some extent on the international estimates. Houle *et al.* (2013) also compared maintenance cycles for a range of low impact development (LID) stormwater approaches including SuDS.

Important decisions in the process of preparing WLC estimates relate to the time period over which the comparison is made and the discount rate chosen to account for the lower perceived value of future maintenance costs when compared to current capital investment. As it is a relatively innovative technology, the opportunity for lifetime evaluations has scarcely existed and their tendency to obsolescence is unknown. Therefore, design life estimates for SuDS are said to be uncertain in comparison to the piped alternatives (HR Wallingford, 2004) and selection of a time horizon may be a critical factor in cross-project comparisons. Selection of the chosen discount rate amounts to a judgement on the relative importance of capital investment, maintenance and operation. The higher the discount rate the more advantageous the low capital investment options appear to be. Different stakeholders typically use discount rates that suit their cost of capital and policy goals. National Government (Treasury) guidance in the UK specifies suitable rates for use in project appraisal. However, if a pluralistic approach to funding is required (e.g. government grant, EU funding, direct private investment, donor agencies, loan financing etc.) or if different stakeholders are responsible for installation vs operations and maintenance, then some conflicts may arise through the use of alternative discount rates for different stakeholders in the funding consortium.

In circumstances where the drainage features of SuDS are most important, cost-effectiveness analysis using a WLC approach can be considered useful and has lower data requirements than a full cost–benefit approach. For new development, this is equivalent to working out the cost of meeting the desired drainage design requirements using different approaches, rather than optimising the benefits. However, it can also be used to consider the alternatives to retrofitting existing infrastructure. In Portland Oregon (USA), for example, it has been estimated that widespread adoption of green roofs could reduce the expenditure needed to upgrade the stormwater system to the tune of $60 million (Bureau of Environmental Services – City of Portland, 2008).

17.2.1 Evidence on WLC from Past Studies

Although it was noted above that authors have found evidence that capital costs can be lower and maintenance costs higher for SuDS than conventional systems, the empirical evidence paints a more complex picture. For example, in their comprehensive report on evidence for costs and benefits of SuDS in the UK, Stevens and Ogunyoye (2012) concluded that

> the cost of SuDS to provide the same performance criteria as traditional drainage are much lower. However conventional drainage systems are not designed to the same requirements as SuDS (Defra, 2009). Defra (2011) noted from case study examples that overall evidence suggests SuDS may be up to 30% cheaper to construct, however for challenging sites it can be 5% more expensive to construct than traditional drainage.

Potential to see variation in costs due to alternative SuDS specifications is high as there are fewer design tools available and a greater range of optional design criteria to specify. Some SuDS systems can be enhanced aesthetically or in other ways to boost the multiple benefits over and above the primary stormwater management purpose. In order to compare like with like on a single benefit appraisal the cheapest specification is preferred. MacMullan and Reich (2007) summarised the evidence and suggested that, whereas LID was often already a cheaper alternative, the sheer novelty of LIDs added to their costs (of design and compliance with regulations) and that with time this would reduce. Shelton and Vogel (2005) compared the cost-effectiveness of infiltration versus detention BMPs for the same performance requirements in different locations and concluded that detention was in general more cost-effective.

For WLC, the variability in estimated costs can be caused by methodology as well as site specific factors that will inevitably increase the cost of installation. The WLC approach was taken by Duffy et al. (2008) for a site in Scotland, concluding that when well designed and maintained, SuDS can cost less to maintain than more traditional drainage. This was reiterated by Wolf et al. (2015), who implemented a whole life costing approach using two different tools for the same site, namely the Water Environment Research Foundation method (2009), and the 'SuDS for roads whole life cost and whole life carbon tool' (Scottish SUDS Working Party, 2009) and under different assumptions, discounting factor and time horizon. This study demonstrated differences between estimates using the alternative methods.

Houle et al. (2013) recorded short-term maintenance costs for stormwater management facilities in New Hampshire, concluding that the variation was quite broad, ranging from an annualised 4–19% of capital costs annually, and that those facilities responding to regular planned maintenance and collecting least pollutants required the lowest resources for maintenance. Gordon-Walker et al. (2007) estimate that permeable block paving has lower replacement and maintenance costs than regular paving and therefore costs less on a life cycle basis. Porsche and Köhler (2003) and Bureau of Environmental Services (2008) argue that the life cycle costs of green roofs are lower than the traditional alternatives because they need to be replaced less often. Green roofs are also argued to protect sensitive membranes from solar damage (Vila et al., 2012; Livingroofs.org, n.d.). The sparsity of, and differences in, the results described here clearly show that more research on the long-term costs of operation and maintenance of SuDS is needed.

17.3   Multiple Benefits of SuDS

The financial case for the use of SuDS to manage stormwater and reduce flood risk can sometimes be made in a straightforward manner, as demonstrated above, based on reduced whole life cost to deliver the required stormwater management benefit. Flood reduction benefits achieved using SuDS can be estimated in the usual way, for example using the Multi-coloured Manual (Flood Hazard Research Centre, 2010). However, SuDS have the potential to contribute more broadly within urban environments and ecosystems so that the stormwater and flood management aspects are only a part of the benefits to be evaluated in a cost–benefit approach or considered in multi-criteria approaches.

The list of potential multiple benefits of SuDS is long, and ranges from specific, scientifically testable, environmental claims around air and water quality to more subjective claims around neighbourhood improvement, wellbeing and amenity. Abbott *et al.* (2013) categorised the potential benefits into types, as listed in Table 17.1 with brief examples. However, it is worth noting that not all SuDS provide all benefits, so that a detailed assessment is required on a case-by-case basis. For example, the claims to improving air quality, carbon and nitrogen sequestration associated with SuDS are highlighted in other chapter of this volume and are largely based on the increase in vegetation in built-up areas. They are most easily provided by street trees and to a limited extent by other vegetated SuDS, and are not provided, for example, by permeable paving. The potential for blue-green installations to mitigate extreme weather is explored by Voskamp and Van de Ven (2015), highlighting those that are mostly of benefit for cooling (such as green facades), others that are best for infiltration (such as porous paving) and some that are mainly tailored to storage (such as rainwater harvesting), while other installations provide a combination of benefits. Voskamp and Van de Ven (2015) propose that guidance and support tools need to recognise multiple options, leaving the local decision-makers to determine the best combination of features to suit the goals of planning within the specific urban environment.

Studies demonstrate that a wide variety of benefits can be provided by SuDS. A comprehensive review of the evidence is provided by CIRIA (CIRIA, 2013), covering, for example, the potential for green roofs to act as an insulation layer, thereby reducing a building's heating and cooling costs and benefits global greenhouse gas emissions (Bamfield, 2005; Bastien *et al.*, 2011). Castleton *et al.* (2010) concluded that this could be significant for retrofit of older UK properties with lower thermal efficiency, with up to 45% savings on heating costs. Wilkinson and Feitosa (2015) found significant cooling benefits (up to 15%) on metal roofs in the context of green roof retrofit in Sydney and Rio.

Table 17.1   List of categories of benefits from SuDS (after Abbott *et al.*, 2013).

Category of benefit	Example within category
Water quality	Filtration devices improving water quality discharging into watercourses, leading to cleaner streams
Flood risk management	Peak flow attenuation and delay as well as infiltration reducing and delaying runoff
Food and urban agriculture	Increased green spaces providing the opportunity for food/crop production
Energy/carbon	Thermal efficiency of green roofs leading to lower heating and cooling costs
Wastewater	Quantity and quality of runoff entering the water treatment cycle improved through retention and filtration
Water supply	Infiltration increasing subsurface flows and groundwater recharge
Health and wellbeing	For example, improvements in air quality leading to better respiratory health
Economy	Linked to other benefits such as uplift in property value due to neighbourhood improvement
Place and community	Improved aesthetics in the urban environment
Habitat and biodiversity	Increased aquatic and terrestrial species including insects, birds, amphibians and plants
Microclimate adaptation	Urban cooling and reduction of urban heat island

Some health benefits are linked to air quality improvement, predicated on the removal of particulates by vegetation that lead to less respiratory damage. The link between air pollution and increased demand on hospital services is well recognised. Quantifying the link between SuDS and air pollution is less well evidenced. However, studies including Clark *et al.* (2008) contend that the consideration of reduction of nitrogen oxides in urban areas should not be ignored. Hoang and Fenner (2014) offer reflections on the differential performance of green roofs to trap pollutants depending on soil moisture content and air temperature. Further health benefits are posited due to increased physical activity, higher levels of walking and wellbeing due to proximity to green space (de Vries *et al.*, 2003; Groenewegen *et al.*, 2006; Maas *et al.*, 2006; Sinnett *et al.*, 2011).

Biodiversity improvements due to provision of habitat within urban areas are potentially significant but problematic to estimate in advance. The assessment of benefit of increase in wildlife within cities calls for subjective judgement regarding the value of biodiversity, although the preservation and encouragement of rare species, as seen in the Barclays Bank 400 m green roof, where extremely rare beetles were recorded (Warwick, 2007), is usually regarded as beneficial. Features can be designed to include rare or locally rare plant species or to attract specific wildlife but are more potent when wildlife connectivity is considered.

Social benefits claimed from SuDS include recreation, educational opportunities and community cohesion. Clearly the implementation and design of public realm facilities that include SuDS can lend themselves to such benefits (Graham *et al.*, 2012). For example, the exemplar work in schools has provided direct educational benefit for the children attending adapted schools (Graham *et al.*, 2012).

17.3.1 Valuing Benefits

The process of monetising benefits is useful for the purpose of cost–benefit analysis (CBA) and is advocated by proponents of CBA as a methodology that allows for comparison of projects on a like-for-like basis. Converting the benefits of SuDS into a monetary equivalent also allows the calculation of net present value (NPV) thereby enabling the justification of capital investment and raising of finance where necessary. Making the business case for SuDS is particularly important when public money is invested for the public good by governmental and non-governmental organisations (CIRIA, 2013). Post hoc measurement of benefits can draw upon a range of valuation tools, including stated and revealed preference methods such as hedonic analysis (Ichihara and Cohen, 2011), contingent valuation methods (Bowman *et al.*, 2012), substitution or cost based methods. A priori estimation of benefits can draw on expected reductions in damages, estimated reductions in air pollution and predictions of reduced heating or cooling costs (Bureau of Environmental Services – City of Portland, 2008).

Tools for valuing benefits are available to help decision-makers assess the relative merits of different schemes, and there are many options available (EFTEC/Cascade, 2013). Assessment of the scope and focus of the various tools is provided in the EFTEC/Cascade review (2013) for Natural England. For example, the Center for Neighbourhood Technology in the USS quantifies 20 benefits provided by SuDS – including flood risk reduction, amenity benefits and environmental improvements – in its valuation tool (Center for Neighborhood Technology (CNT), n.d.). The green infrastructure valuation toolkit is a tool developed for the UK context by Green Infrastructure Northwest and is a prototype open source toolkit available on-line. The latest UK tool is the CIRIA BeST tool (CIRIA, 2015), drawing on the previously published review and using the principles of benefit transfer to allow valuation of multiple types of SuDS.

Benefit transfer is the use of the results of site-specific detailed benefits studies to estimate benefits that might accrue from similar installations in a different location (CIRIA, 2013). The advantage of the benefit transfer approach is that it reduces the need for detailed site studies – thus saving cost and time. Naturally, the trade-off lies in the inadequacy of the data available on comparable installations and site specific factors that will fundamentally alter the benefits achieved. Such tools are potentially most helpful for a feasibility assessment or comparison of options at a broad design stage. More precise studies can then be focused on critical benefits and/or smaller numbers of options.

17.3.2 Distribution of Benefits

Consideration of the benefits needs to be contextualised within an understanding of the spatial scale of benefits and the identity of the beneficiary (Abbott *et al.*, 2013). For example, benefits to commercial property owners are discussed by Lamond *et al.* (2014) and by NRDC (Clements *et al.*, 2013) which range from direct savings in heating costs to hypothesised reputational enhancement, whereas Abbott *et al.* (2013) consider various institutional stakeholders and water companies in relation to lower water processing costs. The multiple benefits of SuDS installation are sometimes spread over a wide populace; for example, flood reduction benefits go beyond the property that installs them, stormwater benefits may be spread among all customers of a given utility company and amenity benefits accrue to local businesses, residents and visitors to an urban area. In the city of Portland, Oregon the *Tabor to the River* programme includes several SuDS features, such as bioswales, to manage the rainwater at source (Church, 2015). Portland has been active in installing such features across the city under a number of initiatives, and private residents are often asked to partially or wholly fund their installation in the paving areas outside their property (Everett *et al.*, 2016). However, the stormwater benefits of bioswales are felt in reduced nuisance flooding (not necessarily in the same street as the bioswale) and reduced costs of stormwater treatment and management that are reaped much more widely. Water quality improvement is another benefit that accrues at a city-wide scale both from directly experiencing cleaner waters and indirectly having lower clean-up costs. Other benefits from the green streets are expected to be more concentrated on the immediate neighbourhood, such as air quality and community liveability (Entrix, 2010). There is some evidence that the neighbourhoods with green street facilities are more desirable in the long term (Netusil *et al.*, 2014), resulting in property value uplift. However, Netusil *et al.* (2014) also suggest that those living closest to the facilities may experience a negative impact on property value compared to those in the neighbourhood but not immediately adjacent to a bioswale, and some regard them as ugly, or dislike the overshadowing of their own gardens by trees in bioswales (Everett *et al.*, 2015). The apportionment and spatial distribution of costs and benefits for SuDS is an area that has received minimal attention but will be increasingly important if widespread adoption and maintenance costs are to be equitably apportioned.

17.4   Conclusions

Sustainable urban drainage systems are advocated as a more natural and flexible, and therefore sustainable, solution to flood and stormwater management for urban and suburban settings. However, if the uptake of SuDS is to be promoted as opposed to more familiar and trusted traditional drainage approaches, it will be necessary to understand the cost-effectiveness of the SuDS alternative, in order to include the approaches within formal options appraisal.

This chapter has explored relevant literature and touched on a variety of tools that can contribute to the estimation of the costs of implementing and maintaining SuDS, and the evaluation of flood damage reduction, stormwater management and other benefits associated with these approaches. It is apparent from this review that the analysis of costs and multiple benefits of SuDS is complicated by the multiplicity of SuDS options, the variety of design choices, the heterogeneity of the urban setting within which they are proposed to be installed and the fact that those schemes and beneficiaries are not necessarily co-located.

Owing to this complexity, in practical terms, it will be important for decision-makers to select evaluation methods that are appropriate and proportionate. This may sometimes be limited to cost-effectiveness calculations when runoff reduction is a major goal or legislative requirement for the SuDS, specifying the lowest cost option to achieve the required stormwater performance. However, in other instances, such as retrofit SuDS during renewal or regeneration, a greater focus on other positive features of SuDS is necessary. The increasing use of benefits transfer via tools such as the CIRIA BeST tool can help in making the case for SuDS, but it is important to continue to add to the evidence base supporting such tools given the scarcity of studies that demonstrate robust estimates of benefits achieved and even of the ongoing cost of maintaining SuDS under differing climatic and locational factors.

Another area requiring greater focus from research and improved evidence is the distribution of costs and benefits across stakeholders, spatially and temporally. This is an emerging area of research and practice that may enable the leveraging of increased funds into SuDS implementation and also provide insights into appropriate incentive schemes to encourage private stakeholder to make sustainable choices.

Acknowledgement

This research was performed as part of an interdisciplinary project programme undertaken by the Blue-Green Cities Research Consortium (www.bluegreencities.ac.uk). The Consortium is funded by the UK Engineering and Physical Sciences Research Council under grant EP/K013661/1, with additional contributions from the Environment Agency, Rivers Agency (Northern Ireland) and the National Science Foundation.

References

Abbott, J., Davies, P., Simkins, P., Morgan, C., Levin, D. and Robinson, P. (2013) Creating water sensitive places – scoping the potential for water sensitive urban design in the UK. London: CIRIA.

Bamfield, B. (2005) Whole Life Costs & Living Roofs – The Springboard Centre, Bridgewater. Second ed. Cheshunt: The Solution Organisation.

Barr Engineering Company (2011) Best Management Practices Construction Costs, Maintenance Costs, and Land Requirements. Minneapolis: Minnesota Pollution Control Agency.

Bastien, N.R.P., Arthur, S., Wallis, S.G. and Scholz, M. (2011) Runoff infiltration, a desktop case study. Water Science and Technology; 2011, 63, 10, 2300–2308.

Bloomberg, M.R. and Strickland, C.H. (2012) Guidelines for the Design and Construction of Stormwater Management Systems New York: New York City Department of Environmental Protection, in consultation with the New York City Department of Buildings.

Bowman, T., Tyndall, J.C., Thompson, J., Kliebenstein, J. and Colletti, J.P. (2012) Multiple approaches to valuation of conservation design and low-impact development features in residential subdivisions. *Journal of Environmental Management*, 104, 101–113.

Bureau of Environmental Services – City of Portland (2008) Cost Benefit Evaluation of Ecoroofs 2008. Portland, Oregon USA: City of Portland, Oregon.

Castleton, H.F., Stovin, V., Beck, S.B.M. and Davison, J.B. (2010) Green roofs; building energy savings and the potential for retrofit. *Energy and Buildings*, 42, 1582–1591.

Center for Neighborhood Technology (CNT). (n.d.) *Green Values Stormwater Toolbox – homepage*. Chicago, USA: CNT. Available: http://greenvalues.cnt.org/.

Church, S.P. (2015) Exploring Green Streets and rain gardens as instances of small scale nature and environmental learning tools. *Landscape and Urban Planning*, 134, 229–240.

CIRIA (2009) Overview of SuDS performance – Information provided to Defra and the EA. London: CIRIA.

CIRIA (2013) Demonstrating the multiple benefits of SuDS – A business case (Phase 2), Draft Literature Review. *Research Project RP993*. London: CIRIA.

CIRIA (2015) New tool assesses the benefits of SuDS. London: CIRIA. Available: https://www.ciria.org/News/CIRIA_news2/New-tool-assesses-the-benefits-of-SuDS.aspx.

Clark, C., Adriaens, P. and Talbot, F.B. (2008) Green Roof Valuation: A Probabilistic Economic Analysis of Environmental Benefits. *Environmental Science Technology*, 42, 2155–2161.

Clements, J., St Juliana, A., Davis, P. and Levine, L. (2013). The Green Edge: How Commercial Property Investment in Green Infrastructure Creates Value. New York: Natural Resources Defense Council.

Constructing Excellence (2006) Wole life costing. Available at: www.constructingexcellence.org.uk.

De Vries, S., Verheij, R.A., Groenewegen, P.P. and Spreeuwenberg, P. (2003) Natural environments – healthy environments? An exploratory analysis of the relationship between greenspace and health. *Environment and Planning*, A35, 1717–1731.

Defra (2009) Appraisal of flood and coastal erosion risk management: A Defra policy statement. London: Defra.

Duffy, A., Jefferies, C., Waddell, G., Shanks, G., Blackwood, D. and Watkins, A. (2008) A cost comparison of traditional drainage and SUDS in Scotland. *Water Science and Technology*, 57, 1451–1459.

EFTEC/Cascade (2013) Green Infrastructure – Valuation Tools Assessment. London: Natural England.

Entrix (2010) Portland's Green Infrastructure: Quantifying the Health, Energy, and Community Livability Benefits. Portland, Oregon: City of Portland, Bureau of Environmental Services.

Environment Agency (2009) Investing for the future. Flood and coastal risk management in England – a long-term investment strategy. Bristol: Environment Agency.

EPA (1999) Preliminary Data Summary of Urban Storm Water Best Management Practices. EPA-821-R-99-012. Washington DC: United States Environmental Protection Agency.

Everett, G., Lamond, J., Morzillo, A., Chan, F.K.S. and Matsler, A.M. (2015) Can sustainable drainage systems help people live with water? *Water Management*, 169, 94–104.

Everett, G., Lamond, J., Morzillo, A., Matsler, A.M. and Chan, F.K.S. (2016) Delivering greenstreets: an exploration of changing perceptions and behaviours over time around bioswales in Portland Oregon. *Journal of Flood Risk Management*.

Flood Hazard Research Centre (2010) The Benefits of Flood and Coastal Risk Management: A Handbook of Assessment Techniques, London, FHRC.

Gordon-Walker, S., Harle, T. and Naismith, I. (2007) Cost-benefit of SUDS retrofit in urban areas – Science Report – SC060024. Bristol: Environment Agency.

Graham, A., Day, J., Bray, B. and Mackenzie, S. (2012) Sustainable drainage systems: Maximising the potential for people and wildlife A guide for local authorities and developers. RSPB/WWT.

Groenewegen, P.P., Van Den Berg, A.E.D.E., Vries, S. and Verheij, R.A. (2006) Vitamin G: effects of green space on health, well-being and social safety. *BioMed Central: Public Health*, 6, 1–9.

HR Wallingford (2004) Whole life costing for sustainable drainage. Oxfordshire: HR Wallingford.

Hoang, L. and Fenner, R.A. (2014) System Interactions of Green Roofs in Blue Green Cities. 13th International Conference on Urban Drainage, 7–12 September 2014 Sarawak, Malaysia. ICUD.

Houle, J., Roseen, R., Ballestero, T., Puls, T. and Sherrard, J. (2013) Comparison of maintenance cost, labour demands, and system performance for LID and conventional stormwater management. *Journal of Environmental Engineering*, 139, 932–938.

Ichihara, K. and Cohen, J. (2011) New York City property values: what is the impact of green roofs on rental pricing? *Letters in Spatial and Resource Sciences*, 4, 21–30.

Lamond, J.E., Wilkinson, S. and Rose, C. (2014) Conceptualising the benefits of green roof technology for commercial real estate owners and occupiers. Resilient Communities, providing for the future, 20th Annual Pacific Rim Real Estate Conference, Christchurch, New Zealand. PRRES.

Lampe, L., Barrett, M.,Woods Ballard, B., Kellagher, R., Martin, P., Jefferies, C. and Hollon, M. (2004) *Post-Project Monitoring of BMPs/SUDS to Determine Performance and Whole-Life Costs*, London, IWA Publishing (UK Water Industry Research).

livingroofs.org. (n.d.) Homepage. Available: http://livingroofs.org/2010030671/green-roof-benefits/waterrunoff.html.

Maas, J., Verheij, R.A., Groenewegen, P.P.D.E., Vries, S. and Spreeuwenberg, P. (2006) Green space, urbanity and health: how strong is the relation? *Journal of Epidemiology and Community Health* 60, 587–592.

MacMullan, E. and Reich, S. (2007) The Economics of Low-Impact Development: A Literature Review. Eugene, OR: ECONorthwest.

Narayanan, A. and Pitt, R. (2005) Costs of Urban Stormwater Control Practices.

Netusil, N.R., Levin, Z., Shandas, V. and Hart, T. (2014) Valuing green infrastructure in Portland, Oregon. *Landscape and Urban Planning*, 124, 14–21.

Porsche, U. and Köhler, M. (2003) Life cycle costs of green roofs – a comparison of Germany, USA, and Brazil. RIO 3 – World Climate & Energy Event. Rio de Janeiro, Brazil.

Scottish SUDS Working Party (2009) SUDS for Roads Guidance Manual. Edinburgh, UK: SUDS Working Party.

Shelton, D.B. and Vogel, R.M. (2005) The value of infiltration- and storage-based BMPs for stormwater management. World Water and Environmental Resources Congress, 15–19 May 2005, Anchorage, AK, USA. American Society of Civil Engineers, 234.

Sinnett, D., Williams, K., Chatterjee, K. and Cavill, N. (2011) Making the case for investment in the walking environment: A review of the evidence. Living Streets. Bristol: University of the West of England (Department of Planning and Architecture, Faculty of Environment and Technology).

Stevens, R. and Ogunyoye, F. (2012) Costs and Benefits of Sustainable Drainage Systems – Final report – 9X1055 – Royal Haskoning. Committee on Climate Change.

Taylor, A. and Wong, T. (2002) Non-structural stormwater quality best management practices – a literature review of their value and life-cycle costs. Monash, Australia: Cooperative Research Centre for Catchment Hydrology.

Vila, A., Pérez, G., Solé, C., Fernández, A.I. and Cabeza, L.F. (2012) Use of rubber crumbs as drainage layer in experimental green roofs. *Building and Environment*, 48, 101–106.

Voskamp, I.M. and Van De Ven, F.H.M. (2015) Planning support system for climate adaptation: Composing effective sets of blue-green measures to reduce urban vulnerability to extreme weather events. *Building and Environment*, 83, 159–167.

Warwick, H. (2007) The garden up above. *Geographical Magazine* 79 (7), 38.

Water Environment Research Foundation (2009) *User's Guide to the BMP and LID Whole Life Cost Models Version 2.0*, Alexandria, VA.

Wilkinson, S. and Feitosa, R.C. (2015) Retrofitting Housing with Lightweight Green Roof Technology in Sydney, Australia, and Rio de Janeiro, Brazil. *Sustainability*, 7, 1081–1098.

Wolf, D.F., Duffy, A.M. and Heal, K.V. (2015) Whole Life Costs and Benefits of Sustainable Urban Drainage Systems in Dunfermline, Scotland.

Green Roof and Permeable Paving Retrofit to Mitigate Pluvial Flooding

Sara Wilkinson, David G. Proverbs and Jessica E. Lamond

18.1   **Introduction**

With increasing urbanisation, predicted climate change and population growth, the potential for intense rainfall events to cause flood damage and disruption within urban settlements including central business districts (CBDs) grows (Met Office Hadley Centre for Climate Research, 2007; Jha *et al.*, 2011). Urban planning has increased the development density and the amount of impermeable surfaces, and population increase will impose further pressure to increase densities ever higher. Stormwater runs off impermeable surfaces, rather than infiltrating into the ground, and in this way increases flooding potential. In many Australian CBDs, piped drainage systems were installed when lower density development existed and the capacity has not increased sufficiently. According to the City of Melbourne flood emergency plan 'Flooding within City of Melbourne catchments is generally caused by short duration thunderstorm events because these produce the highest rates of runoff in hard-lined drainage systems serving relatively small and highly impervious catchments.' (City of Melbourne and VICSES Unit(s) St Kilda and Footscray, 2012).

Retrofitting and replacing below-ground drainage systems is expensive, time-consuming and disruptive. Australia is cited here as an example, but many other cities globally are experiencing similar issues with respect to pluvial flooding overwhelming existing infrastructure and causing physical and economic damage to buildings. Cost of building remediation following the 2010/11 floods in Queensland and Victoria was estimated to be up to A$20 billion (Companies and Markets, 2011). Chetri *et al.* (2012) noted that indirect impacts, such as transport disruption, also affected the local economy during the 2011 event. However, the full impact on local economies can be longer lasting and difficult to measure, given that many businesses fail to recover after flooding (Gissing, 2003).

Sustainable Surface Water Management: A Handbook for SuDS, First Edition.
Edited by Susanne M. Charlesworth and Colin A. Booth.
© 2017 John Wiley & Sons, Ltd. Published 2017 by John Wiley & Sons, Ltd.

Widespread green roof retrofit has been proposed by Charlesworth and Warwick (2011) as a roof treatment that mimics natural infiltration patterns, decreasing runoff and reducing flood risk. Permeable paving is advocated by Gordon-Walker *et al.* (2007) as a treatment for impermeable surfaces, such as car parks, pavements and roads, subject to light traffic. Within CBDs, this could be achieved through retrofit during urban renewal or refurbishment. Policies to encourage green roof retrofit have been introduced, for example, in the USA in Portland, Oregon (Environmental Services – City of Portland, 2011), Chicago and Philadelphia (Bureau of Environmental Services – City of Portland (2008). In adopting such policies it is important to consider whether widespread retrofit of green roofs and permeable paving is a viable option given structural and functional conditions in a given city. Furthermore storm-water management is just one of the functional, social and environmental benefits attributable to green roofs (Hoang *et al.*, 2014) and therefore it has to be considered along with other complementary and conflicting urban planning objectives. In Australia, the pattern of alternating flooding and heatwaves (Chetri *et al.*, 2012) highlights the need for urban renewal that is tolerant to drought and contributes to reducing urban heat islands (UHI).

This chapter uses a recent study within the city of Melbourne as an example to explore the potential of retrofitted green roofs and permeable paving (referred to as sustainable urban drainage systems – SuDS) to attenuate stormwater damage in the context of CBDs. The research highlighted the importance of considering the barriers and drivers to retrofit through three key objectives:

1. Explore the choice of green roof retrofit optimised for flood risk reduction, including the identification of additional social and environmental benefits and trade-offs
2. Evaluate the proportion of buildings that are physically suited to green roof retrofit and of paved areas that are suitable for permeable paving
3. Consider different uptake scenarios in estimating the potential reduction in rainwater runoff from green roof and permeable paving.

18.2 Types of Green Roof for Stormwater Management

The term *green roof,* or *vegetated roof,* describes a roof (or part of a roof) of a building that is covered with vegetation and, in current green roof technology, a substructure supporting the vegetation while protecting the building. Typically, this includes: a roof structure; a waterproof membrane or vapour control layer; insulation (i.e. if the building is heated or cooled); a root barrier to protect the membrane (i.e. made of gravel, impervious concrete, polyvinylchloride (PVC), thermoplastic polyolefin (TPO), high-density polyethylene (HDPE) or copper; a drainage system; a filter cloth (non-biodegradable fabric); a growing medium (soil) consisting of inorganic matter, organic material (straw, peat, wood, grass, sawdust) and air; and plants.

Green roofs can be intensive or extensive. Intensive green roofs, sometimes known as 'roof gardens', typically combine vegetated areas with recreational or social space, planting may include trees and shrubs. Extensive roofs are more common and are usually designed for minimum maintenance, with less emphasis on access for social and recreational purposes. Table 18.1 summarises the attributes of both these common types, but a third type is a hybrid semi-intensive green roof that combines features of intensive and extensive roofs (Czemiel-Berndtsson, 2010). Standard soils are not used for green roofs because they are usually too heavy for roof structures. Alternative growing media vary and should be engineered to suit the particular planting, climate and runoff requirements, but they

Table 18.1 Characteristics of extensive and intensive green roofs.

Intensive green roof	Extensive green roof
Deep growing medium	Shallow growing medium
Higher runoff reduction per m^2	Lower runoff reduction per m^2
Small trees and shrubs feature	Low profile planting such as sedums are common
Often restricted to part of the roof	Cover large expanses of rooftop
Higher carbon sequestration potential per m^2	Lower carbon sequestration potential per m^2
More maintenance required	Requires minimum maintenance
More expensive	Lower capital cost
More common in tropical climates	More common in temperate climates
More often accessible to allow for social and recreational use	Can be accessible or inaccessible, but not usually recreational
Requires irrigation	Does not usually require irrigation
Heavier roof structure required to support roof	Lightweight roof structure needed to support roof
Potentially high structural implications for existing buildings	Minimum structural implications for existing buildings

(Source: Adapted from Wilkinson and Reed (2009), with additions from Getter *et al.*, (2009) and Czemiel-Berndtsson (2010)).

usually include aggregate (e.g. shale, vermiculite), aerated pore space and organic materials. The viable growing medium and the planting need careful consideration, particularly in the highly varied climatic conditions with which Australian cities are faced, from excessive seasonal rainfall in the Northern Territory of Australia to insufficient rainfall in Victoria. Water supply is an issue where watering and irrigation of plants is required, and upgrading to supply water will add further costs. Therefore, in Australia the capacity for rainwater harvesting, perhaps from adjacent roof space, and the use of drought- or heat-tolerant plants is desirable to cope with fluctuations in climate.

Many green roof programmes and regulations do not specify the roof type, resulting in a preponderance of extensive green roofs, which are often sedum based, as these are the least costly to install. However, it is clear from Table 18.1 that choice of the appropriate green roof for optimal stormwater management is complicated by several factors. All else being equal, the stormwater retention performance of a green roof depends on the substrate depth and absorbency of the substrate, meaning that intensive roofs may be preferred for stormwater management. For example, the Beijing Olympic Village (China) had some green roofs included in the design of buildings. However, Jia *et al.* (2012) calculated that improvements could be achieved if the substrate depths were doubled from 0.3 m to 0.6 m. Specific design aspects can be important in regulating prior saturation. Speed of evaporation after a storm is important and depends on external temperatures and humidity (Blanc *et al.*, 2012), but drainage design can increase drying speed, and plant type also affects levels of retention. Furthermore, the proportion of roof covered by an intensive versus an extensive roof may affect total runoff reduction achieved.

18.3 Building Retrofit Characteristics

The potential to retrofit existing structures, as well as the type of green roof considered suitable, depends on factors such as roof type, size and slope and the load-bearing capacity of the structural form. Large Australian commercial buildings tend to have roofs of concrete construction, whereas smaller commercial buildings may have timber roof structures covered with profiled metal sheeting. Green roofs require good drainage and waterproofing; green roofs with <2% slope require additional drainage measures (University of Florida, 2008). If the additional structural load-bearing capacity of the existing roof is low, a light-weight growth medium and/or further structural strengthening may be required. Financial viability of the retrofit is linked to the condition of the existing roof: poor condition and the requirement for upgrading and extra structural support significantly increases costs and can make the project unattractive.

Wilkinson *et al.* (2014) found that numerous building characteristics were important in office retrofits in the Melbourne CBD (Table 18.2). Age was important, with buildings falling within certain ages (i.e. over 25 years old) being more likely to need a retrofit (Fianchini, 2007). Condition was important, with those suffering from wear and tear being prime candidates for upgrade (Kersting, 2006). Physical characteristics of height and depth influenced the likelihood of retrofit (Szarejko and Trocka-Lesczynska, 2007) and smaller buildings were more favoured in retrofits (Gann and Barlow, 1996; Ball, 2002). The building envelope influenced the type, extent and costs of retrofit (Kersting, 2006), and poor acoustic separation leads to less retrofit because the noise transmission issues were perceived as insurmountable (Gann and Barlow, 1996). The age, condition and location of services affected the type and cost of the retrofit (Snyder, 2005). Arge (2005) found that purpose-built buildings have better specifications with greater flexibility for retrofit. Location had an impact, where retrofit provided a good economic return (Remoy and van der Voordt, 2006), and heritage buildings were preferred in some markets (Snyder, 2005). Many found that proactive policy-making and legislation (planning and building codes and fire) influenced the amount of retrofit (Heath, 2001; Ball, 2002; Snyder, 2005; Kersting, 2006). Povell and Eley (cited in Markus, 1979) and Isaac (cited in Baird *et al.*, 1996) noted that the number of site boundaries (i.e. whether a building is adjoined to others) determined the ease of retrofit, with Kincaid (2002) noting that detached buildings are the easiest to adapt because of ease of access and the lack of disturbance caused to neighbours.

Table 18.2   Retrofit characteristics in existing Melbourne office buildings.

Age
Condition
Height
Depth
Envelope and cladding
Structure
Building services
Internal layout
Flexibility for a range of differing uses and functional equipment
Purpose-built buildings (not speculative)
Location
Perceived heritage value
Size
Accessibility
Proactive policy making/legislation (planning and building codes including fire)
Acoustic separation
User demand
Site conditions

Source: Wilkinson *et al.*, 2014.

Table 18.3   Technical features for green roof retrofit.

Position of the building
Location of the building
Orientation of the roof
Amount of overshadowing (if any)
Roof type
Roof size
Roof pitch/slope (2%+)
Load-bearing capacity
Drainage and waterproofing system
Condition of the existing membrane
Access to the roof for construction and user (if accessible to users)
Weight of substrate and planting
Water supply
Preferred planting
Levels of maintenance desired

Furthermore, for the roof area to be successfully retrofitted, it needs to meet certain technical criteria. The rooftop needs a reasonable amount of exposure to sun for the plants to grow, and overshadowing must be considered. The roof pitch affects the type of green roof that can be retrofitted. Clearly, careful consideration of the additional dead and live loads applied in the retrofit need to be considered by a structural engineer. In some cases, additional support may be provided. The type and condition of the existing waterproof membrane needs to be assessed. Where necessary, patch repairs may be needed and, in some cases, replacement may be advisable. Green roofs extend the life cycle of the roof membrane because it is covered and unexposed to wear and tear of the elements and trafficking. Finally, the type of planting is affected by the local climate and rainfall patterns, as well as availability of water for irrigation and the amount of maintenance to be provided. Table 18.3 summarises the technical features to be considered regarding the roof for retrofit.

18.3.1 Melbourne CBD Database

Melbourne CBD was originally laid out in the 1830s, with the 'Hoddle Grid' incorporating heritage architecture now interspersed with modern buildings, sited on sloping land on the banks of the Yarra River. Melbourne has spread out since the 1830s, creating suburbs on previously undeveloped land, so that the catchment surrounding the CBD has also become less permeable over time. Open watercourses have been culverted to create underground drainage channels with the attendant potential for blockages or capacity exceedance. Melbourne has consequently been affected by periodic pluvial 'flash' flooding throughout its history, and this seems to have become more frequent in recent years.

A Melbourne CBD database of 526 buildings (constructed between 1850 and 2005), compiled by the authors, was examined using criteria drawn from the literature above, such as age, position, orientation and location of the building, the roof pitch, weight limitations and ground conditions. Finally, criteria were developed to determine the suitability of roofs for green retrofit.

Analysis revealed an average age of 61 years for buildings in the Melbourne CBD, with the oldest constructed in 1853 and the most recent built in 2005. The top ten years for the construction of new buildings are shown in Table 18.4, and this reflects considerable

Table 18.4   Rank order of year of construction for buildings in Melbourne.

Rank order	Year	Number of buildings constructed
1	1945	38
2	1990	19
3	1972	15
4	1991	14
4	1930	14
4	1920	14
7	1973	12
8	1987	10
8	1969	10
8	1960	10

post-war construction with the majority of commercial buildings in Melbourne (60.4%) constructed after 1940. A large number of buildings (237) have been constructed since the 1960s; therefore, there is a large amount of stock that would be due for major renewal and updating, in which adaptation and retrofitting green roofs could be considered.

Overshadowing may occur if there are significant numbers of high-rise buildings interspersed with low-rise buildings and blocking their sunlight and it is apparent that this pattern may exist in the Melbourne CBD. The Property Council of Australia uses an office building quality matrix that classes buildings from premium (the highest grade) through to A, B, C and D grades (the lowest grade). Part of the grading criteria is net lettable area and not the number of storeys (Property Council of Australia Limited, 2006). According to some definitions (based on converting metres to average storey height), buildings over approximately seven storeys are high-rise and those over twenty storeys are skyscrapers. Figure 18.1 shows the cumulative frequency of buildings in the Melbourne database by number of storeys. The modal number of storeys is three and the median is six, with 67% of the stock up to ten storeys high. However, a significant number (8%) are 21 storeys and over and may be casting shadows over adjoining lower buildings, meaning that most buildings are low-rise but partially, or totally, overshadowed by their high-rise neighbours. Such an arrangement of buildings could mean that existing buildings, which have adequate structural strength to accommodate retrofit, may be unsuitable because overshadowing adversely affects planting.

Orientation has an influence on how much exposure to sunlight the roof gets; in the southern hemisphere north-facing properties are exposed most to direct sun. Orientation was examined for a sample of 72 buildings in the database and revealed that north-facing buildings only represented 12% of the sample. Most faced east (41%) followed by west-facing (31%) and south-facing (16%). Therefore, a large number will have only partial exposure to sunlight, even before overshadowing is considered. These two analyses imply that access to sun should be considered on a case-by-case basis, as this will affect the type of plants specified and/or whether green roof retrofit is viable.

Access factors, for ease of construction, were assessed by categorising the attachment of buildings. Attachment on three sides may cause access issues, requiring main-street disruption for retrofit. However, in the sample, only 18% of the properties were in this category. Almost half (47%) of the properties were bounded on two sides, with 22% bounded on one side only, and 12% detached. Therefore, it was judged that the majority of properties

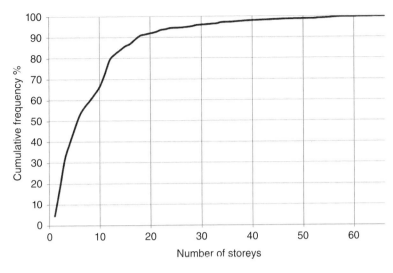

Figure 18.1   Cumulative frequency of Melbourne CBD Buildings by number of storeys.

were not adversely affected by attachment to other buildings or restricted access for construction, and are suited for retrofit.

Structural capacity affects the amount and ease of retrofit, and a full structural appraisal would be required to determine structural suitability for retrofit and strengthening requirements. For the analysis, structural capacity was judged through construction type, and is therefore limited to an indication of whether a building may have the potential to bear the extra loading necessary for a green roof. In Melbourne, most buildings are built of concrete, with 61% of commercial buildings having framed structures. The remaining 39% comprised traditional load-bearing brickwork and/or stone construction, with no timber-framed buildings. The buildings with concrete frames are most suitable for retrofitting with extensive green roof systems; masonry buildings may also be suitable, but timber frames are generally judged unsuitable for retrofit. This analysis, therefore, confirmed good potential for retrofit requiring minimal structural changes to most CBD buildings.

The next stage involved a visual inspection of the roof, using the Google Earth and Google Map software (Google Earth 6.0, 2008), to evaluate the potential of each roof for retrofit. The evaluations called for identification as one of three classifications – (a) yes, (b) no or (c) don't know – with regard to retrofit. The evaluation was based on roof pitch, with those pitched above 30° or below 2% deemed unsuitable. The amount of rooftop plant, especially equipment, which vents air from the building, and the provision of rooftop window cleaning equipment, safety handrails and photovoltaic units was accounted for, where coverage exceeding 40% of area being deemed unsuitable. Lightweight roof construction was unsuitable. Figure 18.2 shows that 15% of buildings were judged suitable for retrofit, 5% were unclassified and 80% were unsuitable, based on the criteria above.

The final stage involved analysis of overshadowing of the stock (Figure 18.3), where orientation and the proximity of taller buildings were considered. This shows that 39.3% of the buildings were overshadowed, 36.3% were partially overshadowed and 24.4% were not overshadowed at all. Therefore, approximately 75% of the existing Melbourne stock was considered unsuitable for retrofit on the basis that insufficient sunlight reaches the rooftop for planting to flourish.

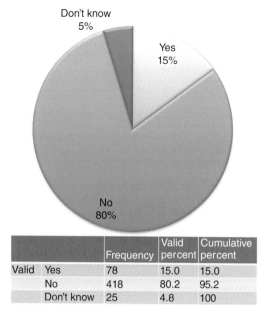

		Frequency	Valid percent	Cumulative percent
Valid	Yes	78	15.0	15.0
	No	418	80.2	95.2
	Don't know	25	4.8	100

Figure 18.2   Proportion of buildings judged suitable for green roof option through roof inspection (Source: Adapted from Wilkinson and Reed (2009)).

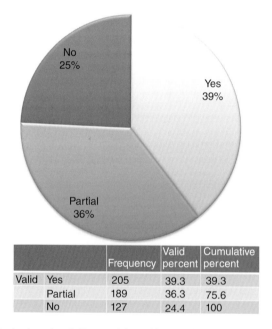

		Frequency	Valid percent	Cumulative percent
Valid	Yes	205	39.3	39.3
	Partial	189	36.3	75.6
	No	127	24.4	100

Figure 18.3   Overshadowing of roofs (Source: Adapted from Wilkinson and Reed, 2009).

Services and service categories	Traditional piped drainage solutions (included here for comparison based on life cycle study in Gouda et al., unpublished data)	Filter drains and pervious pipes	Pervious surfaces	Infiltration basins	Constructed wetlands (optimally designed for access and amenity, biodiversity, etc.)
Provisioning services					
Fresh water	Does not recharge groundwater though may feed into surface water resources albeit likely to aggregate pollutants associated with concentrated stormwater surges	Makes a partial contribution to recharging water resources if there is significant infiltration to groundwater or egress to surface water bodies	Makes a partial contribution to recharging water resources if there is significant infiltration to groundwater or egress to surface water bodies	Supports recharge of groundwater and can attenuate pollutants	Supports recharge of groundwater or surface water resources and can attenuate pollutants
Food (e.g. crops, fruit, fish, etc.)	Heavy infrastructure reduces potential for peri-urban farming	Makes no provision for peri-urban farming	Pervious surfaces make no provision for peri-urban farming	Simple low-sward habitat provides little opportunity for food growing, including peri-urban farming, through grazing animals such as rabbits and geese may be harvested at low frequency	Constructed wetlands may allow for harvesting of limited amounts of and possibly for limited peri-urban farming, though both are rarely realised in practice
Fibre and fuel (e.g. timber, wool, etc.)	Heavy infrastructure reduces potential for peri-urban fibre and fuel production	Makes no provision for peri-urban fibre and fuel production	Pervious surfaces make no provision for peri-urban fibre and fuel production	Simple low-sward habitat provides little opportunity for peri-urban fibre and fuel production, beyond low potential for composting of green waste from management and its potential use in bioenergy production as noted below	Potential for peri-urban fibre and fuel production though rarely realised in practice
Genetic resources (used for crop/stock breeding and biotechnology)	Heavy infrastructure displaces ecosystems potentially hosting genetic resources	Drain and pipe infrastructure displaces ecosystems potentially hosting genetic resources	Pervious paving displaces ecosystems potentially hosting genetic resources	Simple low-sward habitat provides no opportunity for ecosystems potentially hosting genetic resources	Constructed wetlands may host genetic resources, particularly if this is an element of design considerations
Biochemicals, natural medicines, pharmaceuticals	Heavy infrastructure displaces ecosystems potentially hosting medicinal resources	Drain and pipe infrastructure displaces ecosystems potentially hosting medicinal resources	Pervious paving displaces ecosystems potentially hosting medicinal resources	Simple low-sward habitat provides no opportunity for ecosystems potentially hosting medicinal resources	Constructed wetlands may host medicinal resources, particularly if this is an element of design considerations
Ornamental resources (e.g. shells, flowers, etc.)	Heavy infrastructure displaces ecosystems potentially hosting ornamental resources	Drain and pipe infrastructure displaces ecosystems potentially hosting ornamental resources	Pervious paving displaces ecosystems potentially hosting ornamental resources	Simple low-sward habitat provides no opportunity for ecosystems potentially hosting ornamental resources	Constructed wetlands may host ornamental resources, particularly if this is an element of design considerations
Energy harvesting	Whilst there may be opportunities for energy harvesting from water egressing at pace through pipework, this is not seen as a common or significant ecosystem services	Does not provide opportunities for energy harvesting	Does not provide opportunities for energy harvesting	Limited potential for biomass harvesting for energy production during routine management, though little evidence that this practice occurs operationally	Potential for biofuel production based on reuse of green waste from wetland management

Regulatory services

Regulatory services					
Air quality regulation	Heavy drainage infrastructure does not regulate air quality	Artificial infrastructure does not support ecosystems regulating air quality	Artificial infrastructure does not support ecosystems regulating air quality	Simple low sward can influence air quality, though to a limited extent compared to complex habitat	Constructed wetlands can comprise complex vegetation efficient at scrubbing and transforming air quality pollutants
Climate regulation (microclimate)	Heavy drainage infrastructure makes no contribution to modification of local climate	Artificial infrastructure does not support ecosystems modifying local climate	Removal of habitat reduces potential to modify local climate	Simple low sward can influence microclimate, though to a limited extent compared to complex habitat	Constructed wetlands can comprise complex vegetation efficient at modifying microclimate
Climate regulation (global climate)	Heavy drainage infrastructure does not sequester carbon and requires heavy energy consumption during construction and maintenance	Artificial infrastructure does not support ecosystems that sequester carbon and moderate energy consumption required during construction phase, with lower maintenance energy consumption than traditional piped drainage	Removal of habitat reduces potential to sequester carbon and moderate energy consumption during construction phase, with lower maintenance energy consumption than traditional piped drainage	Simple low sward can sequester carbon and transform other climate-active gases, though to a limited extent compared to complex habitat and lower energy consumption during construction phase, and moderate consumption during maintenance stage.	Constructed vegetated systems efficient at sequestering carbon and transforming other climate-active gases, though with a limited risk of generation of some climate-active gases (methane, nitrous oxide) in anaerobic sections and lower energy consumption
Water regulation (timing and scale of run-off, flooding, etc.)	Piped drainage infrastructure addresses local flood relief but may relocate flooding problems elsewhere	Produces local flood relief, and may slow the pace of runoff partially averting the risk of relocating flooding problems elsewhere	Produces local flood relief, and may slow the pace of runoff partially averting the risk of relocating flooding problems elsewhere	Produces local flood relief, and slows the pace of infiltration and runoff averting the risk of flooding problems elsewhere	Constructed wetlands provide mosaic semi-natural habitat that detains floodwater and slows its release to ground and surface flows
Natural hazard regulation (i.e. storm protection)	Makes no contribution to hazard protection, and can displace complex vegetation performing storm buffering services	Artificial infrastructure makes no contribution to hazard protection, and can displace complex vegetation performing storm buffering services	Artificial infrastructure makes no contribution to hazard protection, and can displace complex vegetation performing storm buffering services	Simple low sward makes no contribution to hazard protection, and displaces complex vegetation performing storm buffering services	The complex habitat structure of constructed wetlands contributes to hazard protection, and displaces complex vegetation performing storm buffering services
Pest regulation	Makes no positive contribution to pest regulation, but may eliminate habitat for natural pest predators	Artificial infrastructure makes no positive contribution to pest regulation, but may eliminate habitat for natural pest predators	Artificial infrastructure makes no positive contribution to pest regulation, but may eliminate habitat for natural pest predators	Simple low sward provides only limited habitat for the predators of pest organisms, and very low risk of supporting colonies of pest organisms	The complex habitat structure of constructed wetlands may support the predators of pest organisms, though there is a risk of poor management supporting populations of potential pest organisms such as rats and mosquitoes
Disease regulation	May help to rid contaminated water locally but may displace disease risks	Helps rid contaminated water locally with limited attenuation of potential disease organisms, though may displace risks downstream	Helps rid contaminated water locally with limited attenuation of potential disease organisms, though may displace risks downstream	Simple low sward can eliminate some pathogenic organisms though mixed uses of infiltration basins can expose people to accumulated risks. Little opportunity for hosting potential disease vectors	The complex habitat structure of constructed wetlands attenuates pathogenic organisms, though poor design and management can support populations of potential disease vectors such as mosquitoes
Erosion regulation	More concentrated flows of piped water may exacerbate erosion both upstream and downstream of piped solution	Buffering of flows reduces the risks of erosion on site and downstream, albeit that this solution applies where land surfaces are sealed for development	Buffering of flows reduces the risks of erosion on site and downstream, albeit that this solution applies where land surfaces are sealed for development	Infiltration basins still water allowing for progressive infiltration, so avert local erosion of soil except where overtopping in extreme precipitation	Constructed wetlands still water allowing for progressive runoff and infiltration, averting local erosion of soil except where overtopping in extreme precipitation
Water purification and waste treatment	Habitat performing purification processes is lost, with no substantial within-infrastructure purification	Limited physical filtration but artificial infrastructure performs purification processes suboptimally	Limited physical filtration but artificial infrastructure performs purification processes suboptimally	Simple low sward provides limited physicochemical purification of accumulated water, albeit not as effectively as more complex habitat	The complex, mosaic habitat structure of constructed wetlands can be highly efficient at a range of physical and chemical purification
Pollination	Habitat supporting pollinators is lost	Habitat supporting pollinators is lost, and artificial drainage infrastructure provides no compensatory habitat	Habitat supporting pollinators is lost, and artificial drainage infrastructure provides no compensatory habitat	Simple low sward provides only limited habitat for natural pollinating organisms	The complex habitat structure of constructed wetlands can support populations of pollinating organisms
Salinity regulation (mainly and landscapes)	No impact	Emulation of natural hydrology may help avert risks of salinisation of land downstream through buffered infiltration and runoff rates	Emulation of natural hydrology may help avert risks of salinisation of land downstream through buffered infiltration and runoff rates	Emulation of natural hydrology may help avert risks of salinisation of land downstream through buffered infiltration and runoff rates	Emulation of natural hydrology may help avert risks of salinisation of land downstream through buffered runoff and infiltration rates
Fire hazard regulation (mainly arid and landscapes)	Lack of potentially combustible vegetation reduces overall urban fire risk	Lack of potentially combustible vegetation reduces overall urban fire risk	Lack of potentially combustible vegetation reduces overall urban fire risk	The simple sward structure of managed basins provides only a low biomass of potentially combustible vegetation, reducing overall urban fire risk	Complex habitat may create fire risks in urban settings, albeit that wetland habitat generally tends to be low risk

	Heavy (engineering) pipework	Filter drains and pervious pipes / Artificial infrastructure	Pervious paving	Infiltration basins	Constructed wetland ecosystems
Cultural services					
Cultural heritage	Heavy pipework has little or no cultural value, with some exceptions such as old drainage infrastructure of localised heritage interests, but is more generally perceived as a threat	Artificial infrastructure does not of itself provide cultural heritage, but may be a solution that works sympathetically with heritage structures and landscapes		Infiltration basins contribute minimal cultural heritage, but may work sympathetically with heritage structures and landscapes	Constructed wetland ecosystems can potentially be a source of cultural value in urban settings
Recreation and tourism	Heavy engineering pipework lacks recreation and tourism value	Artificial infrastructure makes no contribution to recreation and tourism		Infiltration basins commonly serve as valuable recreational spaces e xcept when holding water after heavy precipitation events	Constructed wetland ecosystems may be of recreational value, depending on wetland design and conflicts with other values (such as habitat for wildlife, particularly where semi-natural spaces are limited in urban settings
Aesthetic value	Heavy engineering pipework lacks aesthetic value	Artificial infrastructure is not itself of aesthetic value, but may be a solution that works sympathetically with mixed use landscapes		Infiltration basins may form part of the landscaping of public and private spaces	Constructed wetland ecosystems can potentially be of aesthetic value in otherwise impoverished urban environments, with the added benefit of complex vegetation providing visual and noise buffering services
Spiritual and religious value	Heavy engineering pipework lacks, and may positively erode, spiritual and religious value	Artificial infrastructure is not itself of aesthetic value, but may be a solution that works sympathetically with spiritually valued landscapes		Infiltration basins contribute minimally to spiritual values, but may work syrmpathetically with spiritually valued landscapes	Constructed wetland ecosystems can potentially be a limited source of spiritual value in urban settings
Inspiration of art, folklore, architecture, etc.	Heavy engineering pipework generally lacks artistic inspiration value, though there is some potential interest and inspiration of local heritage structures	Artificial infrastructure is not itself of inspirational value, but may be a solution that works sympathetically with landscapes		Infiltration basins may fit sympathetically with landscapes valued for inspirational qualities	Constructed wetland ecosystems can potentially be a limited source of creative value in urban settings
Social relations (e.g. fishing, grazing or cropping communities)	Heavy engineering pipework lacks support for local communities other than those united by an interest in urban drainage	Minimal support for local communities other than those united by an interest in urban drainage		Infiltration basins may provide communal places in dry condition for social gatherings, albeit of lesser value than natural spaces	Constructed wetland ecosystems can be a focus for common interests and community activities in urban areas
Education and research	All drainage solutions may support learning and research activities	All drainage solutions may support learning and research activities		All drainage solutions may support learning and research activities	All drainage solutions may support learning and research activities
Supporting services					
Soil formation	Heavy pipework makes no contribution to soil formation	Filter drains and pervious pipes make no contribution to soil formation	Pervious paving makes no contribution to soil formation	Infiltration basins support only simple low sward which contributes to soil formation, albeit suboptimally	Constructed wetlands can be efficient at soil formation
Primary production	Heavy pipework makes no contribution to primary production	Filter drains and pervious pipes make no contribution to primary production	Pervious paving makes no contribution to primary production	Infiltration basins support only simple low sward which contributes suboptimally to primary production	Constructed wetlands can perform significant primary productivity
Nutrient cycling	Heavy pipework makes no contribution to nutrient cycling	Filter drains and pervious pipes make no contribution to nutrient cycling	Pervious paving makes no contribution to nutrient cycling	Infiltration basins support only simple low sward which contributes suboptimally to nutrient cycling	Constructed wetlands can perform significant nutrient trasnformation and cycling services
Water recycling	Heavy pipework makes no contribution to local-scale water recycling	Filter drains and pervious pipes make no contribution to local-scale water recycling	Pervious paving makes no contribution to local-scale water recycling	Infiltration basins support only simple low sward which contributes suboptimally to local-scale water recycling	Constructed wetlands can perform significant local-scale water recycling and retention services
Photosynthesis (production of atmospheric oxygen)	Heavy pipework makes no contribution to photosynthesis	Filter drains and pervious pipes make no contribution to photosynthesis	Pervious paving makes no contribution to photosynthesis	Infiltration basins support only simple low sward which contributes suboptimally to photosynthesis	Constructed wetlands can perform significant photosynthetic processes
Habitat for wildlife	Heavy pipework provides no useful habitat for wildlife, other than for organisms such as rats that may generally be considered as pest rather than assets	Filter drains and pervious pipes provide no useful habitat for wildlife	Pervious paving provides no useful habitat for wildlife	Infiltration basins support only simple low sward which provides some habitat for wildlife, albeit not supporting diverse biodiversity	Constructed wetlands can provide a diversity of habitats for wildlife, which may be of particular value both directly and as a 'stepping stone' where habitat is scarce in urban settings

18.4   Drivers and Barriers to the Uptake of SuDS in Melbourne

Drivers for green roof retrofit are economic, social, technological and environmental and exist at building, local and regional scales (Rajagopalan and Fuller, 2010). For example, on a building scale, the life-cycle costs of some green roofs are lower than traditional alternatives, such as bitumen and gravel (Porsche and Köhler, 2003), due in part to the protective effect of green roofs on the waterproofing membrane (Vila *et al.*, 2012). For older building stock there may be improved thermal performance, reducing the need for heating in winter and cooling in summer, with associated reduction in energy costs for the occupants (Fioretti *et al.*, 2010).

Green roofs may provide a range of broader environmental and amenity benefits that are not easily measured or valued (MacMullan and Reich, 2007). For example, large-scale green roof retrofit may offset the UHI effect, leading to thermal comfort; this is important when inner city temperatures can reach up to five degrees higher than suburban and rural areas (Williams *et al.*, 2010). Carbon sequestration is another potential environmental benefit of greening roofs. Getter and Rowe (2009) measured the sequestration on extensive roofs and estimated that 55,000 tonnes of carbon could be sequestered in the plants and substrates if the available commercial and industrial rooftops in Detroit (USA) were retrofit (15,000 hectares of rooftops).

In respect of social benefits, green roof retrofit can be positive for owners, the community and the environment. For example, in Malmö (Sweden) a project initially driven by flood risk management (Kaźmierczak and Carter, 2010), found the creation of green infrastructure improved the neighbourhood aesthetically and benefitted overall reputation. Greening in the CBD may generate a socially beneficial closer relationship of residents with the natural world, known as the 'biophilia phenomenon' (Kellert and Wilson, 1993), as illustrated by Wilkinson *et al.* (2013a), where proximity to nature was seen to enhance worker satisfaction and productivity. A further direct benefit is to use the rooftop for urban food production: as urban densities increase and food security issues increase, it creates potential for social engagement and interaction for city dwellers (Wilkinson *et al.*, 2013b). Table 18.5 summarises the potential benefits of green roof retrofit.

Table 18.5   Potential benefits of, and drivers for, green roof retrofit.

- Reputational enhancement for suburbs/areas /projects
- Aesthetic improvements
- Flood mitigation
- Reduced maintenance costs
- Carbon sequestration
- Reduction in urban heat island
- Improved air quality
- Improved biodiversity and nature conservation
- Reduced risk of pollution and stream degradation
- Reduced fees for owners where there is a charge for runoff into streams
- Increased energy efficiency of buildings and lower carbon GHG emissions
- Enhancement of familiar landmarks and buildings
- Closer relationship and access to nature for urban populations (biophilia effect)
- Enhanced user satisfaction and worker productivity for commercial stock
- Rainwater harvesting opportunity can reduce use of potable water
- Cost of finance for retrofit is often cheaper as building remains occupied
- Property values may be enhanced
- Possibility of growing food crops – particularly vegetables and fruit
- Reducing noise pollution
- Water recycling and nutrient stripping

Table 18.6   Potential barriers to green roof retrofit.

- Lack of awareness of the economic, social and environmental benefits
- Lack of experience of green roofs
- Fear of the unknown
- Existence of split incentive re cost and benefits
- Lack of urban and regional planning policy re green roofs
- Lack of incentives at a policy level
- Lack of data on cost benefits from previous projects

The barriers (Table 18.6) exist partly in an erroneous perception among practitioners that installing soil on a roof will lead to building defects and increased maintenance. Given limited experience, practitioners are wary of 'unproven' technologies and are unlikely to have been taught about green roofs (Wilkinson *et al.*, 2015). This perception may change, but the pace is slow, which is a challenge that will reduce the take-up of retrofit on suitable roofs. Planners may also have limited exposure to green roof technology, which may make them hesitant when dealing with applications featuring green roofs. Again, this may change over time, as green roof technology is taught in academic programmes and becomes more widely accepted. Wider benefits are not within the remit of planning committees.

There is also a perception among practitioners that costs are high and green roofs are thus value engineered out of projects (Wilkinson *et al.*, 2015). Better consideration of the full life-cycle costs and multiple benefits may help to alleviate these concerns, and the perception may shift. However, development of example roofs and detailed monitoring of them may be needed to generate the required evidence to support investment in green roofs. For wider benefits, and in tenanted buildings, a complicating factor is the 'split incentive' regarding the balance between who pays for installation and who benefits from the improvement (Abbott *et al.*, 2013). Within the stormwater and flooding aspects of SuDS, the installer reaps minimal direct benefit, except where drainage authorities charge for runoff entering piped systems. Flood mitigation may extend to properties downstream; the benefits for the environment are the reduced pollutant load entering watercourses, and amenity benefits may accrue to other local businesses, CBD residents and visitors to the CBD.

Australian government policy on water-sensitive cities incorporates green roofs and permeable paving (Department of Infrastructure and Transport, 2011), with initiatives at state level in Victoria. For Melbourne, green roof installation is part of the climate adaptation strategy (Department of Industry Innovation Climate Change, 2013). Melbourne City Council promotes green roofs for the purposes of urban greening as part of their 'Blueprint to Green Roof Melbourne'. Examples of green roofs in Melbourne include the Melbourne City Council Building CH2 and Freshwater Place residential tower (Rajagopalan and Fuller, 2010). However, the driver for green roof programmes in Melbourne has much to do with cooling and greenhouse gas mitigation, and less consideration has been given to stormwater. According to Rajagopalan and Fuller (2010), the motivations of installers were even more varied and divorced from the environmental benefits. Melbourne City Council and Melbourne Water have also been exploring SuDS, including permeable paving, to address water quality and peak flow issues (Wong, 2006; Abbott *et al.*, 2013). Within Melbourne, therefore, the uptake of suitable retrofit may be higher for porous paving in municipal spaces (roads and pavements) but lower for commercially installed green roofs.

18.5   Estimation of Runoff Under Different Scenarios

To explore runoff reduction, two approaches were adopted. To calculate total potential percentage permeable surface and applied average runoff performance, assumptions were that 15% of rooftops were suitable for retrofit, as determined from the database analysis. Also, 40% of roads and pavements were considered suitable for permeable retrofit. The runoff retention during peak storm events is assumed to be 60% for green roof with 100% for permeable paving assuming that they are under-drained. The results, as shown in Table 18.7, show that a runoff reduction of 22% may be achievable if high levels of potential retrofits are achieved.

UK research suggests 10% reduction in runoff to the sewer system has the potential to prevent 90% of flood incidents (Gordon-Walker *et al.*, 2007) and Table 18.7 indicates that retrofit could be a significant mitigation factor in Melbourne, if high levels of retrofit could be achieved. The findings imply that further detailed feasibility studies and modelling are warranted.

18.6   Conclusions and Further Research

Green roof retrofit may be an important part of climate adaptation strategies for CBDs generally and in Melbourne. Retrofit of permeable paving is also an important SuDS approach within CBDs. The potential for mitigation of pluvial flooding in CBDs through retrofitting of green roofs on commercial office buildings and retrofitting permeable paving in Melbourne, Australia, was explored. Some 15% of existing office buildings could potentially be retrofitted with extensive green roof technology and the potential reduction in rainwater runoff from green roof and permeable pavement retrofit in Melbourne was estimated to be 22%. This could provide a useful contribution to mitigation of future flood damage and could deliver multiple environmental, social and economic benefits. Further research into the optimal types and positioning of roofs and paving, and study of the wider catchment could confirm and strengthen the case for green roof and permeable paving retrofit. However, the study also revealed the importance of considering barriers and drivers to retrofit and the sensitivity of estimates to assumptions. It will be important to give

Table 18.7   Runoff calculations for Melbourne, showing estimated percentage of total rainfall falling on the CBD managed by potential retrofit under two scenarios.

	Assume all roof and road retrofit	Assume suitable roof and road retrofit	Assume 50% uptake for green roof
Total study area (1000 m²)	2150.0	2150.0	2150.0
Area of roof (1000 m²)	1150.0	172.5 (15%)	86.25 (7.5%)
Area of road (1000 m²)	500.0	200.0 (40%)	200.0 (40%)
Runoff reduction roof % of total rainfall[1]	32.1	4.8	2.4
Runoff reduction road % of total rainfall[2]	23.3	9.3	9.3
Runoff reduction pavement % of total rainfall[2]	7.9	7.9	7.9
Total runoff reduction %	63.3	22.0	19.6

[1] Assume 60% runoff reduction over the area of green roof.
[2] Assume 100% runoff reduction over the area of permeable paving but no drainage from adjacent areas.

further thought to appropriate strategies to increase uptake by removing the institutional, economic and social barriers, as well as continuing to develop technological approaches and improved guidance.

Acknowledgements

The authors are grateful for the financial support and assistance provided by the RICS through the research project entitled 'Retrofit of Sustainable Urban Drainage (SUDS) in CBD for improved flood mitigation'. This chapter draws on the main research findings of this project.

References

Abbott, J., Davies, P., Simkins, P., Morgan, C., Levin, D. and Robinson, P. (2013) *Creating water sensitive places – scoping the potential for water sensitive urban design in the UK C724*. London: CIRIA.

Arge, K. (2005) Adaptable office buildings: theory and practice. *Facilities* 23(3) 119–127.

Baird, G., Gray, J., Isaacs, N., Kernohan, D. and Mcindoe, G. (eds) (1996) *Building Evaluation Techniques*. Wellington, NZ: McGraw Hill.

Ball, R.M. (2002) Re-use potential and vacant industrial premises: revisiting the regeneration issue in Stoke-on-Trent. *Journal of Property Research*. 19, 93–110.

Blanc, J., Arthur, S. and Wright, G. (2012) Natural flood management (NFM) knowledge system. Part 1 – sustainable urban drainage systems (SUDS) and flood management in urban areas – Final report. Edinburgh: CREW – Scotland's Centre of Expertise for Waters.

Bureau of Environmental Services – City of Portland (2008) *Cost Benefit Evaluation of Ecoroofs 2008*. Portland, Oregon USA: City of Portland, Oregon.

Charlesworth, S. and Warwick, F. (2011) Adapting to and mitigating floods using sustainable urban drainage systems. in Lamond, J.E., Booth, C.A., Proverbs, D.G. and Hammond, F.N. (eds) Flood hazards, impacts and responses for the built environment. New York: Taylor CRC press.

Chetri, P., Hashemi, A., Basic, F., Manzoni, A. and Jayatilleke, G. (2012) Bushfire, heat wave and flooding: Case studies from Australia. Report from the International Panel of the WEATHER project funded by the European Commission's 7th framework programme. Melbourne: RMIT University.

City of Melbourne and VICSES Unit(s) St Kilda and Footscray (2012) City of Melbourne flood emergency plan: A sub-plan of the municipal emergency management plan. Melbourne: City of Melbourne and VICSES.

Companies and Markets (2011) Australian Flood Damage Reconstruction Likely to Cost Billions. *Companies and Markets*.

Czemiel-Berndtsson, J. (2010) Green roof performance towards management of runoff water quantity and quality: A review. *Ecological Engineering*, 36, 351–360.

Department of Infrastructure and Transport – Australia (2011) *Our cities, our future: A national urban policy for a productive, sustainable and liveable future*. Canberra: Department of Infrastructure and Transport, Australia.

Environmental Services – City of Portland (2011) *Portland's Ecoroof Program*. Available at: http://www.portlandoregon.gov/bes/article/261074.

Environmental Services – City of Portland (n.d.) *Downspout disconnection program*. [Accessed 21/06/13]. Available at: http://www.portlandoregon.gov/bes/54651.

Fianchini, M. (2007) Fitness for purpose. A performance evaluation methodology for the management of university buildings. *Facilities* 25(3/4) 137–146.

Fioretti, R., Palla, A., Lanza, L.G. and Principi, P. (2010) Green roof energy and water related performance in the Mediterranean climate. *Building and Environment*, 45(8), 1890–1904.

Gann, D.M. and Barlow J. (1996) Flexibility in building use: the technical feasibility of converting redundant offices into flats. *Construction Management & Economics* 14(1) 55–66.

Getter, K.L. and Rowe, D.B. (2009) Carbon sequestration potential of extensive green roofs. Greening Rooftops for Sustainable Communities Conference (Session 3.1: Unravelling the energy/water/carbon sequestration equation). Atlanta, GA, USA.

Getter, K.L., Rowe, D.B., Robertson, G.P., Cregg, B.M. and Andresen, J.A. (2009) Carbon sequestration potential of extensive green roofs. *Environ. Sci. Technol.*, 43, 7564–7570.

Gissing, A. (2003) Flood action plans – making loss reduction more effective in the commercial sector. *Australian Journal of Emergency Management*, 18, 46–54.

Google Earth 6.0 (2008) *Google Earth Australia*. Available at: http://www.google.com.au/maps.

Gordon-Walker, S., Harle, T. and Naismith, I. (2007) *Cost-benefit of SUDS retrofit in urban areas – Science Report – SC060024*, Nov 2007. Bristol: Environment Agency.

Heath, T. (2001) Adaptive reuse of offices for residential use. *Cities* 18(3) 173–184.

Hoang, L. and Fenner, R.A. (2014) System interactions of green roofs in blue green cities. 13th International Conference on Urban Drainage, 7–12 Sept 2014, Sarawak, Malaysia. ICUD.

Jha, A., Lamond, J., Bloch, R., Bhattacharya, N., Lopez, A. *et al.* (2011) *Five Feet High and Rising – Cities and Flooding in the 21st Century*. Washington: The World Bank.

Jia, H., Lu, Y., Yu, S.L. and Chen, Y. (2012) Planning of LID-BMPs for urban runoff control: The case of Beijing Olympic Village. *Separation and Purification Technology*, 84(0), 112–119.

Kaźmierczak, A. and Carter, J. (2010) *Adaptation to climate change using green and blue infrastructure – A database of case studies*. Manchester UK: University of Manchester: Green and Blue Space Adaptation for urban areas and eco-towns (GRaBS); Interreg IVC.

Kellert, S.R. and Wilson, E.O. (1993) *The Biophilia Hypothesis*. Washington DC, USA: Island Press.

Kersting, J.M. (2006) *Integrating past and present: The Story of a Building through Adaptive Reuse*. Master of Architecture Masters, University of Cincinnati.

Kincaid, D. (2002) *Adapting buildings for changing uses – guidelines for change of use refurbishment*. London: Spon Press.

MacMullan, E. and Reich, S. (2007) *The Economics of Low-Impact Development: A Literature Review*, November 2007. Eugene, OR: ECONorthwest.

Markus, T.A. (1979) *Building conversion and rehabilitation – designing for change in building use*. London: Butterworth Group.

Met Office Hadley Centre for Climate Research (2007) *Climate Research at the Met Office Hadley Centre – Informing Government Policy into the Future*. Available at: http://www.metoffice.gov.uk/research/hadleycentre/pubs/brochures/clim_res_had_fut_pol.pdf.

Porsche, U. and Köhler, M. (2003) Life cycle costs of green roofs – A comparison of Germany, USA, and Brazil. *RIO 3 – World Climate and Energy Event*. Rio de Janeiro, Brazil 1–5 Dec 2003.

Property Council of Australia Limited (2006) *A guide to office building quality*. Sydney: Property Council of Australia Limited.

Rajagopalan, P. and Fuller, R.J. (2010) Green Roofs in Melbourne – Potential and Practice. Australia: School of Architecture and Building, Deakin University.

Remoy, H.T. and van der Voordt, T.J.M. (2006) A new life: Transformation of vacant office buildings into housing. CIBW70 Changing user demands on buildings. Needs for life-cycle planning and management, 12–14 June 2006 Trondheim.

Snyder, G.H. (2005) Sustainability through adaptive reuse: the conversion of industrial buildings. College of Design, Architecture and Planning, University of Cincinnati. Master of Architecture.

Szarejko, W. and Trocka-Leszczynska E. (2007). Aspects of functionality in modernization of office buildings. *Facilities* 25(3) 163–170.

University of Florida (2008) Green roofs/Eco-roofs. Available from: http://buildgreen.ufl.edu/Fact_sheet_Green_Roofs_Eco_roofs.pdf.

Vila, A., Pérez, G., Solé, C., Fernández, A.I. and Cabeza, L.F. (2012) Use of rubber crumbs as drainage layer in experimental green roofs. *Building and Environment*, 48(0), 101–106.

Wilkinson, S., Lamond, J., Proverbs, D.G., Sharman, L., Heller, A. and Manion, J. (2015) Technical considerations in green roof retrofit for stormwater attenuation in the Central business district. *Structural Survey*, 33, 36–51.

Wilkinson, S., Rose, C., Glenis, V. and Lamond, J. (2014) Modelling green roof retrofit in the Melbourne Central Business District. *Flood Recovery Innovation and Response IV*. Poznan, Poland 18–20 June 2014.

Wilkinson, S.J., Van Der Kallen, P. and Leong Phui, K. (2013a) The relationship between occupation of green buildings, and pro-environmental behaviour and beliefs. The Journal for Sustainable Real Estate. ISSN 1949 8276. *Vol 5*, 1–22.

Wilkinson, S.J., Ghosh, S. and Page, L. (2013b) Options for green roof retrofit and urban food production in the Sydney CBD. RICS COBRA New Delhi, India. 10–12 Sept 2013. ISSN 978-1-78321-030-5.

Wilkinson, S.J. and Reed, R. (2009) Green roof retrofit potential in the central business district. *Property Management*, 27(5), 284–301.

Williams, N.S., Raynor, J.P. and Raynor, K.J. (2010) Green roofs for a wide brown land: Opportunities and barriers for rooftop greening in Australia. *Urban Forestry and Urban Greening*, 9, 245–251.

Wong, T.H. (2006) Water sensitive urban design – the journey thus far. Australian Journal of Water Resources, 10, 213–222.

Contemporary Landscapes and Buildings of Motorway Service Areas

Colin A. Booth and Anne-Marie McLaughlin

19.1 Introduction

Motorway service areas (MSAs) or highway rest areas provide an essential 24-hour, 365-day service for road traffic users, presenting them with somewhere to rest, relax and recharge. The range of provisions that MSAs offer, such as truck and car parking, toilets, food outlets, shops, picnic areas and refuelling stations, is similar in most countries (e.g. Evgenikos and Strogyloudis, 2006). However, public expectations and standards of MSAs vary between nations (Tunusa, 2015), with Austria, Germany, Switzerland and Spain reported to have the best rated MSAs in Europe (The AA Motoring Trust, 2004).

Most modern architecture and new infrastructure developments are designed and built to more sustainable standards than their predecessors (Beddoes and Booth, 2012; Khatib, 2012). Buildings have previously been built and used as a wasteful enterprise of a throwaway society. With greater awareness of environmental issues and the impacts caused by the built environment (Lamond *et al.*, 2011; Booth *et al.*, 2012), sustainable buildings, sustainable businesses and sustainable behaviours are becoming commonplace (Baird, 2010; Martin and Thompson, 2010; Crocker and Lehmann, 2013). These types of expectations and standards now extend to modern-day MSAs.

This chapter describes a recent shift in the archetype of contemporary MSAs in the UK, towards sustainability-driven businesses, operating in eco-designed buildings that give greater consideration to the natural environment and attention to sustainable drainage as a site precedence.

19.2 Motorway Service Areas in the UK

The first UK motorway (the 8-mile Preston bypass) was opened to traffic in 1958 (Cox, 2004), heralding a new form of efficient high-speed surface transport that would facilitate

Sustainable Surface Water Management: A Handbook for SuDS, First Edition.
Edited by Susanne M. Charlesworth and Colin A. Booth.
© 2017 John Wiley & Sons, Ltd. Published 2017 by John Wiley & Sons, Ltd.

a dramatic improvement in the mobility of people and goods (Bridle and Porter, 2003; Wootton, 2010). Today, there are 75 motorways (including M-designated motorways) forming a national network of 3559 km. These provide 18.2% of passenger traffic (Wootton, 2010).

Stopping on motorways is prohibited, except in emergencies, so it has always been necessary to provide opportunities for road traffic users to rest, obtain fuel and refreshments, and utilise toilet facilities, without leaving the motorway network (Williams and Laugharne, 1980; Charlesworth, 1984). The first MSA in the UK, Watford Gap (M1), opened its doors in 1959 to road traffic users, albeit without catering facilities until 1960. By 1963 a further six MSAs had opened across the expanding of motorway network, at Newport Pagnell (M1), Keele (M6), Charnock Richard (M6), Knutsford (M6), Farthing Corner (M2) and Strensham (M5).

Since the early days of being 'too few, too far apart' (Williams and Laugharne, 1980), the number of MSAs has substantially increased. To accommodate the requirements of today's motorway users, there are over three hundred MSAs now located across the UK (The Motorway Archive Trust). An original intention of the Department of Transport (DoT) was to have MSAs at 30 mile (48 km) intervals with potential infill sites at 15 mile (24 km) intervals (DoT, 2008). So where there are sections of motorway that the distance between MSAs exceeds 40 miles (64 km) these are considered priority areas for where new MSAs need to be built (Highways Agency, 2010). This policy has enabled several new MSAs to be built but it has also meant that there has been a growth in opposition to new MSAs, with some applications being refused due to local pressure groups (e.g. Catherine de Barnes (M42)) and others only granted permission after battles fought out in the court room (e.g. Gloucester (M5)).

For many years the UK was reported to have the worst MSAs in Europe, with Sandbach (southbound) on the M6 rated bottom of all those included in a survey, and the best UK MSAs, located at Cardiff (west) on the M4 and Oxford on the M40, only reaching an overall 'acceptable' score on the survey (The AA Motoring Trust, 2004). The categories used to inform the assessment included access/indoor facilities, catering, shop/kiosk, service, communication, hygiene, prices, road safety and parking, and outdoor facilities, with none of the UK MSAs involved in the survey scoring well for the latter two categories. This has prompted MSA operators to undertake substantial refurbishment and/or rebuild programmes, which now include the addition of take-away food outlets, and the provision of overnight hotel facilities, all located in improved aesthetic surroundings.

The landscaping of modern-day MSAs has seen considerable thought and sizeable investment. Compared to their earlier counterparts, the most recently built MSAs are notably different in their design and surroundings. For instance, most of the original MSAs were considered too small (10–15 acres) and struggled to meet user demands and were often perceived as being overcrowded and outdated places to rest; whereas, the latest MSAs are much larger (35–40 acres) and offer a greater array of products and services to their customers. This attentive concern has also extended to the architectural design of the MSA buildings and their landscapes.

19.3   Exemplar Motorway Service Areas

Two exemplars of modern-day MSAs are now described, together with some of the challenges faced: (1) Hopwood Park (M42) MSA, built 1999, includes sustainable drainage as a major architectural design component of its landscape, but findings indicate that this may now be facing maintenance issues; and (2) Gloucester MSAs, built 2014 and 2015, are considered the greenest MSAs in the UK because of their sustainability-driven farm-shop business model, their eco-designed buildings, and their sustainable drainage designed landscape, but they had to overcome a major obstacle before being built.

19.3.1 Hopwood Park (M42) Motorway Service Area

The Hopwood Park MSA is located in Worcestershire on the M42 (Junction 2) and is designed similarly to the RIBA award winning Wheatley MSA (www.jwa-architects.co.uk), which opened a year earlier (Oxford, M40, Junction 8a). The Hopwood Park MSA cost £25 million to complete and facilitated the first designated wildlife reserve, with an area of 25 ha, included in the design of an MSA in England. This was included as a condition by the Bromsgrove District Council during planning (Graham *et al.*, 2012). The MSA area is a total of 9 ha and comprises parking for HGVs, coaches and cars, a fuel filling area and an amenity building. This is one of the first examples of applying the sustainability paradigm to the construction of an MSA.

One of the most notable features of the Hopwood MSA are the four SuDS management trains surrounding the amenity building, designed by Robert Bray Associates and Baxter Glayster Consulting Ltd. Despite the limited guidance on SuDS design at the time (Woods Ballard *et al.*, 2007), the management trains were implemented to control surface runoff from different sections of the MSA. As one of the first management trains installed in the UK, these devices were established as an Environment Agency SuDS demonstration site. The original arrangement of the four management trains is subdivided to receive runoff from the (1) HGV park; (2) coach park, fuel filling area, service yard and main access road; (3) car park (4) the amenity building roof. The outflow of the SuDS and the MSA area are drained into the adjacent wildlife reserve and the Hopwood stream, a tributary of the river Arrow.

The HGV management train consists of a 10 m wide grass filter strip, a stone-filled trench, a spillage basin (Pond 1), a final attenuation pond (Pond 2), another grass filter strip and a swale to manage overflow in excess of 10 mm first flush. The annual average daily traffic at the HGV park is approximately 400 vehicles (Jefferies and Napier, 2008). The main access road, fuel filling area and coach park management train receive runoff from a conventional gully and pipe system that is initially treated by a silt and oil interceptor prior to the spillage basin (Pond 3) and wetland basin (Pond 4). The service yard runoff is treated by a wetland basin (Pond 5). Discharge from Ponds 3, 4 and 5 are further treated by a shallow ditch directing flow to a balancing pond (Pond 6) or grass swale, installed to control excess runoff. Ponds 3 and 4 are designed with an outlet valve to block any accidental spillages (Graham *et al.*, 2012). The size of the two management trains are larger as a result of the high pollutant loading associated with car park runoff (Revitt *et al.*, 2014). Slotted kerbs direct the car park runoff through gravel-filled collector trenches, which discharge into a balancing pond (Pond 7). The roof water from the amenity building is piped to a decorative balancing pond feature (Figure 19.1). Ponds 6, 7 and 8 are connected by a drainage basin before the discharge enters the Hopwood stream. The size of Ponds 7 and 8 are designed for low pollutant loading in comparison to the first two management trains (Heal *et al.*, 2009).

Research on the flow attenuation, maintenance and pollution control of the Hopwood MSA SuDS management trains was compiled by Heal *et al.* (2009). The management trains were designed to attenuate a 1 in 25 year storm event and provide a greenfield runoff rate of 5 l/s/ha. Between May 2002 and June 2004, Woods Ballard *et al.* (2005) found 70% of

Figure 19.1　Decorative balancing pond outside the main amenity building at Hopwood Park MSA (M42).

peak flows exceeded the greenfield runoff rate in the outlet of the silt and oil interceptor, but significantly reduced to 5% at the inlet of Pond 6. Although the peak flow downstream of the inlet of Pond 6, was not measured, prior to discharge into the Hopwood stream, it has been assumed that the criterion for greenfield runoff rates has been met by the end of the management train. The SuDS have been successful in the attenuation of surface runoff as no flooding occurred after the heavy rainfall event in summer 2007 (Heal *et al.*, 2009).

The water treatment efficiency of the management trains was monitored by the Environment Agency for 2000–2005, and found a reduction in contaminant concentrations in water samples from the first to the last SuDS components of the management train (Heal *et al.*, 2009). Percentage removal efficiency for Cu and Zn concentrations was 70–90%. In 2007, Jefferies *et al.* (2008) assessed the sediment accumulation and composition of the first management train. They found that sediment had been efficiently trapped in the top 10 cm of the grass filter strip. The reduction in contaminant concentrations with depth in the grass filter strip and distance from the car park is evident. The vertical reduction of contaminant concentrations in the filter strip suggests that the potential of groundwater pollution from the downward movement of pollutants is unlikely.

Sediment was excavated in October 2003, removing approximately 25% of the pond vegetation and the attached sediment (Heal *et al.*, 2009). Once the material was dewatered, it was recycled as compost on site. However, the removal of sediment in Pond 1 did not effectively reduce the concentrations of total petroleum hydrocarbons (TPH) to below sediment quality guidelines (Table 19.1). The high TPH and polycyclic aromatic hydrocarbon (PAH) concentrations in Pond 1 are the result of a diesel spillage of approximately 200 litres in 2000 (Heal *et al.*, 2009). Although Pond 1 successfully trapped the majority of contamination from the spillage event, this indicates issues associated with remediating unforeseen spillage events effectively and suitable maintenance techniques to sustain the initial functioning of the SuDS. Furthermore, the HGV park extension in 2007 has had an

Table 19.1 Contaminant concentrations in sediment (mg kg^{-1} dry weight) in the HGV park and coach park management trains.

Location	Cd	Cu	Pb	Zn	Ni	TPH	Total PAHs
HGV park management train							
Filter PF 1 m 0–10 cm[a]	0.4	71	66	351	31.4	398	5.16
Filter PF 1 m 10–20 cm[a]	0.3	51	69	146	48.7	153	1.72
Filter PF 3 m 0–10 cm[a]	0.3	50	52	199	30.3	1199	16.2
Filter PF 3 m 10–20 cm[a]	0.2	30	39	106	50.8	86	1.56
Filter 3 m 0–10 cm[a]	0.3	28	40	145	21.05	277	10.0
Filter 6 m 0–10 cm[a]	0.3	24	36	118	18.9	151	2.61
Filter 9 m 0–10 cm[a]	0.3	26	40	123	20.2	166	3.55
Pond 1[a]	0.7	192	92	733	40.1	3152	19.2
Pond 2[a]	0.6	89	67	393	49.6	629	4.27
Coach park, fuelling area, main access road management train							
Interceptor (2005)[b]	2.16	350	193	2500	—	10660	112
Interceptor (2006)[b]	1.15	224	101	1790	—	26030	64.7
Pond 3[c]	1.78	352	183	2580	—	—	108
Pond 4[c]	0.586	215	136	1290	—	—	
Pond 5[c]	1.03	161	120	1680	—	—	30.1
Pond 6[c]	0.115	23.9	32.1	75.5	—	—	4.29
Standards							
Ontario Ministry of Environment (1993)	10	110	250	820	75	1500	—

[a] Jefferies et al. (2008) sampled the grass filter strip soil sampled at 1, 3, 6 and 9 m from the pavement edge and at a depth of 0–10 cm. In an area of apparent preferential flow (PF) samples were taken at 0–10 cm and 10–20 cm depths.
[b] Sampled in 2005 and 2006 (Faram et al., 2007).
[c] Sampled in 2003 by Willingale (2004).

impact on the direction of flow into the system, potentially bypassing the filter strip and trench entirely, which may increase contaminant concentrations in Ponds 1 and 2 (Heal et al., 2009).

The oil and silt interceptor is at present unable to filter particles <2 mm (R. Bray, pers. comm.) and therefore the concentrations of contaminants could have increased since these samples were collected. This is a concern, as an increase in pollutant concentrations has been associated with finer sediment fractions (Horowitz 1991), and sediment highly contaminated in hydrocarbons (Table 19.1) might be released into the management train, exceeding the treatment capacity. Contributory factors to the difference in concentrations between the HGV management train and the coach park management train include the effective sediment trapping of the grass filter strip as a pre-treatment component and possibly a larger volume of vehicles using the surfaces that are drained in the second management train.

Heal *et al.* (2009) discussed the maintenance routine applied to the SuDS, including contractors who were assigned to remove litter, and cut the grass and wetland vegetation. Separate contractors maintained the oil and silt interceptor every 6 months. The sediment from Ponds 1–7 were removed in 2003 at a cost of £554 (2007 prices), after four years in operation (Heal *et al.*, 2009). The cost for removal was lower than expected, as the equipment hired for excavation was only required for half a day instead of a full day. Currently, the Hopwood Park MSA SuDS have not had the maintenance (e.g. grass cutting) that would have been implemented at the time of these studies. Maintenance of the site slowly declined as the annual budget has been reduced to £300 (Graham *et al.*, 2012), despite the cheaper annual cost to maintain SuDS (£2500) in comparison to conventional drainage structures (£4000). UK non-statutory standards express the necessity to incorporate a maintenance regime throughout the design life of the SuDS devices to reach optimal performance (Defra, 2015). The lack of maintenance may prove to be costly in the long term as the accumulation of sediment, and therefore contaminants, will contribute to increasing costs associated with disposal. As certain parameters exceed severe effect levels to aquatic biota, these values are indicative of how long-term maintenance is important to the biodiversity of SuDS. The structural integrity of the management trains and the efficiency of attenuating the flow at greenfield runoff rates may dwindle as the volume storage of the SuDS devices are reduced.

Both sediment and water quality results of the Hopwood MSA management trains show a horizontal reduction in contamination from the inlet to the outlet of each SuDS component. Although the retention of sediment and contaminants has been clearly demonstrated from the integrated studies, the concern is how the pollutants behave in the SuDS structure once they are trapped. Jefferies and Napier (2008) suggested the insufficient degradation rates of TPH and PAH in submerged pond sediments could potentially classify the sediment as hazardous waste if it was disposed of. This is a serious concern if the oil and silt interceptor is no longer functioning properly, as the excessive concentrations of TPH and PAH in the sediment could be entering the second management train. Furthermore, the continuous accumulation of non-degradable pollutants (e.g. heavy metals) may have a negative impact on the treatment characteristics or biodiversity of the management trains. The Ponds Conservation Trust (2003) completed a study on the ecological performance of the management trains and advised that pollutant loading should be kept to a minimum to enhance the ecology of the area. Additionally, the reduced maintenance budget may impact on the flow attenuation and removal efficiencies of pollutants. It is unknown whether the leachability or bioavailability of contaminants and the long-term efficiency of contaminant removal is affected by long-term pollutant loading in SuDS.

19.3.2 Gloucester (M5) Motorway Service Areas

A partnership between Gloucestershire Gateway Trust and Westmorland Ltd has facilitated the delivery of unique and visionary designed (BREEAM excellent) MSAs (Figure 19.2). Gloucester MSAs (M5 northbound and southbound, between Junctions 11a and 12) are unlike traditional services and appear almost seamless with their surroundings (Pegasus Planning Group, 2010). Founded on sustainable practices and philosophy, the Gloucester MSAs proffer local benefit through an innovative 'farm-shop' business model that donates and invests two per cent of sales to charity and local communities. Dedicated to local food, farming and the surrounding community, the MSAs source local food from over 130 local suppliers within 50-km (30 miles) of the services (Stroud News and Journal, 2015).

Figure 19.2   Views of the newly built Gloucestershire southbound MSA (M5).

The MSAs employ over 400 staff, with many from the Gloucestershire Gateway Trust's Academy that helps the long-term unemployed back into work. With an annual average daily traffic flow of 90,700 vehicles (Bean, 2012), the MSAs hope to serve 4.5 million customers per year (Gloucester Citizen, 2015).

Costing £40 million to build (by Buckingham Group Contracting Ltd.), they are set in a backdrop of undulating escarpments and vales with native indigenous planting. The buildings on each side of the motorway display similar designs, layout and floor space, with a main facilities building (~3300 m²), LGV drivers' building (~30 m²) and a petrol filling station (~230 m²). The main facilities buildings are timber framed with lattice roof structures (~9 m high) convened between solid dry-stone walls made of locally sourced buff-coloured Cotswold limestone. The buildings are draped in a blanket of soil and turf to create bespoke green roofs supporting native grass cover (Figure 19.2), which contributes to the water management and biodiversity of the sites. Water conservation inside and outside the buildings incorporates rainwater harvesting, low flow sanitary appliances, which include dual flush toilets and aerated flow restricted taps, plus leak detection and control (using smart metering), and low water landscaping that reduces the quantity and improves the quality of site runoff.

Drainage of the MSAs follows the SuDS management train and has been designed for storm events in exceedance of 100 years, plus 30% climate change events. Besides the source control measures already mentioned (e.g. green roofs and rainwater harvesting), a suite of devices has been incorporated into the landscape of the MSAs. The parking bays of the car parks (with kerb drains) and pedestrian walkways have permeable paving (Figure 19.3a and b), which allow surface waters to percolate between the blocks to infiltrate and be stored in the stone beneath. The LGV parking and access roads are drained by shedding the flow to filter strips, before entering stone-filled treatment trenches. A series of roadside swales and trenches (both wet and dry) provide an infiltration and conveyance network across the sites (Figures 19.3c and d). These are joined by underground piped (various diameters) inlets/outlets, fitted with silt traps and flow control chambers. Excess waters eventually enter a series of ponds (permanent volume of 154.4 m³, maximum attenuation volume 1004.7 m³) and wetlands (permanent volume of 96.5 m³, maximum attenuation volume 556.4 m³) at the end of the MSA sites (Figure 19.3e and f). As well as providing a visually attractive

(a)   (b)   (c)

(d)   (e)   (f)

Figure 19.3 Devices used for the SuDS management train at the newly built Gloucestershire southbound MSA (M5).

landscape, the sequence of site control devices improves the quality and runoff characteristics, which also contributes to the biodiversity and management of the MSAs.

Without doubt, Gloucester can now boast exemplar MSAs, both as a sustainably driven business and as sensitively designed MSA sites, but it is conceivable they might never have been built if it was not for a legal dispute (17–18/01/2012). In 2010, Stroud District Council (the local planning authority) received an application for planning permission on the same site that was refused in 1994 but decided to grant permission to this latest application. However, this decision was challenged by a collective of neighbouring MSA owners, local parish councils and an association of opponents to MSAs (with a 1089 signature petition against the scheme) (Bean, 2012).

Permission to seek judicial review was granted (09/05/2011) to the claimants on four grounds, which included the impact that the MSAs would have on the landscape. This is because planning policies aim to protect rural landscapes and particularly those sites designated as areas of outstanding natural beauty (AONB). In this case, the MSAs adjoin the Cotswolds AONB and would be visible from the surrounding terrain. It was therefore necessary to scrutinise how adverse the impact would be on the landscape, and an independent landscape assessor was appointed to undertake a review of the landscape assessment prepared as part of the Environmental Statement. This revealed that the proposal to build visionary designed buildings and site layout would, in fact, be sensitive to its setting and would cause only a slight adverse impact on the landscape designation. The court concluded that the regeneration and highway safety benefits of the scheme outweighed concerns over landscape impact (Bean, 2012). The application was considered to comply with planning policies, so the application for judicial review was dismissed in its entirety.

Despite the conceptual challenge in the court, the Gloucester MSAs now hold potential and promise, which others can aspire to replicate. In fact, on the day they were opened by Prince Charles (July 2015), a TV camera crew from the USA was filming the opening of the MSAs and revealed 'we have a lot of motorway services in America but nothing like this. It's all franchises... people will be interested to see what is happening here' (Gloucester Citizen, 2015). However, the MSAs' maintenance strategy, submitted as part of the planning application (Pegasus Planning Group, 2010), fails to mention or consider the needs for maintaining the SuDS devices, so their performance may also be a subject of interest and scrutiny by outside parties. However, depending on the quality of design and construction of the management train, a new concept suggests that it is possible to undertake minimal maintenance but retain performance (Bray, 2015). Let's watch this space to see what happens!

19.4 Conclusions

Modern-day MSA infrastructure has evolved from traditional basic designs to cover amenity requirements for visitors and to manage the demands of increasing numbers of motorway users. The implementation of sustainable practices from construction to operation has become integral to the business strategy of MSAs. The utilisation of the landscape through the installation of SuDS has been successfully demonstrated by the Hopwood Park MSA. The Gloucester MSAs are evidence of progression in MSA design, providing an exemplary sustainable building and landscape design, with an eco-business model. However, a concern for both sites is enforcing the long-term maintenance schedule of the SuDS as the performance of these systems may potentially deteriorate without attention to their maintenance.

References

Baird, G. (2010) *Sustainable Buildings in Practice: What the Users Think*. Routledge Publishers. p. 352.

Bean, J. (2012) Neutral Citation Number: [2012] EWHC 140 (Admin); Case No: CO/1654/2011 http://www.stroud.gov.uk/info/welcome_break_final.pdf

Beddoes, D.W. and Booth, C.A. (2012) Insights and perceptions of sustainable design and construction. In: Booth, C.A., Hammond, F., Lamond, J. and Proverbs, D.G. (eds) *Solutions to Climate Change Challenges in the Built Environment*. Wiley-Blackwell, Oxford, 127–139.

Booth, C.A., Hammond, F.N., Lamond, J.E. and Proverbs, D.G. (2012) *Solutions to Climate Change Challenges in the Built Environment*. Wiley-Blackwell, Oxford.

Bray, R. (2015) Designing SuDS for nominal maintenance: integrating SuDS design with creative management. SUDsnet International Conference, Coventry University, September 3–4 2015. http://tinyurl.com/hcfcezd

Bridle, R.J. and Porter, J. (2003) Engineering the UK motorway system 1950–2000. *Proceedings of the Institution of Civil Engineers: Civil Engineering*, 163, 137–143.

Charlesworth, G. (1984) *A History of British Motorways*. Thomas Telford Ltd., London, pp. 236–265.

Cox, L.J. (2004) Britain's first motorway: the Preston Bypass. In: Baldwin, P. and Baldwin, R. (eds) *The Motorway Achievement (Volume 1) – Visualisation of the British Motorway System: Policy and Administration*, Thomas Telford Ltd., London, 497–520.

Crocker, R. and Lehmann, S. (2013) *Motivating Change: Sustainable Design and Behaviour in the Built Environment*. Routledge Publishers. p. 472.

Defra (2015) *Sustainable Drainage Systems: Non-statutory technical standards for sustainable drainage systems*. PB14308. Crown Copyright, London.

Department of Transport (2008) *Policy on Service Areas and Other Roadside Facilities on Motorways and All-purpose Trunk Roads in England*. The Stationery Office Ltd., Norwich. p.47.

Faram, M.G., Iwugo, K.O. and Andoh, R.Y.G. (2007) Characteristics of urban run-off derived sediments captured by proprietary flow-through stormwater interceptors. *Water Science and Technology*, 56, 21–27.

Graham, A., Day, J., Bray, R. and Mackenzie, S. (2012) *Sustainable Drainage Systems – Maximising the Potential for People and Wildlife: A Guide for Local Authorities and Developers*. Royal Society for the Protection of Birds and Wildfowl & Wetlands Trust.

Gloucester Citizen (2015) *Prince calls into Gloucester Services*. http://tinyurl.com/pg2sqp7

Heal, K.V., Bray, R., Willingale, A.J., Briers, M., Napier, F., Jefferies, C. and Fogg, P. (2009) Medium-term performance and maintenance of SuDS: a case-study of Hopwood Motorway Service Area, UK. *Water Science and Technology*, 59, 2485–2494.

Highways Agency (2010) *Highway Agency: Spatial Planning Framework Review of Strategic Road Network Service Areas – National Report*. Highways Agency Publication Code PR272/09.

Horowitz, A.J. (1991) *A Primer on Sediment-Trace Element Chemistry*, Lewis Publishers, Chelsea.

Jefferies, C. and Napier, F. (2008) *SuDS Pollution Degradation*. Final Report, Project UEUW02, SNIFFER, Edinburgh.

Jefferies, C., Napier, F., Fogg, P. and Nicholson, F. (2008) *Source Control Pollution in Sustainable Drainage*. Final Report, Project UEUW01, SNIFFER, Edinburgh.

Khatib, J.M. (2012) Progress in eco and resilient construction materials development. In: Booth, C.A., Hammond, F., Lamond, J. and Proverbs, D.G. (eds) *Solutions to Climate Change Challenges in the Built Environment*. Wiley-Blackwell, Oxford, 141–151.

Lamond, J.E., Booth, C.A., Hammond, F.N. and Proverbs, D.G. (2011) *Flood Hazards: Impacts and Responses for the Built Environment*. CRC Press – Taylor and Francis Group, London.

Martin, F. and Thompson, M. (2010) *Social Enterprise: Developing Sustainable Businesses*. Palgrave Macmillan Publishers. p. 264.

Ontario Ministry of Environment (1993) *Guidelines for the Protection and Management of Aquatic Sediment Quality in Ontario*. Queen's Printer for Ontario.

Pegasus Planning Group (2010) *Gloucester Gateway Motorway Service Area: Design and Access Statement*. http://tinyurl.com/hvyb6u5

Ponds Conservation Trust (2003) *Maximising the Ecological Benefits of SUDS Schemes*. Report SR 625, Ponds Conservation Trust: Policy and Research, Oxford Brookes University, Oxford.

Revitt, M., Lundy, L., Coulon, F. and Fairley, M. (2014) The sources, impact and management of car park runoff pollution: A review. *Journal of Environmental Management*, 146, 552–567.

Stroud News and Journal (2015) *Gloucester Southbound Services officially opened by HRH The Prince of Wales*. http://tinyurl.com/jjscoq5

The AA Motoring Trust (2004) *Eurotest 2004 – Motorway Service Area Tests: Results of the 2004 Pan-European motorway service area testing programme*. https://www.theaa.com/public_affairs/reports/MSA_Report_2004.pdf

Tsamboulas, D., Evgenikos, P. and Strogyloudis, A. (2006) The financial viability of motorway rest areas. Public Works Management and Policy, 11, 63–77.

Williams, T.E.H. and Laugharne, A. (1980) Motorway usage and operations. In: *Twenty years of British Motorways*. Institution of Civil Engineers, London. p. 71–88.

Willingale, S.A. (2004) *A Study into the Bioremediation of Silt from a Sustainable Drainage System*. Unpublished B.Sc. Thesis, Department of Geography, Swansea University, Wales.

Woods Ballard, B., Dimova, G., Weisgerber, A., Kellagher, R., Abbot, C., Manerio Franco, E., Smith, H. and Stovin, V. (2005) *Benefits and Performance of Sustainable Drainage Systems*. Report SR 677, HR Wallingford.

Woods Ballard, B., Kellagher, R., Martin, P., Jefferies, C., Bray, R. and Shaffer, P. (2007) *The SuDS Manual*. Report C697, CIRIA, London.

Wootton, J. (2010) The history of British motorways and lessons for the future. *Proceedings of the Institution of Civil Engineers: Transport*, 156, 121–130.

Yunusa, M.B. (2015) Physical planning and the development of Dankande Rest Stop Area in Kaduna, Nigeria. *City, Culture and Society*, 6, 53–61.

Modelling for Design

Craig Lashford, Susanne M. Charlesworth and Frank Warwick

20.1 Introduction

Computational modelling is the desk-based analysis of the characteristics of a site (Ellis *et al.*, 2012), allowing the user to model a variety of different scenarios before a site is developed. Rainfall-runoff modelling is an example of simulations that are run to determine areas that are likely to flood as a result of a given storm event (Tramblay *et al.*, 2011). Programs such as SWMM and MicroDrainage® have created SuDS add-ons that can model the role of different devices in reducing runoff. Different models and methods have been developed depending on the amount of data to incorporate in the model and the required output resolution; large-scale regional modelling, strategic catchment scale and local scale. There are therefore three common methods; one-dimensional, two-dimensional and three-dimensional and these are covered in turn in the following sections.

20.2 One-Dimensional Modelling

One-dimensional modelling is reasonably simple, and analyses the environment across one plane (Mahdizadeh *et al.*, 2012). It is typically used as a first pass attempt, requiring limited computational power due to the simplicity of the parameters modelled (FWR and WAPUG, 2002). It provides users with the flood width across a channel, but does not provide depths (Henonin *et al.*, 2013), enabling only an initial outline of the scope of flooding. Mark *et al.* (2004) incorporated a number of factors into their model, including pipes and roads, with the extent of flooding modelled. The research questions the full extent of flooding likely since the depth of the event is not accounted for in the model. Henonin *et al.* (2013) state that one-dimensional modelling is not suitable for measuring overflow, owing to the simplicity of the model, but it can give an indication of potential ponding sites. A further dimension is required to provide a more comprehensive model of the floodplain (Bates and De Roo, 2000).

Sustainable Surface Water Management: A Handbook for SuDS, First Edition.
Edited by Susanne M. Charlesworth and Colin A. Booth.
© 2017 John Wiley & Sons, Ltd. Published 2017 by John Wiley & Sons, Ltd.

20.3   Two-Dimensional Flood Modelling

Two-dimensional modelling acts as the benchmark for fluvial flood simulation (Henonin *et al.*, 2013). Models are typically run using elevation data to compute runoff extent and depth at a site after a storm (Bates and De Roo, 2000). The method is unable to model underground drainage, which is often estimated, causing uncertainty with the method, which is therefore unable to model pluvial flooding (Henonin *et al.*, 2013). Qi and Altinakar (2011) calculated the impact of a flood event in Georgia, USA, offering stakeholders information on the likely damage, and therefore showing what flood-proofing was required. However, to provide a more detailed simulation, incorporating overland flow and pipe channel flow, a combination one-dimensional and two-dimensional method should be used (Mahdizadeh *et al.*, 2012).

20.4   One-Dimensional and Two-Dimensional Modelling

Incorporating both 1D and 2D modelling enables a more detailed analysis of overland flooding in terms of both extent and depth, alongside a simple 1D pipe channel model (Pathirana *et al.*, 2011). The 1D-2D modelling is frequently used for both pluvial and fluvial flood simulation, but is reliant on the factors that are input into the model (Henonin *et al.*, 2013). Ellis *et al.* (2011) suggest that the method identifies critical areas of a site at risk of flooding, enabling stakeholders to evaluate mitigation methods. The major limitation with the method is that a coupled 1D and 2D model assumes that runoff is a result of surcharging of the sewer system (Zhou *et al.*, 2012); as a result, three-dimensional modelling software was developed to provide more accurate data.

20.5   Three-Dimensional Modelling

A three-dimensional model involves more parameters than the previous methods, including geomorphology and site conditions, with more detail of pluvial flooding not associated with sewer surcharge (Poole *et al.*, 2002; Chen and Liu, 2014). Limited research has been undertaken using this method due to the large computational power required in order to use the software, for example MicroDrainage® (MicroDrainage, n.d). However, it provides a detailed simulation of rainfall-runoff, the likely areas of inundation, including depth and extent, and methods of mitigation (Merwade *et al.*, 2008; Lee *et al.*, 2011).

20.6   Modelling Uncertainty

As a result of the number of possible parameters that can impact runoff, including climatic conditions, soil type and state, infiltration rate, underlying lithology, topography of the area and the characteristics of the channel, there is an associated level of uncertainty (Refsgaard *et al.*, 2007). This can be reduced by field-based validation, which involves comparing the model outputs with real-life scenarios to determine the overall accuracy of the model. This adds further confidence in the results and ultimately, the outputs of the model (Nativi *et al.*, 2013).

20.7 Validation of Models: Monitoring of SuDS Management Trains

Much of the field research related to SuDS focuses on the benefits generated from individual SuDS devices with minimal monitoring of the impacts of a combined SuDS management train (González-Angullo *et al.*, 2008; Stovin, 2009; Freni *et al.*, 2009). In the UK, Heal *et al.* (2009) analysed the long-term impacts of installing a SuDS management train, consisting of multiple ponds, filter strips, swales and wetlands at Hopwood Park motorway service area. Results of monitoring (covering 2000–2008) determined that the system was highly effective at improving runoff water quality. SNIFFER (2004) reported on the monitoring of both individual devices and management trains at the Dunfermline Eastern Expansion for water quality improvement and also provided anecdotal evidence of the reduction of flood events as a result of installing SuDS.

The SuDS management train at Lamb Drove, Cambridge, UK, has been monitored since its construction in 2006, focusing on all aspects of the SuDS triangle. The management train consists of a green roof, detention pond, filter strip, swale, water butts, permeable paving and a retention pond. Overall, it was calculated that the SuDS management train reduced runoff in comparison to an impermeable pipe-based control site, and that the retention pond reduced runoff most effectively, bringing it down to 3 l/s/ha. However, the impacts of other devices at the site were not quantified (Cambridgeshire County Council, 2012).

20.8 Scale of Drainage Modelling

Modelling SuDS can replicate reduction in water quantity and improvements in water quality (Bastien *et al.*, 2010). It is an effective way of understanding likely impacts prior to development and can be used to inform policy and best practice procedures (Elliott and Trowsdale, 2007). Modelling is one of the most suitable ways of understanding the abilities of a SuDS management train, as it allows the characteristics of the site to be examined, combined with the necessary SuDS devices, to understand the reduction in runoff (Viavattene *et al.*, 2010). To obtain the full benefits, site characteristics need to be added to the model to acquire more detailed data. There are a series of levels that modelling can occur at: regional, strategic and local (Figure 20.1).

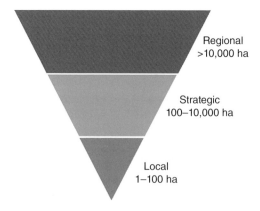

Figure 20.1   Series of levels that modelling can occur at: regional, strategic and local.

20.8.1 Regional Level Modelling

Regional analyses are at the large catchment scale, and as a result are usually associated with a reduction in resolution due to the amount of data required and the coarseness of scale. Consequently, much of the modelling is related to overall flood risk, or the modelling of a single characteristic related to the hydrological cycle, as opposed to the overall impacts of SuDS (Wheater 2002). For example, Bell *et al.* (2012) modelled the impact that climate change would have on flooding in the Thames Basin. In their study, the resolution for climatic data was 25 km, while the river flow was 1 km. By reducing the scale of modelling, increased resolution can be achieved.

20.8.2 Modelling at the Strategic Level

At the strategic or catchment level, better resolution is achieved; at this scale, Warwick (2013) was able to create a SuDS feasibility map to enable decisions to be made concerning choice of SuDS device to be used in Coventry, UK, and accounted for various site characteristics, such as topography and geology. Mitchell (2005) was able to identify sites for SuDS retrofit that could improve water quality in the Aire Basin, Yorkshire, UK, with the following three outcomes:

1. production of a map locating diffuse pollution hotspots
2. identification of areas most at risk of pollution
3. Assessment of the impact of land-use change on runoff quality

Even at the strategic level, however, the scale is still too coarse to yield sufficient information to assess any likely impacts (Moore *et al.*, 2012), and significant amounts of data are still required.

20.8.3 Local Level Modelling

The local scale can provide two separate but related levels: site and building. Site modelling involves much smaller areas than the strategic level, but can use information generated from strategic level modelling in order to design a drainage system and predict the impacts. This scale requires more detail in comparison to the previous two levels and requires the information at a high enough quality (Bastien *et al.*, 2010). Site level simulations have been run both for specific devices and for the wider combined management train. For example, Petrucci *et al.* (2012) modelled the potential impact on runoff of installing RWH butts to a neighbourhood in Paris, using the Storm Water Management Model (SWMM). They calculated that by installing RWH on 157 out of the 450 houses in the area it was possible to limit runoff for all events up to the 1 in 5 year return period.

Both Bastien *et al.* (2010) and Hubert *et al.* (2013) addressed the whole of the SuDS triangle related to management trains, but this breadth leads to a lack of detail. Furthermore, Bastien *et al.* (2010) did not include site characteristics, such as topography, which are required to model runoff routes and potential ponding sites. However, Viavattene *et al.* (2010) modelled the local water quantity impacts of individual SuDS devices, such as PPS and green roofs, applied to a site in Birmingham, UK. The study was part of the SuDS selection and location tool (SUDSLOC), and led to the conclusion that the installation of PPS and green roofs had the potential to reduce runoff by 28% and 26%, respectively.

At the building level, Ellis *et al.* (2011) were able to utilise high-resolution ground-based light detection and ranging (LiDAR) data to map small-scale topographic changes. This information is useful to enable modelling of how certain structures impact flood flow routes and it has the potential to be applied to the estimation of the benefits to water quantity reduction of standalone devices. Other single devices include Freni *et al.* (2009) who modelled infiltration trenches, concluding that they are more efficient at improving water quality rather than reducing water quantity. Using the Model for Urban Stormwater Improvement Conceptualisation (MUSIC), Khastagir and Jayasuriya (2010) found that even a single RwH device was able to improve overall water quality.

20.9   Issues with SuDS Modelling

Modelling can supply information about the impacts of development on local drainage, or the installation of a single device, thereby allowing for the optimisation of space (Moore *et al.*, 2012). However, there are some limitations associated with modelling, not least of which is to what extent the model's output bears any relationship to the real world (Wheater, 2002; Merwade *et al.*, 2008). A model is only as good as the data used to construct it and if there are inaccuracies, the output will be inaccurate. There are also specific uncertainties surrounding modelling SuDS. The type and density of vegetation may vary over the chosen area, which is typically complex to model and can produce variable results (Elliott and Trowsdale, 2007; Burszta-Adamiak and Mrowiec, 2013). Additionally, the results from a modelled system are of a 'perfect' scenario, whereas models are based on consistently maintained and non-clogged drainage plans, which are often not available and this therefore reduces their impact (González-Angullo *et al.*, 2008; Bergman *et al.*, 2011).

The selection of software to be used is critical; some packages are more effective than others and more appropriate to specific modelled scenarios. The following sections consider some computer models that have been used to model drainage, have the potential to model SuDS or have been used for that purpose.

20.9.1 Choice of Drainage Modelling Software

Typically, modelling software is used to simulate the impacts of imposing a predetermined event on a catchment at the scales (Ellis *et al.*, 2012). Arguably, there are five commercially available models that are widely used to model drainage and storm events: SWMM, MUSIC, MOUSE, Infoworks and MicroDrainage®.

20.9.2 Stormwater Management Model

The Stormwater Management Model (SWMM) is a rainfall-runoff model designed by the United States Environmental Protection Agency (EPA), which estimates potential water quantity and quality improvements (Rossman, 2010). It has become a widely used freeware model that can simulate both single and continuous rainfall scenarios (Burszta-Adamiak and Mrowiec, 2013) and has a limited range of SuDS devices that can be incorporated, including green roofs, PPS, swales, infiltration trenches, bioretention zones and rain barrels (Liao *et al.*, 2013). Research by Lee *et al.* (2012) used infiltration trenches and rain barrels in the SWMM to estimate that a potential reduction of runoff of 7–15% could be achieved for the 50-year return period in Korea. Upon validation of the model, they found error margins of up to 13.3%. The accuracy of the software for measuring the reduction in water quantity has been questioned by researchers such as Burszta-Adamiak and Mrowiec (2013), who concluded that the software underestimated the outflow from green roofs for over half of their experiments. This conclusion coincides with that from previous studies, carried out by Elliott and Trowsdale (2007), who found that water which would be able to percolate into the ground under swales and infiltration trenches is not accounted for in SWMM, and is therefore not added to groundwater flow.

20.9.3 Model for Urban Stormwater Improvement Conceptualisation

The Model for Urban Stormwater Improvement Conceptualisation (MUSIC), developed by the Cooperative Research Centre for Catchment Hydrology in Australia, is an urban stormwater modelling tool (Wong *et al.*, 2002; Ellis *et al.*, 2012). It is widely used because it has a variety of SuDS integrated into the package and has the ability to combine whole life costing (WLC), with the hydraulic modelling tools, as well as an assessment of water quality (Elliott and Trowsdale, 2007). It was constructed so that devices with known impacts could be input and has been used in research projects to assess the effects of SuDS on stormwater flows (e.g. Khastagir and Jayasuriya, 2010; Beck and Birch, 2013). Bastien *et al.* (2011) used MUSIC, for example, to determine the impacts of installing a SuDS management train at the local scale at the Clyde Gateway, Glasgow, Scotland, on the full breadth of the SuDS triangle. However, while considerable improvements in water quality were predicted, and the modelling was able to size the management train to deal with an up to 1 in 30 year storm, nonetheless WLC, in terms of land-take and maintenance costs, appeared to undermine the will of the local authority to implement a SuDS train at the site.

MUSIC can be used to design at the catchment scale; for example, Khastagir and Jayasuriya (2010) used the model to estimate the impacts of installing RwH, finding that hydraulic loading reduced by 13–75%, but there was also a 72–80% reduction in total nitrogen, and total suspended solids export reduced by >90%. While MUSIC has therefore been used to some effect, nonetheless there have been doubts as to its accuracy, with Dotto *et al.* (2011) casting doubt over MUSIC's rainfall/runoff modules' ability to accurately predict stormwater flows in a highly urbanised catchment. This is echoed by Imteaz *et al.* (2013), who completed a series of tests to validate the software, concluding that MUSIC over-estimated several of the results.

20.9.4 Model for Urban Sewers

The Model for Urban Sewers (MOUSE) was developed by the Danish Institute of Hydrology (2002), and represents urban runoff well, but it is not particularly user-friendly (Viavattene *et al.*, 2008), resulting in it not being commonly used in the UK (Defra and Environment Agency, 2005). In terms of integration of SuDS, MOUSE is limited to PPS, bioretention, rain tanks, swales and infiltration trenches. However, like SWMM, it is does not incorporate groundwater flows (Elliott and Trowsdale, 2007). For this reason, the use of the package in research is restricted to modelling surface flow from impermeable surfaces, ignoring the need to measure groundwater characteristics (Semadeni-Davies *et al.*, 2008). A review by Elliott and Trowsdale (2007) concluded that the model was more successful than others at simulating the improvement of water quality, but less effective with regards to water quantity.

20.9.5 Infoworks

Infoworks is a hydrodynamic package that focuses on a series of hydraulic structures, primarily to model flow and runoff routes (Salarpour *et al.*, 2011; Moore *et al.*, 2012). However, it is also capable of modelling reduction in runoff that is possible through implementing SuDS (Bastien *et al.*, 2010). Regarded as the industry standard tool for modelling sewer flows, it is generally used to model existing structures (Atkins, 2008), although Moore *et al.* (2012) used it to investigate the impact of installing retrofit SuDS, and the effect of disconnecting of conventional drainage on runoff.

20.9.6 MicroDrainage®

MicroDrainage® by XP Solutions is a commercially available urban stormwater drainage design model (MicroDrainage, 2009) and is the UK flood and drainage industry standard system (Mott Macdonald Ltd. and Medway Council, 2009; RPS Group, 2012; Hubert *et al.*, 2013). The software enables interaction in the design procedure, since data is input by drawings as opposed to numerically in a spreadsheet; this provides visual animations and easier transfer of data between GIS packages (Afshar, 2007; MicroDrainage, n.d). It is generally used to develop new designs, but it has the capability of incorporating SuDS retrofit (Moore, 2006; Atkins, 2008) and to produce outflow hydrographs based on a predetermined rainfall event, accounting for topographical features, and the presence of housing (Bassett *et al.*, 2007). MicroDrainage® is often used by stakeholders and consultancies (Moore, 2006) when creating flood risk assessments (e.g. Mott Macdonald Ltd and Medway Council, 2009). Hubert *et al.* (2013) used MicroDrainage® to compare the overall site benefits of installing a SuDS management train to an office site, in comparison to conventional pipe-based drainage.

20.10 Case Study: Modelling the Impacts of a SuDS Management Train at Prior Deram Park, Coventry, UK, Using Microdrainage®

This case study is based in the Canley Regeneration Zone (CRZ), 6 km south-west of Coventry city centre, and covers 5 ha in total, part of which was brownfield (Charlesworth *et al.*, 2013; Lashford *et al.*, 2014). Outline planning permission had been given for a total of 250 dwellings, at a density of 50 houses/ha, with new community services and improvements to the open space.

Using ArcGIS, a plan for the housing was designed, using an access road layout that was provided by Coventry City Council. Two drainage scenarios were then developed: a pipe-based conventional system and a SuDS management train, with all runoff routed into the local Canley Brook. The conventional system was modelled to address the usual 1 in 30 year storm, with the SuDS system designed for the 1 in 100 year event, as is required by

Table 20.1   Devices used in the SuDS management train at Prior Deram Park, Coventry, UK, their purpose and modelled volumes.

Device	Location	Purpose	Modelled Volume (m³)
Swale	Alongside pavements	Conveyance	729
Porous paving	Every house driveway	Source control	761
Green roofs	Every house		2014
Detention ponds	Five located around the site	Detention	6890

Defra (2015). Table 20.1 lists the devices used and their purpose in the train, as well as the modelled volume and location.

A throttle, such as a weir plate, was added to the outlet of the ponds in order to slow water flow from the pond and ensure compliance with Defra (2015) to constrain the run-off to greenfield runoff. The results of the modelling showed the pipe-based system would have resulted in the equivalent of 40 of the 200 houses being flooded, amounting to 858 m³ of excess water, but the SuDS management train would be able to successfully deal with the 1 in 100 year (±30 years to account for climate change) storm easily, and would in fact have coped with a 1 in 275 year storm (Lashford *et al.*, 2014).

20.10.1 Decision Support Tools

Decision support tools aid practitioners early in the decision-making process in their choice of suitable SuDS devices; they do not make the final decision, but are a first pass, taking account of some of the characteristics of the area to be drained, such as the degree to which water can infiltrate, if the area is brownfield, or what the underlying lithology is (Stovin and Swan, 2007; Scholz and Uzomah, 2013; Newton *et al.*, 2014). For example, Todini (1999) improved flood mapping by simulating flood flows and was then able to present potential management options across Europe. The support system required high computing power, due to the complexity of the area being modelled, but nonetheless it was able to provide reasonably accurate results. Other systems, such as the support tool created by Shim *et al.* (2002), have also attempted to enhance the flood management selection process, with a focus on river basin catchment systems in South Korea. However, the models used by both Shim *et al.* (2002) and Todini (1999) were complicated, particularly when some SuDS devices were included.

Owing to the complexity of the different SuDS devices, a decision support tool factoring in all their requirements and roles does not exist. However, there have been limited attempts at producing a system that supports one facet of the triangle. Stovin and Swan (2007) were able to quantify hydraulically efficient solutions for SuDS retrofit with a system whose primary aim was to ensure cost-effectiveness and to provide stakeholders with a quick understanding of eventual costs and cost savings, in an attempt to further incentivise the implementation of SuDS. However, although the system was successful, aspects such as the potential for high density housing were not accounted for, thus making it difficult sometimes to estimate the number of devices required.

There are also decision support tools for determining the benefits of specific devices; for example, Scholz and Uzomah (2013)'s rapid assessment system was able to quantify any improvements to the local ecosystem by installing trees in close proximity to PPS. The overall aim of the tool was to increase the implementation of PPS and enhance the ecology of the urban landscape.

Kahinda *et al.* (2009) developed a tool for assessing RwH, called RHADESS (Rainwater Harvesting Decision Support System), to indicate site suitability for RwH in South Africa, using a combination of ArcView 3.3 and Microsoft Excel. The project was driven by the inclusion of the concept of water security in the Millennium Development Goals, and hence promotion of the wider application of RwH. A successful SuDS decision support tool can ensure a more resilient site, whether the design is to tackle flooding or pollution, or to provide more amenity potential, assisting in the design of a site.

20.11 Case Study: Decision Support Tool for Coventry, UK

A decision support tool for Coventry, developed by Warwick (2013), determined the spatial distribution of each factor driving SuDS device choice across Coventry in the West Midlands, UK. The information required was accessed using data from a number of sources, including the British Geological Survey, Coventry City Council, Ordnance Survey, National Soil Resources Institute and the Environment Agency. Rules were created for each of the factors; for example, different rock types were assessed in relation to their capacity for infiltration or detention of runoff. Figure 20.2 shows a flowchart of this process using rules based on the ability of the geology to infiltrate of detain excess stormwater. The sites deemed to be suitable present the best options for above-ground vegetated devices, whereas those that are less suitable would need extra effort and/or expense to install suitable SuDS.

These rules were agreed in collaboration with local government, environmental regulators and the responsible water utility, all of whom had local knowledge, and the were rules coded so they could be applied spatially. These spatial relationships were then analysed in a GIS, to determine appropriate locations for the SuDS devices across the city. An example of one of the set of maps it produces as a result of this exercise is given in Figure 20.3, which shows that detention is possible over most of the city apart from where there are already lakes or streams. The 'engineered' detention would be located on brownfield sites where it would be necessary to avoid infiltration, and therefore tanked porous paving could be used, lined ponds or hard infrastructure bioretention, and also the Nottingham rain gardens as shown in Figure 20.4. As shown in Figure 20.3, in Coventry, approximately

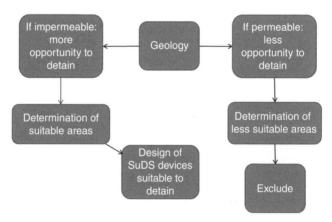

Figure 20.2 The map creation process in the GIS system using geological data for the detention SuDS map as an example. More and less suitable areas were determined independently, and their geographical overlaps removed.

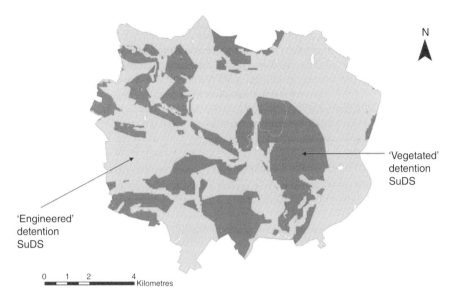

Figure 20.3 The output map for Coventry, UK, with potential locations for both vegetated (darker colours) and engineered (lighter colours) detention SuDS. Unshaded locations (in white) are regarded as unsuitable for any form of detention or retention SuDS.

Figure 20.4 Daybrook rain garden, Nottingham, UK.

one-third of the city's area is suitable for vegetated SuDS, while some form of engineered solution is likely to be necessary in two-thirds of the area.

By using a GIS approach, the maps were scalable; they could be viewed at different resolutions from full city scale to that of individual development and regeneration sites. They were intended to support local government officers to assess the use of SuDS in new developments and areas of regeneration during early discussions with developers. However, it is acknowledged that in order to submit a more detailed planning application, more technical tests and modelling would be required. However, the decision support tool is able to provide readily understandable information, which can support initial discussions between planning officers and developers, using a form that local governments are already familiar with. Consequently, it might be able to reduce some of the barriers currently limiting SuDS uptake.

20.12 Site Design

Designing a site properly that effectively integrates SuDS to achieve the requirements of the SuDS triangle is critical (Charlesworth, 2010). Ensuring that they are designed successfully reduces the likelihood for large future maintenance costs (Jefferies *et al.*, 2009), and ensures that they do not deteriorate too quickly (Wilson *et al.*, 2004) and that they meet site requirements (Woods Ballard *et al.*, 2007). Factors that need to be considered to ensure that the site is designed effectively are the optimal rainfall scenarios that will be modelled, taking account of the potential impacts of climate change and the overall site characteristics.

20.12.1 Designing for the Optimal Rainfall Event

The storm event that will have the greatest impact on the site, termed the critical storm duration, is the event that produces the largest amount of discharge (Kanga *et al.*, 2009). In the UK, this can be broken down into two events: summer and winter. The winter event provides the greatest volume and velocity of runoff, due to changes in ground conditions, which further promote runoff. The duration of the event is also a key factor to be considered; Scholz (2004) discovered that it was shorter events, of about one hour, that usually triggered the critical storm duration. In addition, the return period needs to be selected which decides the magnitude of event that will be modelled, and therefore what the SuDS device of management train needs to be able to withstand. The Defra (2015) non-statutory technical standards suggest that a 1 in 100 year event should be designed for, with runoff less than greenfield runoff. As well as the critical storm duration, climate change must be factored into the design, since it is accepted that it will have an effect on the climate of the UK for the design life of the management train (IPCC, 2007). For this reason, a 30% increase should be added to any storm event to provide resilience against climate change (Environment Agency, 2009).

In terms of site-specific characteristics, infiltration is an important consideration. This has a large bearing on what type and how many SuDS can be used (Kirby, 2005), as different soils have different infiltration rates (Ward and Robinson, 2000). Furthermore, Merwade *et al.* (2008) suggest that topographic and site elevation data is required for detailed flow route modelling to provide accurate outputs.

20.13 Conclusions

There are a number of techniques that can be applied to drainage modelling, each presenting different benefits dependent on the output requirements and available resources. Models can integrate SuDS at various scales to identify their role at reducing runoff. However, due to the complexity of modelling and their data requirements, there are a number of uncertainties with the outputs.

While previous research has utilised MUSIC and Infoworks (Bastien *et al.*, 2010), MicroDrainage® is more widely used by practitioners for new-build sites (Atkins, 2008; Mott Macdonald Ltd and Medway Council, 2009), and is the industry standard for UK drainage and flood systems (Hubert *et al.*, 2013). Attempts have been made to model management trains, but accounting for water quantity has been problematic (Bastien *et al.*, 2010; Hubert *et al.*, 2013). These few studies indicate that overall, a SuDS management train is an effective strategy for mitigating flood risk (Hubert *et al.*, 2013), but the relationship between this effectiveness and the devices that make up the train is not known.

References

Afshar, M.H. (2007) Partially constrained ant colony optimization algorithm for the solution of constrained optimization problems: Application to storm water network design. *Advances in Water Resources*, 30, 954–965.

Atkins (2008) *Development of guidance for sewerage undertakers on the implementation of drainage standards*. Available at: http://tinyurl.com/px9maeh

Bassett, D., Pettit, A., Anderson, C. and Grace, P. (2007) *Scottish Flood Defence Asset Database Final Report*. Available at: http://tinyurl.com/ppp5tso

Bastien, N., Arthur, S., Wallis, S. and Scholz, M. (2011) Runoff infiltration, a desktop case study. *Water, Science and Technology*, 63(10) 2300–2308.

Bastien, N., Arthur, S., Wallis, S. and Scholz, M. (2010) The best management of SuDS treatment trains: a holistic approach. *Water Science and Technology*, 61(1) 263–272.

Bates, P.D. and De Roo, A.P.J. (2000) A simple raster-based model for flood inundation simulation. *Journal of Hydrology*, 236, 54–77.

Beck, H.J. and Birch, G.F. (2013) The magnitude of variability produced by methods used to estimate annual stormwater contaminant loads for highly urbanised catchments. *Environmental Monitoring Assessment*, 185, 5209–5220.

Bell, V.A., Kaya, A.L., Cole, S.J., Jones, R.G., Moorea, R.J. and Reynard, N.S. (2012) How might climate change affect river flows across the Thames Basin? An area-wide analysis using the UKCP09 Regional Climate Model ensemble. *Journal of Hydrology*, 442–443, 89–104.

Bergman, M., Hedegaard, M.R., Peterson, M.F., Binning, P., Mark, O. and Mikkelsen, P.S. (2011) Evaluation of two stormwater trenches in Central Copenhagen after 15 years of operation. *Water, Science and Technology*, 63(10) 2279–2286.

Burszta-Adamiak, E. and Mrowiec, M. (2013) Modelling of green roofs hydrological performance using EPA's SWMM. *Water, Science and Technology*, 68(1) 36–42.

Cambridgeshire County Council (2012) Lamb Drove Sustainable Drainage (SUDS) Monitoring.

Charlesworth, S. (2010) A review of the adaption and mitigation of global climate using sustainable drainage cities. *Journal of Water and Climate Change*, 1(3) 165–180.

Charlesworth, S.M., Perales-Momparler, S., Lashford, C. and Warwick, F. (2013) The sustainable management of surface water at the building scale: preliminary results of case studies in the UK and Spain. *Journal of Water Supply: Research and Technology-AQUA*, 62, 8, 534–544.

Chen, W.-B. and Liu, W.-C. (2014) Modelling flood inundation induced by river flow and storm surges over a river basin. *Water*, 6, 3182–3199.

Danish Institute of Hydrology (2002) *MOUSE surface water runoff models reference manual.* Hørsholm, Denmark.

Defra (2015) *Non-statutory technical standards for sustainable drainage.* Available at: http://tinyurl.com/qem92y4.

Defra, Environment Agency (2005) *Defra/Environment Agency Flood and Coastal Defence RandD Programme* W5-074/A/TR/1.

Dotto, C., Deletic, A., McCarthy, D. and Fletcher, T. (2011) Calibration and sensitivity analysis of urban drainage models: MUSIC rainfall/runoff module and a simple stormwater quality model. *Australian Journal of Water Resources*, 15(1) 85–94.

Elliott, A. and Trowsdale, S. (2007) A review of models of low impact urban stormwater drainage. *Environmental Modelling and Software*, 22, 394–405.

Ellis, B., Revitt, M. and Lundy, L. (2012) An impact assessment methodology for urban surface runoff quality following best practice treatment. *Science of the Total Environment*, 416, 172–179.

Ellis, B., Viavattene, C. and Chlebek, J. (2011) *A GIS-based integrated modelling approach for the identification and mitigation of pluvial urban flooding.* WAPUG Spring Meeting held 18 May 2011.

Environment Agency (2009) Managing flood risk. Available at: http://tinyurl.com/plejuh3

Freni, G., Mannina, G. and Viviani, G. (2009) Stormwater infiltration trenches: a conceptual modelling approach. *Water Science and Technology*, 60(1) 185–199.

FWR and WAPUG (2002) Urban flood route prediction-can we do it? Report by the Foundation for Water Research, Wastewater Research and Industry Support Forum. Workshop held 26 Sept 2002, Renewal Conference Centre, Solihull, UK. Available at: http://www.fwr.org/wapug/fldpred.pdf

González-Angullo, N., Castro, D., Rodríguez-Hernández, J. and Davies, J.W. (2008) Runoff infiltration to permeable paving in clogged conditions. *Urban Water Journal*, 5(2) 117–124.

Heal, K.V., Bray, R., Willingale, S.A.J., Briers, M., Napier, F., Jefferies, C. and Fogg, P. (2009) Medium-Term performance and maintenance of SUDS: a case-study of Hopwood Park Motorway Service Area, UK. *Water, Science and Technology*, 59(12) 2485–2494.

Henonin, J., Russo, B., Mark, O. and Gourbesville, P. (2013) Real-time urban flood forecasting and modelling-a state of the art. *Journal of Hydroinformatics*, 15(3) 717–733.

Hubert, J., Edwards, T. and Jahromi, B.A. (2013) Comparative study of sustainable drainage systems. *Engineering Sustainability*, 166 (ES3) 138–149.

Imteaz, M.A., Ahsan, A., Rahman, A. and Mekanik, F. (2013) Modelling stormwater treatment systems using MUSIC: accuracy. *Resources, Conservation and Recycling*, 71, 15–21.

IPCC (2007) *Climate Change 2007: Impacts, Adaption and Vulnerability.* Available at: http://tinyurl.com/ywywt8

Jefferies, C., Duffy, A., Berwick, N., McLean, N. and Hemingway, A. (2009) Sustainable urban drainage systems (SUDS) treatment train assessment tool. *Water Science and Technology*, 60(5) 1233–1240.

Kahinda, J.M., Taigbenu, A.E., Sejamoholo, B.B.P., Lillie, E.S.B. and Boroto, R.J. (2009) A GIS-based decision support system for rainwater harvesting (RHADESS). *Physics and Chemistry of the Earth*, 34, 767–775.

Kanga, M.S., Koob, J.H., Chunc, J.A., Herd, Y.G., Parka, S.W. and Yooe, K. (2009) Design of drainage culverts considering critical storm duration. *Biosystems Engineering*, 104, 425–434.

Khastagir, A. and Jayasuriya, L.N.N. (2010) Impacts of using rainwater tanks on stormwater harvesting and runoff quality. *Water Science and Technology*, 62(2) 324–329.

Kirby, A. (2005) SuDS-innovation or a tried and tested practice? *Municipal Engineer*, 158, 115–122.

Lashford, C., Charlesworth, S., Warwick, F. and Blackett, M. (2014) Deconstructing the sustainable drainage management train in terms of water quantity-preliminary results for Coventry, UK. *CLEAN-Soil, Air, Water*, 42(2) 187–192.

Lee, J.-M., Hyun, K.-H., Choi, J.-S., Yoon, Y.-J. and Geronimo, F. (2012) Flood reduction analysis on watershed of LID design demonstration district using SWMM5. *Desalination and Water Treatment*, 38, 326–332.

Lee, S., Birch, G. and Lemckert, C. (2011) Field and modelling investigations of fresh-water plume behaviour in response to infrequent high-precipitation events, Sydney Estuary, Australia. *Estuarine, Coastal Shelf Science*, 92, 389–402.

Liao, Z.L., He, Y., Huang, F., Wang, S. and Li, H.Z. (2013) Analysis on LID for highly urbanized areas' waterlogging control: demonstrated on the example of Caohejing in Shanghai. *Water, Science and Technology*, 68(12) 2559–2567.

Mahdizadeh, H., Stansby, P. and Rogers, B. (2012) Flood wave modelling based on a two-dimensional modified wave propagation algorithm coupled to a full-pipe network solver. *Journal of Hydraulic Engineering*, 138(3) 247–259.

Mark, O., Weesakul, S., Apirumanekul, C., Aroonnet, S.B. and Djordjević, S. (2004) Potential and limitations of 1D modelling of urban flooding. *Journal of Hydrology*, 299, 284–299.

Merwade, V., Cook, A. and Coonrod, J. (2008) GIS techniques for creating river terrain models for hydrodynamic modeling and flood inundation mapping. *Environmental Modelling and Software*, 23 (10-11) 1300–1311.

MicroDrainage (2009) *Working with WinDes®: An example-led instruction to the Windows-based Micro Drainage Suite*. Unpublished handbook. MicroDrainage: Newbury.

MicroDrainage (n.d) WinDes DrawNet 3D model build for drainage engineers. *Waste and Water Magazine*, 1–5.

Mitchell, G. (2005) Mapping hazard from urban non-point pollution: a screening model to support sustainable urban drainage planning. *Journal of Environmental Management*, 74, 1–9.

Moore, S., Stovin, V., Wall, M. and Ashley, R. (2012) A GIS-based methodology for selecting stormwater disconnection opportunities. *Water Science and Technology*, 66(2) 275–283.

Moore, S. (2006) *Modelling of a Small UK Out of Town Retail Park Catchment using WinDes and MUSIC*. 2nd SUDSnet Student Conference held Coventry University 7 Sept 2006.

Mott Macdonald Ltd., Medway Council (2009) *Waterfront and Town Centre Development, Chatham, Kent*. Proposed Dynamic Bus Facility. Flood Risk Assessment. Available at: http://tinyurl.com/o7ve4sl

Nativi, S., Mazzetti, P. and Geller, G.N. (2013) Environmental Model Access and Interoperability: The GEO Model Web Initiative. *Environmental Modelling and Software*, 39, 214–228.

Newton, C., Jarman, D., Memon, F.A., Andoh, R. and Butler, D. (2014) Developing a decision support tool for the positioning and sizing of vortex flow controls in existing sewer systems. *Procedia Engineering*, 70, 1230–1241.

Pathirana, A., Tsegaye, S., Gersonius, B. and Vairavamoorthy, K. (2011) A simple 2-D inundation model for incorporating flood damage in urban drainage planning. *Hydrology and Earth System Sciences*, 15, 2747–2761.

Petrucci, G., Deroubaix, J.-F., de Gouvello, B., Deutsch, J.-C., Bompard, P. and Tassin, B. (2012) Rainwater harvesting to control stormwater runoff in suburban areas. An experimental case-study. *Urban Water Journal*, 9(1) 45–55.

Poole, G., Stanford, J.A., Frissell, C.A. and Running, S.W. (2002) Three-dimensional mapping of geomorphic controls on flood-plain hydrology and connectivity from aerial photos. *Geomorphology*, 48, 329–347.

Qi, H. and Altinakar, M.S. (2011) A GIS-based decision support system for integrated flood management under uncertainty with two dimensional numerical simulations. *Environmental Modelling and Software*, 26, 817–821.

Refsgaard, J.C., van der Sluijs, J.P., Højberg, A.L. and Vanrolleghem, P. (2007) Uncertainty in the environmental modelling process – a framework and guidance. *Environmental Modelling and Software*, 22(11) 1543–1556.

Rossman, L. (2010) *Storm Water Management Model User's Manual-Version 5.0*. USEPA. EPA/600/R-05/040.

RPS Group (2012) *Flood Risk Assessment: Prior Deram Walk, Canley, Coventry*. On behalf of Taylor Wimpy Midlands. Available at: http://tinyurl.com/oekjgu6

Salarpour, M., Rahman, N.A. and Yusop, Z.G. (2011) Simulation of flood extent mapping by Infoworks RS-Case study for tropical catchment. *Journal of Software Engineering*, 5(4) 127–135.

Scholz, M. and Uzomah, V.C. (2013) Rapid decision support tool based on novel ecosystem service variables for retrofitting of permeable pavement systems in the presence of trees. *Science of the Total Environment*, 458–460, 486–498.

Scholz, M. (2004) Case study: design, operation, maintenance and water quality management of sustainable storm water ponds for roof runoff. *Bioresource Technology*, 95, 269–279.

Semadeni-Davies, A., Hernebring, C., Svensson, G. and Gustafsson, L. (2008) The impacts of climate change and urbanisation on drainage in Helsingborg, Sweden: suburban stormwater. *Journal of Hydrology*, 350(1–2) 114–125.

Shim, K.-C., Fontane, D. and Labadie, J. (2002) Spatial decision support system for integrated river basin flood control. *Journal of Water Resources Planning and Management*, 3, 190–201.

SNIFFER (2004) *SUDS in Scotland-The Monitoring Programme.* Available at: http://tinyurl.com/oanggfp

Stovin, V.R. (2009) The potential of green roofs to manage urban stormwater. *Water and Environment Journal*, 24, 192–199.

Stovin, V.R. and Swan, A.D. (2007) Retrofit SuDS-cost estimates and decision-support tools. *Water Management*, 160, 207–214.

Todini, E. (1999) An operational decision support system for flood risk mapping, forecasting and management. *Urban water*, 1, 131–143.

Tramblay, Y., Bouvier, C., Ayral, P. and Marchandise, A. (2011) Impact of rainfall spatial distribution on rainfall-runoff modelling efficiency and initial soil moisture conditions estimation. *Natural Hazards and Earth System Science*, 11(1) 157–170.

Viavattene, C., Ellis, B., Revitt, M., Seiker, H. and Peters, C. (2010) The Application of a GIS-Based BMP Selection Tool for the Evaluation of Hydrologic Performance and Storm Flow Reduction. NovaTech 2010, 7th International Conference on Sustainable Techniques and Strategies in Urban Water Management held 27 June – 1 July 2010 at Lyon, France.

Viavattene, C., Scholes, L., Revitt, D.M. and Ellis, J.B. (2008) A GIS based decision support system for the implementation of stormwater best management practices. *Pollution Research*, 1–9.

Ward, R.C. and Robinson, M. (2000) *Principles of Hydrology* (4th edition) London: McGraw-Hill Publishing Company.

Warwick, F. (2013) *A GIS-based decision support methodology at local planning authority scale for the implementation of sustainable drainage.* Unpublished PhD thesis. Coventry University, UK.

Wheater, H.S. (2002) Progress in and prospects for fluvial flood modelling. *Philosophical transactions of the Royal Society*, 360, 1409–1431.

Wilson, S., Bray, R. and Cooper, P. (2004) *Sustainable Drainage Systems: Hydraulic, Structural and Water Quality Advice.* C609. London: Construction Industry Research and Information Association.

Woods Ballard, B., Kellagher, R., Martin, P., Jefferies, C., Bray, R. and Shaffer, P. (2007) *The SuDS Manual.* Available at: http://tinyurl.com/od75lo3.

Wong, T.H.F., Fletcher, T.D., Duncan, H.P., Coleman, J.R. and Jenkins, G.A. (2002) A model for urban stormwater improvement conceptualisation. In: Rizzoli, A. Jakeman, A. (ed.) Proceedings of the First Biennial meeting of the International Environmental Modelling and Software society, Integrated Assessment and Decision Support held 24–22 June 2012 at Lugano, Switzerland, 1, 48–53.

Zhou, Q., Mikkelsen, P.S., Halsnæs, K. and Arnbjerg-Nielsen, K. (2012) Framework for economic pluvial flood risk assessment considering climate change effects and adaptation benefits. *Journal of Hydrology*, 414, 539–549.

Public Perceptions of Sustainable Drainage Devices

Glyn Everett

21.1 Introduction

This chapter will argue that people's perceptions and understandings of the purpose, function and wider potential benefits of sustainable drainage systems (SuDS) are central to their *performed* sustainability. These perceptions and understandings will affect people's understanding of, and desire to perform, good and bad behaviours that will encourage or discourage function, and thereby impact upon performance, expected product life cycle and the accrual of associated benefits. These will, in turn, cycle back to influence perceptions and so the development and mainstreaming of individual inclinations and social norms to perform good behaviours and contribute to maintenance.

The chapter considers how people are engaging with flood risk as an issue. It presents a literature review of works that have focused on public preferences and behaviours regarding approaches to sustainable flood risk management (FRM), before it then highlights a prominent case study of sustainable FRM, which considers perceptions and behaviour towards 'bioswales'. It then considers public awareness and understanding of bioswale functions and how these may affect perceptions of amenity, costs and benefits, and so impact upon behaviour. Finally, the chapter argues that maximising opportunities for involving citizens in the development and tailoring of SuDS could help encourage the take-up of more appropriate – sustainable – behaviours. Without such efforts, comprehension of the purpose and nature of installations might rest at such a level that inappropriate behaviours remain commonplace, reducing the efficacy, cost-effectiveness and *sustainability* of nominally more 'sustainable' approaches.

Sustainable Surface Water Management: A Handbook for SuDS, First Edition.
Edited by Susanne M. Charlesworth and Colin A. Booth.
© 2017 John Wiley & Sons, Ltd. Published 2017 by John Wiley & Sons, Ltd.

21.2   Public Preferences and Understanding of Flood Risk Management

Around 5.5 million properties currently stand in areas at risk of flooding from rivers, the sea and surface water in England and Wales alone (Environment Agency, 2009a, 2009b), yet action to install flood protection measures remains surprisingly low. Only around 1 in 4 of those who have been flooded have since taken action, while for those that have not experienced anything this drops to 6% (Thurston *et al.*, 2008; Harries, 2012). Some forecasts predict that with climate change, UK flooding may increase dramatically over the next 75 years. If so, this could cost tens of billions of pounds every year in repairs and protection work (King, 2004). This will be mirrored across the world as climate change impacts upon built environments.

As Lamond and Proverbs (2009) have argued, people must go through a number of stages of thinking before they can accept the realities of possible flooding and begin to engage with this. That is, there needs to be a developing of the *desire* to act through awareness, perception and ownership of the problem, as well as of the *ability* to act in developing the knowledge, the available capital and a belief that acting will alter the situation. Public ownership of flood risk is apparently low; respondents sometimes demonstrate aversion to acknowledging the scale of risks faced (Speller, 2005; Defra, 2011). This may stem from what Harries (2010) has referred to as preferences for *feeling* secure over and above actually *being* more secure, and being constantly reminded of potential risk by measures put in place (seeing flood doors and so forth).

This may be no less true regarding the perceived utility of municipal defences such as household adaptations; communities might not accept labelling and actions that acknowledge and work to reduce the risk, for fear they could negatively affect property prices (Burningham *et al.*, 2008). Nonetheless, trends have been observed of people perceiving the responsibility for installing flood protections as being a government-level one rather than that of households (Correia *et al.*, 1998; Werritty *et al.*, 2007), with people remaining passive and expecting government or insurance to cover the costs (Brilly and Polič, 2005; Wedawatta *et al.*, 2011; Ludy and Kondolf, 2012). Other studies have found that publics tend to recognise at least joint responsibility for managing flood risk with designated authorities (Laska, 1986), but findings have, in turn, shown that people's willingness to pay for mitigations can be as low as one-off payments of less than £100 (Kaźmierczak and Bichard, 2010), which would not cover the cost of effective measures, and this strongly implies that the problem still remains.

It could be, however, that there are some in-built preferences within society towards more sustainable approaches to managing flood risk. A number of studies have shown that increasing green space and biodiversity, or wildlife corridors, within the built environment is generally perceived positively (Coley *et al.*, 1997; Dunnett and Muhammad, 2000; Chiesura, 2004; Fuller *et al.*, 2007). Thus, it *could* be that sustainable approaches to FRM will be perceived more positively due to their increasing available green space and contributing to biodiversity, while not presenting, in the first instance, as flood-risk defences; this is something that is considered in the next section.

21.3   The Sustainability of SuDS

Policy in both the UK and the USA now favours employing more sustainable approaches to FRM (Scottish Government, 2003; Defra, 2005; EPA, 2013). Implementing this shift away from hard 'grey' infrastructure will require the involvement of all stakeholders, including local publics who will be affected, in developing new practices and behaviours to ensure functionality and sustainability. This raises questions around where public preferences lie, and whether – and if so, how – they might develop, positively or negatively, with the wider adoption of SuDS.

In contrast to generally more hidden grey infrastructure, SuDS will frequently alter the visible urban environment: 'green' SuDS, such as green roofs, swales and rain gardens, will involve locating green spaces within or atop the built environment, while rain barrels will alter aspects of home aesthetics, and permeable paving may change aesthetics and the 'feel' of the ground. All will, therefore, involve developments in thinking with regard to what flood risk management should involve and look like (Shandas *et al.*, 2010), and will necessitate shifts in behaviour to enable them to carry out their function over the medium to long term. Furthermore, perceptions of SuDS could influence homebuyer preferences, house values and so developer practices (Netusil *et al.*, 2014; Bolitzer and Netusil, 2000; HR Wallingford, 2003).

For this reason, understanding public perceptions and behaviours is vitally important. Crucially, SuDS will only ever be as sustainable as the behaviour surrounding them. Mistreated permeable paving or swales might last only a few years if people allow their cars to leak oil onto paving or use swales as convenient places to dispose of refuse, thereby blocking water flows.

Looking back to the SuDS triangle (Chapter 1), water quality and quantity make up a substantial part of the research that has been conducted around SuDS. Yet, as Singleton (2012) acknowledges, the third arm of the triangle, amenity (and biodiversity or wildlife), is frequently less considered. Indeed, 'sometimes it is sidelined, or even forgotten completely' (Singleton, 2012). This is possibly because, as Singleton (2012) acknowledges, targets for amenity can be hard to set, and outcomes in turn vague. Biodiversity is furthermore a quite separate consideration without overall agreed measurement metrics, scales of assessment (Purvis and Hector, 2000; Franklin, 2008) or formulae for connecting this back to how it would benefit amenity (Hanley *et al.*, 1995).

Yet the amenity arm of the SuDS triangle is arguably the most important from both social and sustainability perspectives. People need to understand the direct functions of SuDS (reducing flooding and improving water quality), as well as their more indirect benefits, such as adding to the urban environment's green infrastructure, in order to be cognisant of how they contribute to amenity (such as reducing water consumption and enabling more access in times of drought, improving aesthetics and air quality, providing wildlife corridors to encourage biodiversity and creating leisure and recreation spaces that frequently benefit mental and physical health, etc.). If they do not feel the devices contribute to their lives, people may be more unwilling to alter behaviour to encourage longer-term functioning, and to pay for the wider rollout as well as maintenance of such approaches.

Amenity is a frequently referenced benefit of using SuDS (Defra, 2011; Anglian Water, 2011; Graham *et al.*, 2012), yet the preferences and perceptions of those who live around devices are under-researched. A few studies have produced findings indicating that publics prefer structural defences to SuDS. Werritty *et al.* (2007) found over 90% of respondents preferred structural defences to proposed alternatives, these being viewed as 'the first line in flood defence'. In looking at the potential benefits of SuDS, Johnson and Priest (2008) also concluded that the public, media and insurance industry remained heavily focused upon structural defences.

In contrast, Kenyon (2007) found participants preferred rural SuDS approaches such as regeneration of woodlands, with structural defences the least favoured option. Three other studies also noted public preferences lying with more sustainable approaches to FRM; HR

Wallingford (2003), Apostolaki and Jefferies (2005) and Bastien *et al.* (2011) all found that SuDS ponds were valued by local residents for their aesthetics, amenity and contributions to wildlife, with wildlife being rated as the most important factor, but aesthetics being a deciding factor.

Apostolaki and Jefferies (2005) found low levels of awareness of local schemes' functions, with many respondents unaware of either the term 'SuDS' or the ponds' contributions to flood-control. It was observed that people's views about SuDS ponds related, at least in part, to their awareness of functions and services. Bastien *et al.* (2011) found that public awareness of ponds' functions was much higher than in Apostolaki and Jefferies' (2005) research, with almost 75% of those surveyed having an understanding. However, safety was a major concern of residents, and large differences were observed between perceived and actual safety levels (McKissock *et al.*, 1999). Tunstall *et al.* (2000), however, evaluated several flood risk and amenity improvement river restoration projects, concluding that 'well-presented' schemes could be implemented, alongside consultation and awareness-raising, without raising safety concerns. The overriding conclusion of these studies is, therefore, that education and consultation are vital to the effective pursuit of sustainable strategies.

Studies from the USA were often more around green infrastructure (GI) generally than SuDS specifically, but with findings of central relevance, and these indicate similarly that awareness and understanding can be quite low (Barnhill and Smardon, 2012; Everett *et al.*, 2015, 2016). Barnhill and Smardon (2012) provide a concise but extensive literature review of the situation in the USA. They cite LaBadie's (2010) findings of poor knowledge regarding the design, construction, maintenance and funding of such techniques in Albuquerque, New Mexico, as an example of how core understanding is generally lacking, and how this can negatively impact upon willingness to consider SuDS alternatives. Similarly, Shandas *et al.* (2010) stress the need to improve knowledge of stormwater management techniques, having observed some significant variance among neighbourhoods in their studies in Portland. Others have observed misconceptions regarding SuDS harbouring increased populations of mosquitoes (Traver, 2009; Everett, 2016), which could in turn negatively affect perceptions.

Barnhill and Smardon (2012) point to related potential issues to be acknowledged and dealt with, such as how interventions might affect the socio-economic profile or the felt safety and security of areas (Seymour *et al.*, 2010; Pincetl and Gearin, 2013). However, they also point to studies that reflect upon the potentially positive social equity impacts of increasing access to green spaces (Floyd *et al.*, 2009; Pincetl and Gearin, 2013) and how this could develop safer and healthier neighbourhoods (Abrahams, 2010; Qureshi *et al.*, 2010; Shandas *et al.*, 2010). Similarly, Dill *et al.* (2010) observed in their Portland study that residents saw children playing outside more on green streets, felt they were better places to live and found walking in their neighbourhoods more pleasant. The overall outcome is a sense of the significant potential positive or negative impact of designing GI SuDS into urban environments. In the next section, we will look to a series of case studies that have been conducted around the use of sustainable approaches towards flood risk management in Portland, Oregon.

21.4   Attitudes and Behaviour: Portland, Oregon, USA

In the USA, green infrastructure (Benedict and McMahon 2006) has been promoted for around 20 years for environmental, economic and social reasons. Portland, Oregon has a history of flood events with an expected 10-year return period known as 'nuisance flooding' – relatively minor floods, which nonetheless cause road blockages and basement and house flooding, and contribute to worsening water quality through runoff from roads and industry (BES, 2001). As a result, the Portland Government's Bureau of Environmental Services (BES) has been developing more sustainable approaches to managing stormwater throughout this time (Reinhardt, 2011).

The BES's *Grey to Green* initiative (2008–2013) focused on expanding the use of stormwater management techniques that mimic natural systems, to restore and protect existing natural areas, improve water quality and reduce problems with street and basement flooding (BES, 2010). This has included, firstly, a 'willing seller land acquisition program', targeting three specific areas that experienced regular nuisance flooding, to buy up houses and return the land to a more natural state: restoring wetlands, improving flood storage for surrounding areas and benefitting wildlife and leisure activity opportunities (BES, 2015a). Secondly, the 'clean river rewards' program has offered households up to a 100% reduction in their stormwater utility fees when stormwater is managed at property level rather than feeding into the drainage system, as well as the city offering free workshops on how to register and how to manage stormwater at a household level (BES, 2014a).

This has previously included offering reductions if households disconnected their downspout, (roof drainpipe), so that rainwater fed directly into their garden or rain bucket rather than entering the drainage system (Wise, 2010; BES, 2014b), as well as reductions for using green roofs (BCIT, 2006). Portland has a mandatory policy of installing green roofs on city-owned buildings, unless this would be impractical (BPS, 2009). Further, planning policy allows for increases in building density where green roofs are used (BCIT, 2006). Although the city no longer offers free work or such incentives, they do still proudly assert that these programmes have led to 56,000 downspouts being disconnected, which has removed 1.3 billion gallons of stormwater from the combined sewer overflow systems each year within the city (BES, 2014).

Portland is now considered one of the leading cities in the USA in its pursuit of using green infrastructure to improve many aspects of city life (improving liveability, promoting sustainable development practices and helping to prepare for climate change – see Slavin and Snyder, 2011 and Mayer and Provo, 2004). Portland, for example, receives a high score for sustainability endeavours in Portney's (2013) review of US cities.

The city adopted its first stormwater management manual in 1999 (SWMM) (BES, 2005) and then officially assumed a green streets policy in 2007 (BES, 2007). As a result of this, one further key element within their approaches to dealing with stormwater runoff, bioswales, have been being installed on city streets for over ten years. Bioswales, or bioretention gardens, are highly engineered SuDS stormwater management facilities similar to rain gardens, but with drainage installed underneath to transport the filtered water, using native plants to extract pollutants before the water returns to the main watercourse (Figure 21.1). These have been used extensively in Portland for reducing street and basement flooding and for improving water quality, both as city retrofits to developed areas and through changes to legislation that require developers to undertake GI SuDS work wherever more than 500 sq. ft (46 m^2) of hardstanding is to be laid down (BES, 2014).

In Portland, Shandas *et al.* (2010, 2015) researched the 'Tabor to the River' (T2R) programme, a series of works involving extensive tree-planting, bioswale installation, habitat improvements and sewer pipe restoration, to improve the area's ability to cope with the limits of a historic combined sewer overflow pipe system in the face of increases in urbanisation, hardstanding and climate change. Shandas *et al.*'s (2010, 2015) work looked at resident understandings and attitudes in areas within the T2R programme, where bioswales had been installed, and compared this with areas where no closer engagement work had yet been undertaken.

Figure 21.1   Bioswales in Portland, Oregon and preplanting information.

The surveys they conducted found respondents, in general, to be well informed about the nature of the programme. They found that people in areas with bioswales rated their surroundings more highly on every variable considered (walkability, safety, aesthetics and green space), pointing to a positive relationship between resident satisfaction and green infrastructure SuDS, as posited in the previous section. In terms of willingness to engage with maintenance of devices, Shandas *et al.* (2010) found that, as a rule, higher income households were more likely to engage, and more likely still if (a) they were already involved with the other environmental projects, (b) they had developed social interactions with others in their neighbourhood or (c) they rated the neighbourhood lower regarding the presence of parks and open space. Lower income households were more likely to engage when younger or with a graduate education.

Church (2015) also studied T2R, finding strong awareness of and support for the use of bioswales, crediting this to outreach work undertaken by the BES. Church (2015) found support for the statement that bioswales were 'a good idea' (82%), but weaker support for the notion that bioswales improve aesthetics and act as an amenity (32%). A large proportion of the sample (63%) understood the function of bioswales. Church's (2015) work did, however, discover mixed views of bioswales as 'nature'; around two-fifths felt that they were, and the same number felt they were not, the rest considering them a purposive 'manufactured' nature – highly engineered city interventions, rather than natural, or providing green space or wildlife corridors.

Dill *et al.* (2010) looked across several sites in Portland to assess whether green streets impacted upon 'active ageing'. They found green streets residents walked more than in other areas, even controlling for demographics, attitudes and nearby destinations, and were more likely to concur that walking in their neighbourhood was more pleasant since facilities were installed. It was further found that green streets residents stopped and talked with their neighbours more often than on other streets. Concurring to some extent with Shandas *et al.*'s (2010) work, Dill *et al.* (2010) found that older residents tended to hold more negative opinions about facilities.

Everett *et al.* (2015, 2016) also researched public perceptions and behaviour regarding Portland bioswales. They looked outside of the T2R programme area, and possibly due to demographic differences and a different methodological approach, findings differed somewhat from those of the authors detailed above. Everett *et al.* (2015, 2016) adopted a point of opportunity interaction (POI) approach, talking without prior notification with people on the street or in their gardens, to avoid self-selection bias, whereby residents might respond only if they were already aware of the installations and had strong opinions about them (Whitehead, 1991; Hudson *et al.*, 2004). The interactions produced valuable insights from people who may not otherwise have volunteered for more formal engagements.

Everett *et al.* (2016) found a lot of low awareness of the purpose and function of devices. Those with some awareness spoke much more about reducing flood risk and cleaning the water than they did about possible wider benefits of the devices, such as providing wildlife corridors or helping with adaptation to climate change. Importantly, a significant minority of residents in areas not at direct flood risk did not connect with how devices local to them might help mitigate risks elsewhere, or the city-wide economic benefits of avoiding flooding. Others were rather cynical about city claims for flood reduction and water cleaning, indicating lower awareness. With regard to maintenance, some respondents took part in basic litter clearing, but very few were aware of the existence of the green street steward programme. The city publishes materials advising on how to clear facilities (BES, 2012, 2013), and encourages members of the public to sign up as stewards, where they gain training and then 'adopt' bioswales. These points again reemphasise the importance of engagement and awareness-raising taking place prior to, during and following the installation of devices.

Everett *et al.*'s (2016) study also demonstrated some pronounced dissatisfaction with plant choice and maintenance on certain streets, with some residents thinking plants looked like weeds, others that they looked overgrown and unkempt and a final group asking why they could not plant edible produce in the bioswales. Dialogue in such cases might allow for local aesthetic adaptations and negotiation as well as awareness raising. Finally, as a result of poor bioswale understandings and perceptions, Everett *et al.* (2016) heard stories of people emptying their trash into the devices, cutting back or removing plants that had been placed there for a reason, and diverting water *away* from bioswales so that it remained on the street. While such stories were in a minority, they were frequent enough to cause some concern regarding levels of awareness and buy-in to city strategy, and how this might affect longer-term performance and sustainability of devices.

21.5   Co-development and Co-ownership

An increasing number of authors advocate adopting what might be termed a knowledge *co-construction* approach, where all interested parties can discuss and learn from each other in developing together solutions that all might be more satisfied with, over a 'deficit model', expert–public knowledge-transfer approach (Fielding *et al.*, 2007; White and Richards, 2008; Evers *et al.*, 2012; O'Sullivan *et al.*, 2012). Engaging and involving locally affected communities should be a fundamental first step in looking to encourage the community buy-in needed for device longevity. 'Deliberative participation strategies' need to be employed alongside efforts at community education in order to effectively empower and involve publics (Ryan and Brown, 2000). As Tunstall *et al.* (2000) noted, publics expect to be consulted about changes to their local environment, especially ones that will negatively alter aesthetics in the short term and that some may regard as negatively affecting their flood risk.

Dialogue and consultation around engagement work could help to bring in local knowledge, concerns and preferences, with the aim of constructing devices that local people feel greater ownership of and investment in, improving awareness to improve acceptability (HR Wallingford, 2003; DTI, 2006; Hostetler *et al.*, 2011). Consultation could also allow people to input to modifications that could potentially improve preferences and give them a sense that these were 'their' spaces, thereby hopefully encouraging interest in adopting stewardship roles (Larson and Lach, 2008; Dill *et al.*, 2010; Shandas *et al.*, 2010; Everett, 2016).

Conducting engagement could be challenging, and costly; it would be important to try to get beyond the 'usual suspects', as Larson and Lach (2008) and Shandas *et al.* (2010) found with their studies that higher income and more highly educated respondents were more likely to engage with consultation exercises and other city interaction efforts. Henning (2015) presents an interesting approach to thinking about engagement in looking to break down the catchall of 'homeowners' into a more textured analysis of motivations. Henning (2015) arrives at a six-point typology relating to people's concerns, or lack of, with the adoption of green infrastructure and stormwater management techniques. This ranges from those more concerned with maintaining clean aesthetics, through 'the greens' concerned to do what they believe is good for 'the environment', to 'early adopters' of stormwater management techniques, such as rain barrels. This more textured and nuanced attempt at understanding 'the public' could, Henning (2015) argues, allow for more targeted communications pitched at top-level preferences (reducing flooding, increasing green space, improving biodiversity or aesthetics, and so forth). This could, in turn, work to bring more people in to conversations around SuDS and GI.

21.6 Conclusions

This chapter has looked at what we know so far of public understanding, preferences and behaviour around sustainable drainage systems. It has been shown that understanding of the purpose, function and wider benefits of such systems does exist, but that it appears to be far from mainstream. Published literature supports this belief that a strong majority of the public are unaware, or insufficiently aware, of the reasons why these systems are put in place, unless they are engaged with, early and in an ongoing manner. As a result, preferences will frequently be developed based upon the aesthetics and perceived amenity or disbenefits of systems. Yet if people are unaware of the wider potential benefits, while costs such as reduced parking space or perceived reductions in safety are more obvious, this will feed back towards negative preferences regarding SuDS.

As a result of low awareness, further, people have been argued to be often poorly informed about behaviours required to ensure continued functioning and the development of the multiple potential benefits from established devices, as seen in Portland, USA. This will tend again to feed back to reduced functioning, worsened aesthetics and further negative preferences.

We have acknowledged and agreed with arguments from the literature that to encourage more positive preferences it will be important to bring potentially affected, or concerned, publics in to conversations around SuDS as early as possible. Where people are involved, they can express their personal preferences, share their local expert knowledge, learn from professional stakeholders and negotiate towards maximally preferred solutions for all parties. In so doing, members of the public will hopefully become more disposed to assume ownership over devices, and therefore be more willing to engage with both good behaviour and maintenance practices.

Engagement efforts around the Tabor to the River programme in Portland serve as an example of best practice, where local voices have been listened to, awareness is high and behaviour generally good. However, the further Portland research that was cited demonstrated that where such engagement work was not undertaken, awareness remained low.

Engagement will cost in the short term, but engagement of publics with the development and implementation of their local devices might save money over the longer term. A greater desire to have devices that have been co-developed and people feel ownership over alongside improved awareness and appreciation of the devices' multiple benefits could encourage more widespread appropriate behaviour (and community-level disapproval of inappropriate behaviours). Such desire could further help inculcate community-level endeavours at low- to medium-level maintenance work in the manner of Portland's green street stewards.

Engaging communities as early on as possible, and in an ongoing manner, in the co-development and implementation of SuDS solutions is perhaps the best approach for ensuring that sustainable drainage systems truly are *sustainable*, as well as more cost-effective in the long term.

References

Abrahams, P.M. (2010) *Stakeholders' Perceptions of Pedestrian Accessibility to Green Infrastructure: Fort Worth's Urban Villages*. University of Texas, Arlington.

Anglian Water (2011) *Guidance on the use of sustainable drainage systems (SuDS) and an overview of the adoption policy introduced by Anglian Water*. Anglian Water, Huntingdon.

Apostolaki, S. and Jefferies, C. (2005) *Social impacts of stormwater management techniques including river management and SuDS*. Final report, SUDS01. SNIFFER, Edinburgh.

Barnhill, K. and Smardon, R. (2012) Gaining ground: Green infrastructure attitudes and perceptions from stakeholders in Syracuse, New York. *Environmental Practice* 14(01) 6–16.

Bastien, N.R.P., Arthur, S. and McLoughlin, M.J. (2011) Valuing amenity: public perceptions of sustainable drainage systems ponds. *Water and Environment Journal* 26(1) 19–29.

BCIT (2006) *Case Studies of Green Roof Regulations in North America 2006*. British Colombia Institute of Technology, Burnaby, Canada. Available from: http://commons.bcit.ca/greenroof/files/2012/01/2006_regulations.pdf (May 2014).

Benedict, M.A. and McMahon, E. (2006) *Green Infrastructure: Linking Landscapes and Communities*. Island Press, Washington, DC.

BES (2001) *Johnson Creek Restoration Plan (June 2001)*. Bureau of Environmental Services, Portland, Oregon. Available from: www.portlandoregon.gov/bes/article/214367 (August 2015).

BES (2005) *Portland watershed management plan*. Bureau of Environmental Services, Portland, Oregon. Available from: www.portlandoregon.gov/bes/article/107808 (May 2015).

BES (2007) *Portland Green Streets Program*. Bureau of Environmental Services, Portland, Oregon. Available from: www.portlandoregon.gov/bes/45386 (June 2014).

BES (2010) *Tabor to the River*. Bureau of Environmental Services, Portland, Oregon. Available from: www.portlandoregon.gov/bes/47591 [February 2015].

BES (2012) *The Green Street Steward's Maintenance Guide*. Bureau of Environmental Services, Portland, Oregon. Available from: www.portlandoregon.gov/bes/article/319879 [May 2014].

BES (2013) *Green Street Steward Program*. Bureau of Environmental Services, Portland, Oregon. Available from: www.portlandoregon.gov/ bes/52501 [May 2014].

BES (2014) *The Stormwater Management Manual (SWMM)*. Bureau of Environmental Services, Portland, Oregon. Available from: www.portlandoregon.gov/bes/64040 [May 2014].

BES (2015a) *Willing Seller Program*. Bureau of Environmental Services, Portland, Oregon. Available from: www.portlandoregon.gov/bes/article/106234 [May 2015].

BES (2015b) *'Stormwater for Challenging Sites' workshops begin this weekend*. Available from: www.portlandoregon.gov/bes/article/546473 [June 2015].

Bolitzer, B. and Netusil, N.R. (2000) The impact of open spaces on the property values in Portland, Oregon. *Journal of Environmental Management* 59(3) 185–193.

BPS (2009) *Green Building Resolution – The City of Portland, Oregon*. Bureau of Planning and Sustainability, Portland, Oregon. Available from: www.portlandoregon.gov/bps/article/243213 (May 2015).

Brilly, M. and Polič, M. (2005) Public perception of flood risks, flood forecasting and mitigation. *Natural Hazards and Earth System Sciences* 5, 345–355.

Burningham, K., Fielding, J. and Thrush, D. (2008) 'It'll never happen to me': understanding public awareness of local flood risk. *Disasters* 32(2) 216–238.

Chiesura, A. (2004) The role of urban parks for the sustainable city. *Landscape and Urban Planning* 68, 129–138.

Church, S.P. (2015) Exploring green streets and rain gardens as instances of small scale nature and environmental learning tools. *Landscape and Urban Planning* 134, 229–240.

Coley, R.L., Sullivan, W.C. and Kuo, F.E. (1997) Where does community grow? The social context created by nature in urban public housing. *Environment and Behavior* 29, 468–494.

Correia, F.N., Fordham, M., Da Graca Saraiva, M. and Bernardo, F. (1998) Flood hazard assessment and management: interface with the public. *Water Resources Management* 12(3) 209–227.

Defra (2011) *Flood Risk and Insurance: A roadmap to 2013 and beyond*. Defra, London.

Defra (2005) *Making Space for Water: Taking forward a new government strategy for flood and coastal erosion risk management in England*. Defra, London.

Dill, J., Neal, M., Shandas, V., Luhr, G., Adkins, A. and Lund, D. (2010) *Demonstrating the benefits of green streets for active aging: final report to EPA*. Final Report submitted to US Environmental Protection Agency Agreement. Number: CH-83421301. Portland State University, Portland.

DTI (2006) *Sustainable drainage systems: A mission to the USA*. Department of Trade and Industry, London.

Dunnett, N. and Muhammad, Q. (2000) Perceived benefits to human wellbeing of urban gardens. *HortTechnology* 10(1) 40–45.

Environment Agency (2009a) *Flooding in England: A National Assessment of Flood Risk*. Environment Agency, Bristol.

Environment Agency (2009b) *Flooding in Wales: A National Assessment of Flood Risk*. Environment Agency, Bristol.

EPA (2013) *Stormwater to Street Trees: Engineering Urban Forests for Stormwater Management*. United States Environmental Protection Agency, Washington DC.

Everett, G., Lamond, J., Morzillo, A., Chan, F.K.S. and Matsler, A.M. (2015) Sustainable drainage systems: helping people live with water. *Proceedings of the ICE – Water Management*. In press: dx.doi.org/10.1680/wama.14.00076

Everett, G., Lamond, J., Morzillo, A., Chan, F.K.S. and Matsler, A.M. (2016) Delivering green streets: An exploration of changing perceptions and behaviours over time around bioswales in Portland, Oregon. *Journal of Flood Risk Management*. In press: on-linelibrary.wiley.com/doi/10.1111/jfr3.12225/full

Evers, M., Jonoski, A., Maksimović, C., Lange, L., Ochoa Rodríguez, S. *et al.* (2012) Collaborative modelling for active involvement of stakeholders in urban flood risk management. *Natural Hazards and Earth System Sciences* 12(9) 2821–2842.

Fielding, J., Burningham, K. and Thrush, D. (2007) *Public Response to Flood Warning*. Environment Agency, Bristol.

Floyd, M.F., Gramann, J.H. and Saenz, R. (2009) Ethnic factors and the use of outdoor recreation areas: the case of Mexican Americans. *Leisure Sciences* 15, 83–98.

Franklin, J.F. (2008) Preserving biodiversity: Species, ecosystems, or landscapes? *Ecological Applications* 3(2) 202–205.

Fuller, R.A., Irvine, K.N., Devine-Wright, P., Warren, P.H. and Gaston, K.J. (2007) Psychological benefits of greenspace increase with biodiversity. *Biology Letters* 3(4), 390–394.

Graham, A., Day, J., Bray, B. and Mackenzie, Z. (2012) *Sustainable Drainage Systems: Maximising the Potential for People and Wildlife. A Guide for Local Authorities and Developers*. RSPB, London.

Hanley, N., Spash, C. and Walker, L. (1995) Problems in valuing the benefits of biodiversity protection. *Environmental and Resource Economics* 5(30) 249–272.

Harries, T. (2012) The anticipated emotional consequences of adaptive behavior – impacts on the take-up of household flood-protection measures. *Environmental Planning A* 44(3) 649–668.

Harries, T. (2010) Household Flood Protection Grants: The householder perspective. Defra and Environment Agency Flood and Coastal Risk Management Conference, Telford.

Henning, D. (2015) *Social Dynamics of Stormwater Management: Private Lands in the Alley Creek Watershed*. Yale School of Forestry and Environmental Studies, Yale.

Hostetler, M., Allen, W. and Meurk, C. (2011) Conserving urban biodiversity? Creating green infrastructure Is only the first step. *Landscape and Urban Planning* 100(4) 369–371.

HR Wallingford (2003) *An Assessment of the Social Impacts of Sustainable Drainage Systems in the UK*. Report SR 622. HR Wallingford, Wallingford.

Hudson, D., Seah, L.-H., Hite, D. and Haab, T. (2004) Telephone presurveys, self-selection, and non-response bias to mail and internet surveys in economic research. *Applied Economics Letters* 11 (4) 237–240.

Johnson, C.L. and Priest, S.J. (2008) Flood risk management in England: A changing landscape of risk responsibility? *International Journal of Water Resources Development* 24(4) 513–525.

Kaźmierczak, A. and Bichard, E. (2010) Investigating homeowners' interest in property-level flood protection. *International Journal of Disaster Resilience in the Built Environment* 1(2) 157–172.

Kenyon, W. (2007) Evaluating flood risk management options in Scotland: A participant-led multi-criteria approach. *Ecological Economics* 64(1) 70–81.

King, D.A. (2004) Climate change science: adapt, mitigate, or ignore? *Science* 303(5655) 176–177.

LaBadie, K. (2010) *Identifying Barriers to Low Impact Development and Green Infrastructure in the Albuquerque Area*. University of New Mexico, Albuquerque.

Lamond, J.E. and Proverbs, D.G. (2009) Resilience to flooding: lessons from international comparison. *Urban Design and Planning* 162(2) 63–70.

Larson, K.L. and Lach, D. (2008) Participants and non-participants of place-based groups: An assessment of attitudes and implications for public participation in water resource management. *Journal of Environmental Management* 88(4) 817–830.

Laska, S.B. (1986) Involving homeowners in flood mitigation. *Journal of the American Planning Association* 52(4) 452–466.

Ludy, J. and Kondolf, G.M. (2012) Flood risk perception in lands 'protected' by 100-year levees. *Natural Hazards* 61(2) 829–842, 2012.

Mayer, H. and Provo, J. (2004) The Portland Edge in Context. In Ozawa, C. (ed.) *The Portland Edge: Challenges and Successes in Growing Communities*. Island Press, Washington, DC.

McKissock, G., Jefferies, C. and D'Arcy, B.J. (1999) An assessment of drainage best management practices in Scotland. *Water and Environment Journal* 13(1) 47–51.

Netusil, N.R., Levin, Z., Shandas, V. and Hart, T. (2014) Valuing green infrastructure in Portland, Oregon. *Landscape and Urban Planning* 124, 14–21.

O'Sullivan, J.J., Bradford, R.A., Bonaiuto, M., De Dominicis, S., Rotko, P., Aaltonen, J., Waylen, K. and Langen, S.J. (2012) Enhancing flood resilience through improved risk communications. *Natural Hazards and Earth System Sciences* 12, 2271–2282.

Pincetl, S. and Gearin, E. (2013) The reinvention of public green space. *Urban Geography* 26(5) 365–384.

Portney, K.E. (2013) [2003] *Taking Sustainable Cities Seriously: Economic Development, the Environment, and Quality of Life in American Cities* (2nd Edn). MIT Press, Cambridge, MA.

Purvis, A. and Hector, A. (2000) Getting the measure of biodiversity. *Nature* 405, 212–219.

Qureshi, S., Brueste, J.H. and Lindley, S.J. (2010) Green space functionality along an urban gradient in Karachi, Pakistan. A socio-ecological study. *Human Ecology* 38, 283–294.

Ryan, R. and Brown, R.R. (2000) The value of participation in urban watershed management. *Watershed 2000 Conference*, Vancouver, Canada.

Reinhardt, G. (2011) Portland advances green stormwater management practices. In: Kemp, R.L. and Stephani, C.J. (eds.) *Cities Going Green: A Handbook of Best Practices*. Shutterstock, Jefferson, North Carolina, USA.

Scottish Government (2003) *The Water Environment and Water Services (Scotland) Act 2003* (Commencement No 8) Order 2008. Edinburgh, The Scottish Government.

Seymour, M., Wolch, J., Reynolds, K.D. and Bradbury, H. (2010) Resident perceptions of urban alleys and alley greening. *Applied Geography* 30(3) 380–393.

Shandas, V. (2015) Neightborhood change and the role of environmental stewardship: A case study of green infrastructure for stormwater in the city of Portland, Oregon, USA. *Ecology and Society* 20(3): 16, 10.5751/ES-07736-200316.

Shandas, V., Nelson, A., Arendes, C. and Cibor, C. (2010) *Tabor to the River: An Evaluation of Outreach Efforts and Opportunities for Engaging Residents in Stormwater Management*. Bureau of Environmental Services, City of Portland.

Singleton, D. (2012) *SuDS in the Community: A Suitable Case for Treatment?* Susdrain, CIRIA, London.

Slavin, M.I. and Snyder, K. (2011) Strategic climate action planning in Portland. In Slavin, M.I. (ed.) *Sustainability in America's Cities: Creating the Green Metropolis*, Washington DC, Island Press.

Speller, G. (2005) *Improving Community and Citizen Engagement in Flood Risk Management Decision Making, Delivery and Flood Response*. Environment Agency, Bristol.

Thurston, N., Finlinson, B. and Breakspear, R. (2008) *Developing the Evidence Base for Flood Resistance and Resilience: Summary Report*. Environment Agency, Bristol.

Traver, R.G. (2009) *Efforts to Address Urban Stormwater Run-Off. Address Before the Subcommittee on Water Resources and Environment Committee on Transportation and Infrastructure US House of Representatives, March 19, 2009*. US Government Printing Service, Washington, DC. Available from: www.gpo.gov/fdsys/pkg/CHRG-111hhrg48237/html/CHRG-111hhrg48237.htm [May 2014].

Tunstall, S.M., Penning-Rowsell, E.C., Tapsell, S.M. and Eden, S.E. (2000) River restoration: Public attitudes and expectations. *Journal of CIWEM* (14) 363–370.

Wedawatta, G.S.D., Ingirige, M. and Proverbs, D. (2011) Adaptation to flood risk: the case of businesses in the UK. International Conference on Building Resilience. Dambulla, Sri Lanka.

Werritty, A., Houston, D., Ball, T., Tavendale, A. and Black, A. (2007) *Exploring the Social Impacts of Flood Risk and Flooding in Scotland*. Scottish Executive, Edinburgh.

White, I. and Richards, J. (2008) Stakeholder and community engagement in flood risk management and the role of AAPs. *Flood Risk Management Research Consortium, July*. University of Manchester, Manchester.

Whitehead, J.C. (1991) Environmental interest group behavior and self-selection bias in contingent valuation mail surveys. *Growth and Change* 22(1) 10–20.

Wise, S., Braden, J., Ghalayini, D., Grant, J., Kloss, C., *et al.* (2010) Integrating valuation methods to recognize green infrastructure's multiple benefits. In: Struck, S. and Lichten, K. (eds) *Low Impact Development 2010: Redefining Water In The City*, April 11–14 2010, pp.1123–1143. American Society of Civil Engineers, Reston, VA, USA.

Section 6 Global Sustainable Surface Water Management

Sustainable Drainage Out of the Temperate Zone: The Humid Tropics

Susanne M. Charlesworth and Margaret Mezue

22.1 Introduction

Use of sustainable drainage systems (SuDS) in temperate areas is reasonably well researched and established as a surface water management strategy. However, this is not the case in those regions of the world that are classified as having a tropical climate (i.e. those 30° either side of the equator). In these regions, there are two seasons: wet and dry, whereas temperate areas tend to have rainfall throughout the year. Maksimović *et al.* (1993) detail the specific climate-related problems associated with urban drainage in the humid tropics; these are mostly associated with the intense rainfall and high temperatures experienced (Table 22.1).

Associated with climate are other factors that need to be taken into consideration when designing drainage in the tropics, and probably the most important of these are disease vectors such as mosquitoes, which can carry diseases like malaria, and nuisance animals such as snakes. While it is unlikely that snakes would be present in urban areas, the use of snake repellents or 'pest proofing' of some sort should be encouraged. It is important that native vegetation is used in any device, and thus choosing the plantings for use in the tropics is no different. Regardless of climate, the multiple benefits of the sustainable drainage approach are the same, addressing water quantity and quality, biodiversity and amenity, as well as urban heat island reduction and climate change adaptation and mitigation.

Chapters in this volume already discuss the many problems associated with designing SuDS in countries such as Brazil and South Africa; this chapter also presents progress in India, Colombia, Chile and particularly Malaysia, where SuDS have been embraced and redesigned to take account of their specific climatic and ecological problems. Problems associated with many developing countries include the lack of drainage infrastructure, or if there is any, it is degraded, and no longer functions (Figure 22.1). However, this does

Sustainable Surface Water Management: A Handbook for SuDS, First Edition.
Edited by Susanne M. Charlesworth and Colin A. Booth.
© 2017 John Wiley & Sons, Ltd. Published 2017 by John Wiley & Sons, Ltd.

Table 22.1   Urban drainage problems associated with the humid tropics (after da Silveira *et al.*, 2001).

Climate-driven factor	Result	Impact
Rainfall is high intensity	High volumes of runoff in a short time Increased erosivity Increased energy of environment to carry larger sediment loads	Peak flows with higher volumes More sediment produced High capacity to transport solids
Rainfall volumes large; more days with rainfall per annum	Residence time in storm sewer systems longer due to high volumes Residence times of material carried in stormwater longer Few dry days	High volumes of stormwater need to be managed with associated larger volumes of pollutants, waste and sewage
Temperatures high	Disease-carrying vectors thrive and proliferate	Risk of disease

Figure 22.1   Blocked drainage system in Ijora, Lagos State, Nigeria.

provide a clean slate on which SuDS could be installed as a first option, rather than the problems associated with lack of capacity in the storm sewer systems in developed countries around the world. A further problem is associated with informal settlements, which present additional complications. Where SuDS are implemented, there is generally a lack of data on the efficiency of the devices used; thus, examples are rather limited.

22.2 Modification of the Urban Hydrological Cycle by Urbanisation in Tropical Countries

The process of urbanisation in the tropics is no different from elsewhere, but countries have approached their mitigation strategies in a variety of ways to suit their specific conditions. Thus, Chile and India have both revisited their building and development legislation to encourage the implementation of SuDS (Parkinson and Mark, 2005). After suffering increasing problems with flooding due to rapid urbanisation of its major cities, in 1997, the Chilean government brought in the Stormwater Act, which required the installation of sustainable stormwater mitigating devices in all new developments (Parkinson and Mark, 2005). In Malaysia, similar negative impacts led to reduced infiltration and associated lack of groundwater recharge, as well as flooding and polluted waterways (Sidek *et al.*, 2002). Malaysia, therefore, embraced the use of SuDS to manage runoff, installing various SuDS devices and management trains throughout the country, based on the Malaysian drainage manual (Ghani *et al.*, 2008). In Rio de Janeiro, Brazil, green roofs have been encouraged in order to reduce flooding due to its overloaded storm sewer system, and in Colombia, Campuzano Ochoa *et al.* (2015) observed an increase in the frequency and intensity of rainfall with unprecedented rainy seasons during 2010 and 2011, when there were 1734 floods reported, making up 45% of those occurring between 1998 and 2008. There were hundreds of deaths, and more than three million inhabitants were affected (Hoyos *et al.*, 2013). While sustainable drainage is being considered to address these issues in Colombia, it is relatively recent, is not used nationwide and, as with Brazil, is focused on storm attenuation rather than improving water quality. This is reflected in the study by Ávila and Díaz (2012) in Colombia, which investigated reduction of peak volumes when using SuDS techniques.

22.3 Vegetated Devices

As mentioned earlier, native vegetation is always best when designing any SuDS device or management train, whatever the climate. However, while information is available regarding the properties and uses of plantings in temperate regions (e.g. Woods Ballard, 2015; Charlesworth *et al.*, 2016), little is known about, for example, the pollutant retention capabilities of tropical vegetation.

22.3.1 Green Roofs

Köhler *et al.* (2001) list the following differences between the use of green roofs in tropical climates compared with temperate ones:

1. Intense storms are far more frequent, what is a 100-year storm in temperate climates may occur annually in the tropics. This will result in the green roof becoming saturated very quickly, and also erosion may be a problem.
2. Temperatures are much higher, all year round, resulting in constant vegetation growth and year-round evapotranspiration. This may have implications for the increased biomass produced, impacting on drainage through the substrate, interception of rainfall due to the vigorous growth of the plants and the need for maintenance.
3. Dense vegetation may encourage nuisance biota, such as disease vectors – mosquitoes in particular. Plants such as bromeliads, which trap water, should therefore be avoided.

Extensive, rather than intensive, green roofs would be suitable for the tropics, as they are cheaper and easier to maintain, while not offering the recreational benefits of intensive roofs, or roof gardens. Characteristic features of plants suitable for green roofs are that they should be hardy, low-growing, drought-resistant and fire-resistant. They should provide dense cover, and be able to withstand heat, cold and high winds, requiring minimal maintenance, with shallow roots to avoid penetrating the roof membrane and causing leaks (Getter and Rowe, 2008). Popular plants used for this technique include *Sedum* and *Delosperma*, but both of these are from temperate areas, and are not native to the tropics.

Green roofs are fairly widely used; for example, they have been installed in low-income areas of Colombia's capital, Bogotá, which has a sub-tropical highland climate (Forero *et al.*, 2011; Forero Cortés and Devia-Castillo, 2012). There have been rather limited studies of green roofs in Nigeria (Ezema *et al.*, 2015), but they would appear to be useful, since they do not take up any land, and have a variety of environmental benefits, not least their ability to reduce the temperature in urban areas. According to Getter and Rowe (2008), the aloe, which is a Nigerian native species (Figure 22.2a), can work well on green roofs since it has the necessary characteristics. Furthermore, it was found that *Tectorum*, a grass that grows spontaneously on roofs in Nigeria (Figure 22.2b), was thought viable (Köhler *et al.*, 2001, 2004). However, according to Ezema *et al.* (2015), there are a number of barriers to the implementation of green roofs in Nigeria, particularly in Lagos. These include the costs of construction and maintenance, lack of regulation by government or understanding of their role, no incentives to encourage their use and minimal technical expertise in terms of their installation. These are all problems found in other areas of the world too, so Nigeria is not exceptional in this regard. The authorities are, however, driving a green agenda, which promotes the use of green infrastructure in the provision of parks and gardens (Ezema and Oluwatayo, 2014), and green roofs and walls could be included in this approach, but currently they are not.

22.3.2 Wetlands and Swales

Open waters, such as ponds and basins, are not suitable in the tropics due to issues around disease vectors, as mentioned. However, it is possible to control mosquitoes in artificial wetlands by means of appropriate design and suitable management strategies (Knight *et al.*, 2003), such that there is minimal difference in comparison with natural wetlands. Hence, in Colombia, constructed wetlands were built alongside existing natural wetlands in order to control polluted urban runoff (Lara-Borrero, 2010). Linear parks, with large vegetation-lined swales, have also been used; designed to mimic streams, they contain measures such as rock-built leaky dams designed to slow the flow of water, and hence should reduce flooding and erosion. Unfortunately, these systems were not monitored, so their benefits are largely unknown.

Swales would also be suitable in countries, such as Nigeria, due to their relatively low cost of construction and maintenance, but they would need to be designed to convey water in the subsurface, to avoid encouraging disease vectors associated with slow-moving open water. Swales are also effective for reasonably small areas, less than 2 ha, and would therefore have a role as a pre-treatment phase in any SuDS management train. Swales need to be designed and constructed correctly, incorporating suitable vegetation that has deep root systems, the ability to grow under extreme conditions, vigorous growing habit, high stem density to enhance the reduction of flow rate, while facilitating sedimentation, tolerance to

(a)

(b)

Figure 22.2   (a) (top) Aloe sp. (freely available: Erin Silversmith); (b) (bottom) Tectorum sp. By Andrew massyn – Own work, Public Domain, https://commons.wikimedia.org/w/index.php?curid=2091770.

flooding and the ability to take up pollutants (Woods Ballard, 2015). There are, for example, several species of grasses native to the tropics that fulfil most of these criteria, including *Vetiveria fulvibarbis* (Trin.), Stapf or vetiver grass. This is a herbaceous plant, occurring on coastal plains, which grows up to about 2 m, with 2 m deep fibrous roots, which spread vertically forming a dense mat helping to bind the soil together. Due to its growing habit (Figure 22.3) excessive flows of water spread out around the plants, thus slowing the flow

Figure 22.3   Vetifer grass (*Vetiveria fulvibarbis*) (Wikimedia Commons).

and promoting infiltration. They are able to survive inundation and harsh conditions and can grow in water. It has been used in several countries to prevent soil erosion (NRC, 1993; Maffei, 2002).

A second suitable plant native to the tropics is Cynodon, commonly called Bermuda grass (Figure 22.4a). It is a creeping grass, forming a dense mat wherever a node grows; it has a deep root system and can survive extreme drought, thrives in poor soil and, furthermore, has the potential to grow back after fire. *Cymbopogon*, commonly known as lemon grass (Figure 22.4b), has been used to stabilise embankments (Watkins and Fiddes, 1984), and shares similar characteristics to those plants required for swales. Although these plants may prove to be suitable for their use in swales, there has been no empirical information on the pollution removal performance of these grasses, which stems from a lack of research.

22.3.3   Using Green Infrastructure to Mitigate and Adapt to Climate Change

Govindarajulu (2014) acknowledges the higher susceptibility of tropical cities to climate change, particularly cyclones and floods, highlighting urban green infrastructure as a cost-effective and ecosystem-based means of climate adaptation in cities in India. The recommendation is made that a strategy of planning for green space is considered, specifically for the Indian context. The urban heat island (UHI) effect in cities, such as Kolkata, Mumbai and Bangalore, shows a substantial rise in temperature; for example, in Bangalore the rise was found to be about 2 °C. In Mumbai, due to extreme rainfall and resultant flooding in 2005, flood waters rose by 0.5–1.5 m in low-lying areas. It was therefore considered that green infrastructure could assist in addressing these concerns but other approaches, such as rainwater harvesting, could also be utilised to give multiple benefits.

(a)

(b)

Figure 22.4 (a) Bermuda grass (top). (Mike (Own work) [CC BY-SA 3.0 (http://creativecommons.org/licenses/by-sa/3.0) or GFDL (http://www.gnu.org/copyleft/fdl.html)], via Wikimedia Commons; (b) Lemon grass (below). (Vaikoovery – Own work) [GFDL (http://www.gnu.org/copyleft/fdl.html) or CC BY 3.0 (http://creativecommons.org/licenses/by/3.0)], via Wikimedia Commons).

The most information available on the application of vegetated SuDS in tropical regions is from Malaysia, where management trains have been tested in the laboratory and also installed and monitored at the field scale. These systems have been designed with all the concerns in mind expressed over using SuDS in the tropics, and have also been monitored, and their efficiency analysed. The following sections therefore describe the systems themselves and also the results of the experiments and field testing.

22.4   Case Study: Sustainable Drainage in Malaysia

In order to address specific concerns associated with the Malaysian climate and ecology, Bio-Ecological drainage system or BIOECODS was developed by the River Engineering and Urban Drainage Research Centre (REDAC) and Universiti Sains Malaysia (USM) (Zakaria *et al.*, 2003; Parkinson and Mark, 2005; Ghani *et al.*, 2008). They are management trains made up of three main components: bioecological swales, a biofiltration stage and an ecological pond, based on dealing with surface water at source. Native plants were used in the vegetated devices, such as cow grass (*Axonopus compressus*) (Figure 22.5).

The use of BIOECODS promotes:

- the infiltration of stormwater runoff from impermeable areas
- storage of excess stormwater and its gradual release, thus attenuating the storm peak
- improvement in water quality as the water passes through the management train (Sidek *et al.*, 2002; Parkinson and Mark, 2005).

Preliminary testing of the drainage module at the laboratory scale was encouraging, with the flow found to be between that of an open channel and a pipe. It was thought that the

Figure 22.5 Cow grass (*Axonopus compressus*) as used in the bioecological swale component of BIOECODS (freely available: Harry Rose, South West Rocks, Australia).

turbulent flow that developed helped dissolved oxygen concentrations to increase with distance in the drainage module to 4.5–7.5 mg/l. However, reduction of pollutants was less successful, although those tested did decrease: Zn by 10.9%, Cu 38.7%, Ni 33.3% and Pb 15.9% (Sidek *et al.*, 2002).

Several of these BIOECODS management trains have been implemented in Malaysia, but the following three case studies, Universiti Sains Malaysia, the Taiping Health Clinic and the rehabilitation of previous mining ponds, have been the most thoroughly described.

22.4.1 USM Engineering Campus

This served as a pilot study for the BIOECODS concept and was designed to provide an attractive landscape with integrated flood resilience. Due to site conditions, stormwater had to be infiltrated on site where possible (Sidek *et al.*, 2002). The train itself was made up of a perimeter and ecological swale, a dry pond, with detention on-site, a wet pond, a detention pond and also a wetland (Figure 22.6).

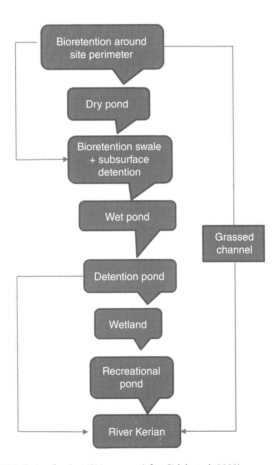

Figure 22.6   BIOECOD design for the USM campus (after Sidek *et al.*, 2002).

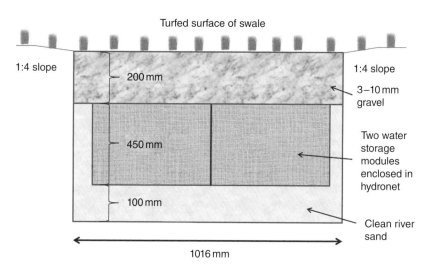

Figure 22.7   Cross-section of a biological swale with underground storage (after Ghani *et al.*, 2008).

The system combined infiltration, delayed flow, storage and treatment of runoff, all designed specifically for the Malaysian climate and environment, beginning with the biological swale which was designed to address a ten-year flood. It had a longitudinal slope of 1 in 1000 and a lateral slope of 1 in 4 and was underlain by a drainage module enclosed with permeable hydronet filter fabric, the function of which was to prevent fines clogging the structure (Lai *et al.*, 2009). The drainage module was then set within a clean river sand bed to further encourage infiltration and provide some treatment (Figure 22.7). Thus, there were two means of infiltration: through the surface of the swale and into the subsurface. Source control was not the only approach, since larger regional controls were used in the form of detention ponds, which encouraged the settling of solids and also provided biological treatment; constructed wetlands were incorporated after the pre-treatment phase for further control and treatment, before runoff was conveyed to a recreational pond, the last phase in the design before discharge to the receiving watercourse (Sidek *et al.*, 2002; Ghani *et al.*, 2008).

Water quality assessment of the management train at USM found that average pH, dissolved oxygen, biological oxygen demand (BOD), chemical oxygen demand (COD), turbidity, total suspended solids and NH_3-N were all well within Malaysian Standards. Most sites tested were below the limits of detection for both oil and grease, and total solids (Ghani *et al.*, 2008). In terms of water quantity, the discharge at the outflow of the management train during a 1 in 100 year event was virtually unmeasurable for the majority of the storm.

22.4.2 Taiping Health Clinic

A chain of SuDS techniques was installed on the 3 ha site, where the Taiping Health Clinic is located in the Larut and Matang district in Perak; they comprise one major and a few smaller control facilities to manage runoff (Ghani *et al.*, 2008); the design is shown in Figure 22.8. In detail, the design included a grassed swale to manage excess runoff from the

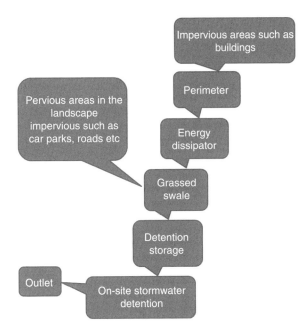

Figure 22.8   Design concept of the drainage system for the Taiping Health Centre (after Ghani *et al.*, 2008).

perimeter of the site, as well as the management of any flows from surrounding pervious and impermeable surfaces. Excess stormwater was stored in subsurface detention located at the connecting points, junctions and critical areas throughout the site; these detention systems were designed to regulate flow velocity, reduce runoff quantity via detention and also increase water quality via sedimentation and filtration processes. A detention pond, designed to blend into the existing landscape, stored excess runoff, up to 600 mm of excess rainfall for an average recurrence interval of ten years. The pond would be emptied in less than 24 hours by various orifices in order to prepare the system to receive the next storm. The whole design allowed for multiple uses of the landscape, such as aesthetics and recreation, with runoff controlled at pre-construction rates (Ghani *et al.*, 2008).

22.4.3 Rehabilitation of Ex-Mining Ponds and Wetlands

The third project was the rehabilitation of some ex-mining ponds and existing wetlands to provide an integrated stormwater management facility, with multi-functional uses for recreation, water reuse and stormwater retention. The site covered 36 ha in the Kinta district, situated on former mining land with a predominantly sandy soil and sparse vegetation. Two ex-mining ponds were located on the site; in general, the whole area was swampy, but particularly along its eastern border and at some points spreading outside the ex-mining area. Any existing drainage on site was conventional, designed to remove runoff quickly to a concrete roadside drain, before channelling it to the River Kinta via a pumping station. Stormwater from the area also flowed into the two ponds that were connected to the river by an earth drain (Ghani *et al.*, 2008).

The first pond was used as a regional stormwater device whose purpose was to control the quantity and quality of water on the site itself and from surrounding areas. The second pond was used to control and treat runoff from only the site itself. The conveyance system, while largely engineered, mimicked the features of natural rivers with connecting swales to transport runoff. Further treatment of runoff from areas outside the site was provided by a constructed wetland connected to the first pond, flowing into the engineered section and into the river via a controlled tidal gate (Ghani *et al.*, 2008).

22.4.4 Lessons Learnt from the Malaysian Experience

Implementation of SuDS in Malaysia has demonstrated that SuDS are a suitable approach for use in developing countries, and could be designed specifically for those with tropical climates (Parkinson and Mark, 2005). However, the Malysian experience of implementing SuDS did produce some problems, and recommendations to address these. Sidek *et al.* (2002), highlight that some, such as ownership, operation and maintenance, are common regardless of the landscape, political, environmental or climatic setting.

Suggestions included:

- there needs to be engagement with the public on the SuDS concept and benefits
- stakeholders need to be engaged as early as possible during design and planning
- long-term maintenance issues should be addressed as soon as possible
- the developer and concerned authority should liaise throughout
- the developer should provide the land, capital and landscaping costs.

From the point of view of constructing the management trains, Sidek *et al.* (2002) recommend:

- grassed channels are not efficient for areas larger than 2 ha
- swales should be installed after construction of the building and associated landscaping
- swales in particular need to be constructed correctly; if the slope is too steep, for example, there would be insufficient time for water quality improvements to take place
- the use of dry swales is recommended over wet swales in tropical climates because of the nuisance of breeding mosquitoes
- thick vegetation cover is required for the system to function efficiently; the grass should be at least 5 cm higher than the designed flow depth.

22.5 Conclusions

Designing and installing sustainable drainage in tropical countries has particular require-ments, since they are mainly located in developing countries, where the population is increasing quickly, and informal settlements house many of the previous rural poor who migrate there seeking employment. In many of these countries, SuDS is a new concept, and the lack of understanding and of guidelines, which would explain the multiple benefits it brings, means that it is unlikely to be embraced in the short term. Since urban development in cities is one of the current major global challenges (Tucci, 2002), the introduction of a sustainable means of addressing excess surface water is key to reducing flood risk, improving both residents' quality of life and the surrounding environment. Frequently, existing drainage is not fit for purpose, so to some extent it is a clean slate, and thus introducing SuDS is made more accessible.

Design of SuDS, however, needs to take account of the climate, whereby regular intense short-duration storms, allied with the high temperatures extant in these regions, mean that short-term storage of large volumes of water is a key factor. However, care needs to be taken in the design of such systems to take account of both disease vectors and nuisance animals, and therefore the use of open water in ponds is to be discouraged. Wherever a SuDS design is to be made, regardless of climate, there needs to be a clear drainage route through the proposed development established at the earliest phase of planning. Particularly in countries unfamiliar with SuDS, the design should:

- be simple enough to enable residents, developers and engineers understanding, easy to construct, using existing materials, skills and technology
- be robust, to allow simple maintenance, repairs or replacement to be carried out
- have funding in place for costs, such as construction and maintenance, as soon as possible
- use native vegetation in green SuDS.

This chapter, however, has shown that countries in the tropics are beginning to see the value of utilising the SuDS approach and the multiple benefits that ensue. With suitable design, construction and maintenance, these devices and trains can improve the environment in general, and the quality of life and human health, in particular, of societies in tropical countries.

References

Ávila, H. and Díaz, K.S. (2012) Disminución del volume de escorrentía en cuencas urbanas mediante tecnologías de drenaje sostenibles. XX National Seminar on Hydraulics and Hydrology, Barranquilla, Colombia, 8–10 August, 2012.

Campuzano Ochoa, C.P., Roldán, G., TorresAbello, A.E., LaraBorrero, J.A., Galarza Molina, S., et al. (2015) Urban Water in Colombia. In: Urban water challenges in the Americas: A perspective from the Academies of Sciences. Published by The Inter-American Network of Academies of Sciences (IANAS).

Charlesworth, S.M., Bennett, J. and Waite, A. (2016) An evaluation of the use of individual grass species in retaining polluted soil and dust particulates in vegetated sustainable drainage devices. Environmental Geochemistry and Health, 1–13. DOI 10.1007/s10653-016-9791-7.

da Silveira, A.L.L., Goldenfum, J.A. and Fendrich, R. (2001) Urban drainage control. In: C.E.M. Tucci (ed.) Urban drainage in specific climates: Urban drainage in humid tropics. IHP-V, Technical Documents in Hydrology, No 40. UNESCO, Paris.

Ezema, I.C., Ediae, O.J. and Ekhaese, E.N. (2015) Opportunities for and barriers to the adoption of green roofs in Lagos, Nigeria. International Conference on African Development Issues (CU-ICADI) 2015: Renewable Energy Track.

Ezema, I.C. and Oluwatayo, A. (2014) Densification as sustainable urban policy: the case of Ikoyi, Lagos, Nigeria. Proceedings, International Council for Research and Innovation in Building and Construction (CIB) Conference, University of Lagos, 28–30 Jan 2014.

Forero, C., Devia, C., Torres, A. and Méndez-Fajardo, S. (2011) Diseño de ecotechos productivos para poblaciones vulnerables. Revista Acodal. 229, 28–36.

Forero Cortés, C. and Devia-Castillo, C.-A. (2012) Green Roof Productive System in Vulnerable Communities: Case Study in La Isla Neighborhood, Altos de Cazucá, Soacha, Cundinamarca. Ambiente y Desarrollo, Bogotá (Colombia) XVI, 30, 21–35.

Getter, K. and Rowe, B. (2008) Selecting Plants for Extensive Green Roofs in the United States. Extension Bulletin E-3047. Available at: http://tinyurl.com/o6llg9o.

Ghani, A.Ab., Zakaria, N.A., Chang, C.K. and Ainan, A. (2008) Sustainable Urban Drainage System (SUDS) – Malaysian Experiences. 11th International Conference on Urban Drainage, Edinburgh, Scotland, UK, 2008 pp. 1–10.

Govindarajulu, D. (2014) Urban green space planning for climate adaptation in Indian cities. *Urban Climate* 10, 35–41.

Hoyos, N., Escobar, J., Restrepo, J., Arango, A. and Ortiz, J. (2013) Impact of the 2010–2011 La Niña phenomenon in Colombia, South America: The human toll of an extreme weather event. *Applied Geography*, 39, 16–25.

Knight, R.L., Walton, W.E., O'Meara, G.F., Reisen, W.K. and Wass, R. (2003) Strategies for effective mosquito control in constructed treatment wetlands. *Ecological Engineering* 21, 211–232.

Köhler, M., Schmidt, M., Grimme, F.W., Laar, M. and Gusmão, F. (2001) Urban Water Retention by Greened Roofs in Temperate and Tropical Climate. In: 38th IFLA Congress, Singapore.

Köhler, M., Schmidt, M. and Laar, M. (2004) Roof gardens in Brazil. Available at: http://www.gruen dach-mv.de/en/RIO3_455_M_Koehler.pdf

Lai, S.H., Kee, L.C., Zakaria, N.A., Ghani, A.Ab., Chang, C.K. and Leow, C.S. (2009) Flow Pattern and Hydraulic Characteristic for Sub-surface Drainage Module. International Conference on Water Resources, 26–27 May 2009, Langkawi, Kedah, Malaysia.

Lara-Borrero, J. (2010) Humedales construidos para el control de la contaminación proveniente de la escorrentía urbana. *Acodal magazine* 226, 1, 19–27.

Maksimović, C., Todorovic, Z. and Braga KJr, B.P.F. (1993) Urban drainage problems in the humid tropics Hydrology of Warm Humid Regions (Proceedings of the Yokohama Symposium. July 1993). IAHSP iSl. No 216, 377–401.

Maffei, M. (2002) *Vetiveria: The Genus Vetiveria*. Taylor and Francis, London.

National Research Council (NRC). (1993) Vetiver grass: a thin green line against erosion. Board on Science and Technology for International Development. National Academies Press. Available at: http://tinyurl.com/nnympl6.

Parkinson, J. and Mark, O. (2005) Urban Stormwater Management in Developing Countries, London: IWA publishing.

Sidek, L.H., Takara, K., Ghani, A., Zakaria, A. and Abdullah, R. (2002) Bio-Ecological Drainage Systems (BIOECODS): An integrated approach for urban water environmental planning. Seminar on water environmental planning: technologies of water resources management. Available at: http://redac.eng.usm.my/html/publish/2002_11.pdf.

Tucci, C.E.M. (2002) Improving Flood Management Practices in South America: Workshop for Decision Makers. Available at: http://www.wmo.int/pages/prog/hwrp/documents/FLOODS_IN_SA.pdf.

Watkins, L.H. and Fiddes, D. (1984) Highway and urban hydrology in the tropics. Pentech Press, Devon, UK.

Woods Ballard, B., Wilson, S., Udale-Clarke, H., Illman, S., Ashley, R. and Kellagher, R. (2015) The SuDS Manual. CIRIA. London.

Zakaria, N.A., Ab Ghani, A., Abdullah, R., Mohd Sidek, L. and Ainan, A. (2003) Bio-ecological drainage system (BIOECODS) for water quantity and quality control, *International Journal of River Basin Management*, 1, 3, 237–251.

Sustainable Drainage Systems in Brazil
Marcelo Gomes Miguez and Aline Pires Veról

23.1 Introduction

The concept of sustainable drainage systems (SuDS) has developed over the past decades, with its origins (for many people) dating back to the 1970s. Nowadays, it is a well-known and reasonably worldwide concept, but it is very far from being considered simple to implement. SuDS encompasses a broad approach, addressing flood control, water quality improvement, urban revitalisation and increasing amenity value, while adding biodiversity benefits. It is not just a matter of a technical approach, but also requires community involvement, a legal and institutional framework and viable economic arrangements in order to have a functional SuDS solution in place. The process of urbanisation and urban flood control has to work together.

The impacts caused by floods have largely increased due to human activities impacting the natural environment, with progressive change to natural surfaces and human occupation of areas that would naturally flood. The fast urbanisation process that took place from the beginning of the Industrial City period, associated with inadequate land use control, led to a great increase in impervious areas, a reduction in the natural retention capacity of the soil and, consequently, to higher peak discharges, flow velocities and runoff volumes. It is also from the period of the Industrial City that the traditional approach for designing urban drainage systems arose. Canalisation and end-of-pipe solutions were proposed as an effective way of avoiding water-borne diseases that greatly affected cities.

In developing countries, as in the case of Brazil, the combination of urbanisation and urban flooding is even worse, due to late industrialisation that concentrated fast urban growth into the final half of the 20th century. Taking this into consideration, Miguez *et al.* (2007) suggest that the following result:

- large population growth in a short period of time
- unplanned and/or non-controlled urbanisation

Sustainable Surface Water Management: A Handbook for SuDS, First Edition.
Edited by Susanne M. Charlesworth and Colin A. Booth.
© 2017 John Wiley & Sons, Ltd. Published 2017 by John Wiley & Sons, Ltd.

- housing policies unable to prevent and avoid illegal occupation, accompanied by a large number of substandard dwellings in informal city settlements
- occupation of flood risk areas, both legal (due to lack of information about flood zoning) and illegal (due to social pressures)
- weak coverage of sanitation infrastructure
- low qualification of municipal technical staff
- low level of environmental education

In this context, this chapter will make a brief overview of SuDS in Brazil, in response to this challenge of fast and (mostly) uncontrolled urban growth. This is a great challenge, because Brazil is a very large country composed of a mosaic of different realities, ranging from physical to socio-economically diverse aspects. Factors include: climate varying from semi-arid to tropical; highly urbanised cities with millions of people and poor municipalities with a few thousands of people; uneven and unequal sanitation conditions; formal and informal cities growing side by side; a highly unequal income distribution. Considering this situation, this chapter will present SuDS in Brazil from a historic perspective, from an academic point of view, and the evolution of the federal legal framework. At the end, some case examples will be presented, focusing on actions developed in the south-east region of Brazil, referring to those in the states of Minas Gerais, São Paulo and Rio de Janeiro.

23.2   The History of SuDS in Brazil – an Academic Perspective

Traditionally, basic concerns about urban drainage systems concentrate on the fast conveyance of storm and wastewaters. However, this approach was rather unsustainable, once it focused on the consequences (the discharge of high runoff generated by urban surfaces) rather than on the causes (the transformation of rainfall into runoff). This approach started to be discussed (and changed) from the 1970s, on the international scene. In fact, Leopold (1968) stated that urbanisation was responsible for modifying hydrology and consequently the increase of runoff volumes and peak discharges. He also envisioned that the impacts of urbanisation over the watershed manifested into three groups: impacts on quantity, quality and basin environmental value.

In parallel, from the 1970s on, concerns about the environment increased, in general terms, considering nature and human development as linked subjects. The creation of the Club of Rome in 1968 (Club of Rome, 1968) was a first step, and its reports prompted several other worldwide studies that sought a better understanding of the relationship between human society and nature. In June 1972, the United Nations Conference on the Human Environment (UNEP, 1972) was held, bringing together more than 110 countries in Stockholm, Sweden. From this meeting, better known as the Stockholm Conference, a new concept of sustainable development was proposed, which would subsequently be included in the development agenda of many countries. The concept related to sustainable development and implied that the fulfilment of present needs should not overcome the needs of future generations. In drainage system terms, a sustainable approach, for instance, should avoid transferring floods in space and time.

In Brazil, this discussion started a little bit late, and an important milestone in his process was the United Nations Conference on Environment and Development, also known as the

Rio Summit, Rio Conference or Earth Summit, held in Rio de Janeiro in 1992 (UNCDE, 1992). The establishment of Agenda 21 (UN, 1993), defined the objectives related to promoting the sustainable development of human settlements:

- providing adequate housing for all
- improving the management of human settlements
- promoting sustainable planning and management of land use
- promoting the integrated provision of environmental infrastructure: water, sewage, drainage and solid waste management.

During the 1990s, different academic groups from Brazilian universities started to work on urban drainage concepts. In 1995, Tucci, Porto and Barros (Tucci *et al.*, 1995) edited the first Brazilian book called *Urban Drainage* (in Portuguese), introducing concepts of sustainable stormwater management. In that same year, Genz and Tucci (1995) started to study on site storage measures to control runoff generation at its source, yet still inside the urban area. Nascimento *et al.* (1997) discussed the importance of floodplains as a matter of planning, to avoid their occupation, guaranteeing space for temporary storage and a more natural environment. However, in the cases where urbanisation had already made important changes in the landscape, these authors also discussed the need to introduce compensatory measures in the drainage system, based on infiltration and storage processes. This concept stressed the proposition of acting to compensate water cycle changes introduced by the urbanisation process, restoring as much as possible the original hydrological functions.

Pompêo (1999) furthered these discussions, by evidencing the collapse of traditional technical solutions for designing drainage systems, and highlighting that a new approach was required, joining technical solutions with social dynamics and multisectoral integrated planning. Even drainage systems that appear technically sound will tend to fail in the long run, if not supported by planned, controlled and equilibrated urban growth; this was also supported by Miguez *et al.* (2014). In this context, Pompêo (2000) commented on typical Brazilian behaviour: several times flood control interventions were guided by isolated actions in response to a critical event or disaster. Pompêo (2000) therefore emphasised the need to think and act preventively, managing the natural and built environment as interdependent and integrated components of the same system. Pompêo (2000) also enunciated six basic principles regarding urban drainage solutions:

1. There is no pure technical–economic solution.
2. There is no simple solution.
3. There is no fast solution.
4. No solution should be the responsibility of just one sector of society.
5. It is not possible to 'copy' a solution from another watershed.
6. The solutions are always tied to a specific context.

Baptista *et al.* (2005) reviewed the experience of SuDS in Brazil up to that point, coining the term 'compensatory techniques' as the basis of Brazilian SuDS. Table 23.1 details the classification of these proposed compensatory techniques in Brazilian urban drainage design.

Table 23.1   Types of compensatory techniques (adapted from Baptista *et al.*, 2005).

Non-structural compensatory techniques	Legislation	
	Flood zoning	
	Urban land use rationalisation	
	Environmental education	
	Preservation of valley bottoms	
Structural compensatory techniques	**In the watershed**	Retention and detention ponds
		Infiltration
	Linear structures	Infiltration trenches
		Swales
	Source structures	Rain gardens
		Green roofs
		Rain barrels
		Permeable pavements

Considering that urbanisation is one of the processes that most affects drainage patterns, Souza *et al.* (2005), discussed and proposed a methodology to apply to new (and sustainable) urban land subdivisions. Briefly, this methodology proposed the following steps:

1. Identify the relevant regulation in terms of master plans, urban zoning, land use and others.
2. Identify natural areas to be protected and desired conditions that will guide future developments.
3. Minimise land surface change.
4. Make use of blue–green fingerprints of the site – use natural drainage paths, minimise vegetation removal, minimise the use of hard engineering structures; disconnect impervious areas from the drainage system, favouring all infiltration opportunities, among others.
5. Minimise imperviousness, preserving natural hydrology characteristics, controlling runoff generation – green roofs, porous pavements, ponds, etc.
6. Integrate hydrological solutions into the urban landscape.
7. Develop integrated management practices.

Miguez *et al.* (2009) called attention to the fact that SuDS should recognise the particular temporal and spatial responses of each basin. It is necessary to evaluate the potential use of different combinations of the different possible interventions, in such a way that their efficacy in terms of flood mitigation can be optimised. Combined effects of the proposed measures may produce a result that is not equivalent to what would be expected from the summation of individual effects. This observation corresponds with Pompêo's principles, stressing that SuDS is dependent on the watershed's configuration and responses.

Righetto *et al.* (2009) stated that present stormwater management aggregates both structural and non-structural actions, involving infrastructure of different magnitudes (from the local to the watershed scale) and co-related with planning and management of land use and occupation of urban space. It is important to note that Righetto *et al.* (2009) was the result of a research network of Brazilian universities, funded by a federal agency called FINEP (Portuguese acronym for 'Funding of Studies and Projects Agency'), in the context of the PROSAB (Portuguese acronym for 'Research Programme on Basic

Sanitation'). This programme had the objective of developing technologies in the themes related to drinking water, wastewaters, urban drainage and solid waste. This programme has been renewed, with 16 public universities joining a research network specifically working with sustainable stormwater management.

23.3 Legal Framework

Stormwater management in the urban environment is the responsibility of municipalities. However, it is not uncommon that they do not have the necessary technical capacity to adequately deal with this issue, allowing practices that produce negative environmental impacts and that often transfer flood problems downstream in the watershed. On the other hand, Brazilian cities count on an adequate federal legal framework to support sustainable urban development, but it is not sufficient. As a federative republic, the federal legal framework acts as a general guideline with detail at the municipal level, but which is, in fact, at the operational level. Nonetheless, frequently this is not what occurs. The 'Federal Urban Land Parcelling Act' (Brazil, 1979), for example, establishes minimum standards for urban development. In terms of urban drainage, this Act indicates that no new construction may take place in a floodable area, unless mitigation measures are in place. However, the majority of the municipalities just repeat the Federal Act text in their local legislation, without providing a flood zoning map to guide urbanisation. As a consequence, cities tend to suffer from flooding, and environmental and urban degradation are relatively common problems.

Another Federal Act, known as the City Statute (Brazil, 2001), was established to regulate two previous articles of the Brazilian Constitution, in order to provide full development of the social functions of the city, guaranteeing the wellbeing of its inhabitants. The City Statute established detailed rules to regulate the use of urban property in favour of communities as a whole, their security and wellbeing as well as environmental balance. Several important urban management tools were made available in the context of the City Statute with the aim of setting proper conditions for the municipalities to act in their management tasks. Some of the basic guidelines proposed in the City Statute, which can be related to urban drainage aspects, are:

- guaranteeing the right to sustainable cities, meaning the right to urban land, housing, environmental sanitation, urban infrastructure, transport and public services, work and leisure for present and future generations
- planning the development of cities to prevent and address the impacts of urban growth and its negative effects on the environment
- supplying urban infrastructure, transport and public services that are capable of fulfilling the interests and needs of the population
- ordering and controlling land use to avoid pollution, environmental degradation and excessive or inadequate use of urban infrastructure
- protection, preservation and restoration of the natural and built environment, and cultural, historical, artistic and landscape heritages.

In 2003, the Ministry of the Cities was created by the Federal Government, with the objective of minimising social inequalities, and turning cities into better spaces, maximising the population access to housing, sanitation and transport. They published a Sustainable Urban Drainage Manual (Brasil, 2004) and launched a funding programme for sustainable stormwater management projects.

The main principles in the Sustainable Urban Drainage Manual state that:

- the watershed (not the urban perimeter) is the basic unit for planning and designing drainage systems and flood control projects
- pre-development behaviour should be taken as reference
- new urban construction should not amplify natural floods
- urban development should produce low hydrological impacts and preserve the natural water cycle
- runoff control should be as near to the source as possible
- design solutions should prioritise infiltration and storage.

The Ministry also stated that a city may request funds to finance actions for urban drainage system only if a 'stormwater management plan' is presented. This plan should be considered a component of the 'urban master plan'. As drainage is part of the urban infrastructure, it should, therefore, be planned in an integrated way. A stormwater management plan should be able to adequately manage stormwater in space and time, based on urban spatial distribution, at the same time being able to improve population and urban environmental health, taking account of economic, social and environmental perspectives. The stormwater system should also be integrated into the sanitation system, with proposals for the control of solid waste alongside reduction of stormwater pollution.

In fact, another Federal Act, known as Basic Sanitation Act (Brasil, 2007) defined basic sanitation as an integration of: public potable water supply; collection, transportation, treatment and adequate final disposal of sewage, as well as household waste and garbage originating from public streets and open areas; drainage and urban stormwater management, in terms of the conduction, detention or retention of flood flows their treatment and final disposal. This definition reinforced a new role for urban drainage systems since it explicitly included detention, retention and water quality treatment as formal components.

23.4   Case Examples

In this section, some examples of SuDS applications in Brazil will be briefly presented to illustrate some important actions that took place in the south-east region of the country, particularly focused on the states of Rio de Janeiro, Minas Gerais and São Paulo. It is also important to stress that these examples do not exhaust the subject and that there are other examples that could have been cited.

23.4.1 The Iguaçu Project – Metropolitan Region of Rio de Janeiro

The Iguaçu-Sarapuí river basin is situated in the Baixada Fluminense lowlands. It has a drainage area of 727 km², all of which is situated in the Rio de Janeiro Metropolitan Region. The Iguaçu river has its source in the mountains of Serra do Tinguá, at an altitude of 1600 m. It runs south-east for approximately 43 km, until it reaches its outfall at Guanabara Bay. Its main tributaries are the Tinguá, Pati and Capivari rivers from the west and the Botas and Sarapuí rivers from the east.

The Baixada Fluminense lowland is located in the western portion of the Guanabara Bay basin, in one of the most critical regions of Rio de Janeiro State in terms of urban flooding. Originally, this was marshland, but during the 1930s the Federal Government supported several drainage interventions to mitigate flooded areas, improving sanitation and minimising epidemics to allow further agricultural development of the region. After agriculture lost its importance, the city of Rio de Janeiro started to attract in-migration, a process which began during the 1950s, accelerating in the 1970s. At the beginning of the 1990s, more than 2 million inhabitants in six counties had settled in the Baixada Fluminense lowlands, of which more than 350,000 suffered the effects of significant floods.

In the 1990s, the first attempt was made to mitigate flooding in the region. However, traditional drainage measures had already been implemented, and this heritage shaped the first Iguaçu Project (LABHID, 1996). Basically, the project proposed the redesign of levees, some temporary reservoirs and canals, and the definition and zoning of a minimum terrain level for new urban development. Unfortunately, the lack of urban growth control led these measures to fail, and 10 years after its implementation, the reservoirs had lost their storage capacity due to all the new construction that had taken place.

Between 2007 and 2009, a review of the Iguaçu Project was carried out by the Federal University of Rio de Janeiro to support the State Government, represented by the State Environmental Institute (INEA is the Portuguese acronym), assessing the actions taken to mitigate flooding. This new version of the Iguaçu Project (LABHID, 2009) intended to control urban flooding and act as an opportunity for the environmental restoration of the Iguaçu, Sarapuí and Botas rivers, while also driving urban revitalisation along their courses (Figure 23.1). Structural and non-structural measures were proposed, in the main:

- the revision of land use zoning to include flood mapping
- the proposal of three environmental protection areas, to preserve important natural storage areas such as rural/low density/green buffers between forests in the upper basin reaches and cities in the lowlands; areas defined in this way were controlled, so that all future developments had to be regulated by the State of Rio de Janeiro
- the proposal to design a set of urban parks, in the built-up areas, to prevent further paving, and along the rivers, both to avoid informal occupation of the river banks and to provide storage capacity for attenuating flood peaks
- the redesign of all levee systems and the proposal to remove one of the levees to recover its floodplain connection and natural storage.

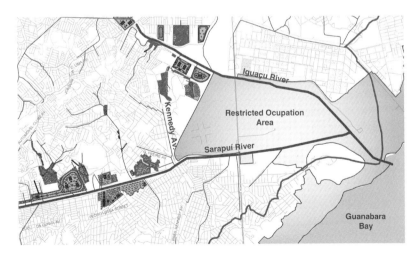

Figure 23.1 Illustrative plan view from Iguaçu Project, showing the designed urban floodable parks working as multifunctional landscapes and the proposed new urban zoning defining a restricted occupation area at the downstream portion of the watershed.

This project has yet to be fully implemented; its construction has been divided into successive phases, funded by the Brazilian Acceleration Growth Program (PAC is the Portuguese acronym), with the first phase already completed. According to INEA, 2200 families were relocated from areas at risk, mainly from houses on river banks, to new housing developments where they created parks and recreational areas and planted trees along the rivers. About 56 km of the watercourses were dredged to restore their original characteristics as far as possible, removing 5 million m³ of sediment and garbage in the process. In order to inform and mobilise local communities of the social and environmental problems due to urbanisation and the importance of participation and social control, INEA has been developing the monitoring and evaluation of this project through local committees and regional forums. They also initiated environmental education campaigns, one of them focused on waste disposal.

23.4.2 DRENURBS – Belo Horizonte/Minas Gerais

The Environment Restoration Program of Belo Horizonte City – DRENURBS (Portuguese acronym) was created by the Municipality of Belo Horizonte to improve the environment by protecting 200 km of urban watercourses still flowing in their natural riverbeds, distributed over 47 watersheds. DRENURBS started in 2001 and received an honourable mention in the Metropolis Awards, 2010, a prize given by the World Association of Major Metropolises.

One of the main features of DRENURBS was its comprehensiveness, acting to intervene in the physical space, but also trying to change socio-economic and environmental realities of the communities settled in the areas covered by the programme actions.

The main objectives of the programme involved reducing flood risks, controlling sediment production, integrating natural water resources with the cityscape, minimising watercourse pollution and institutional strengthening of the Municipality. Important partners of DRENURBS were the Federal University of Minas Gerais (UFMG) and the SWITCH Integrated Project (Sustainable Water Management Improves Tomorrow's Cities' Health) from UNESCO. Among the practical actions introduced by DRENURBS, were:

- the implementation of parks and permanent preservation areas to protect river banks and conserve riverine vegetation
- the implementation of detention reservoirs and the integration of the proposed solutions in the urban landscape
- the promotion of actions to valorise water resources as main components of good environmental quality
- the involvement of the community in the decision-making processes of the rehabilitated spaces (AROEIRA, 2010).

Figures 23.2 and 23.3 show some of the results obtained on completion.

23.4.3 Piscinões – São Paulo Metropolitan Area

In the 1990s, due to a critical problem of urban flooding in the city of São Paulo, the civil engineer Aluísio Canholi designed the first detention reservoir in the neighbourhood of Pacaembu, called a *piscinã*' in Portuguese, the Piscinão Pacaembu has been in operation since 1994.

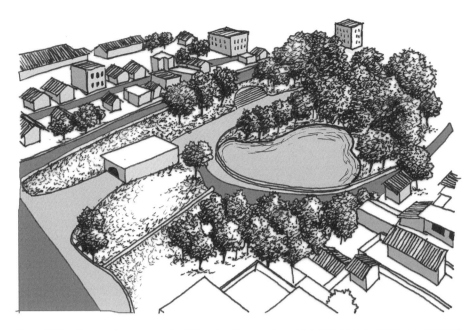

Figure 23.2 Concept image from the 1ˢᵗ of May Creek watershed: a detention reservoir in a new park integrated with the urban landscape.

Figure 23.3 Concept image for Nossa Senhora da Piedade Park, showing a retention reservoir with a permanent lake.

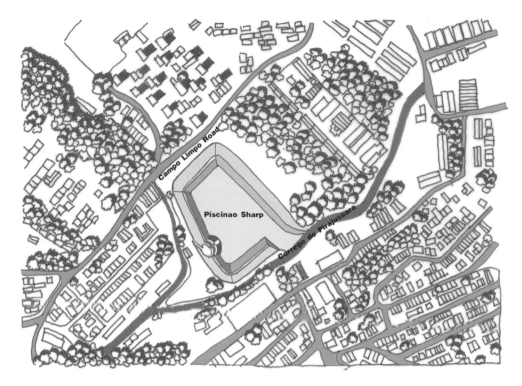

Figure 23.4   "Piscinão" Sharp.

It was excavated under Charles Miller Square with a total capacity of 74,000 m³. The drainage master plan produced in 1998 identified areas in the São Paulo Metropolitan region suitable for the construction of huge new detention reservoirs, which were designed to address patterns of flooding in the area and to avoid inundation of the city. Today, there are 19 piscinões in place, most of them built in concrete, some built underground. The individual average storage capacity of these reservoirs is approximately 200,000 m³.

The construction of these reservoirs was made possible due to different institutional partnerships. Piscinão Sharp, for example, was completed at the end of 2010 with a storage capacity of 500,000 m³; the partnership enabling its construction was between the Municipality and the State Government. The Municipality of São Paulo made the area available, and is responsible for its maintenance, and the State Department of Waters and Energy built the reservoir itself. Figure 23.4 shows an aerial view of Piscinão Sharp, built as an off-line reservoir for the overflowing waters from the Pirajuçara creek, a tributary of the Pinheiros River.

23.4.4   Protijuco – São Carlos/ São Paulo

Protijuco was an environmental recovery project implemented in the valley of Tijuco Preto Creek, in the city of São Carlos, São Paulo State. This creek is totally surrounded by the urban area. Before the project began, Tijuco Preto Creek had been partially canalised and

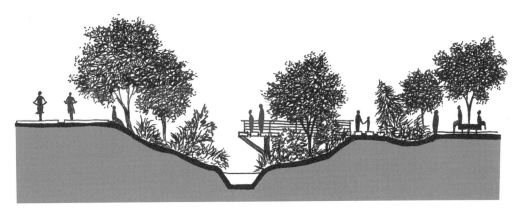

Figure 23.5   Typical cross section designed for Tijuco Preto Creek, São Carlos.

covered, it had poor water quality, had lost its original riparian vegetation and frequent floods used to occur. Solid waste and sewage disposal aggravated local degradation and there was no provision for recreational areas.

The project aimed to design, implement and monitor the restoration of the creek and its floodplains, using sustainable flood control measures. It also intended to promote the revitalisation and redevelopment of urban spaces, valuing historical and environmental aspects. The conceptual design of the proposed interventions intended to recover the natural functions of the system. The first part of the project was to reinstate river flow, and it began by demolishing those stretches that flowed in storm drains, and an open channel was constructed. In this process, urban drainage solutions were coconstructed with landscaping. Mendiondo (2008) discussed the concept of ecosystem services provided by freshwater biodiversity, using this project as a case study. Figure 23.5 shows the design concept in plan view with a cross-section of the project to reopen the creek.

23.5   Concluding Remarks

The SuDS triangle, which considers water quantity, water quality and amenities and biodiversity equally, forms the basis for a better urban environment, opening the possibility to introduce water as a value to be preserved and combined in the cityscape. This trend is seen worldwide, with different levels of engagement, but the focus is changing from end-of-pipe measures to a more sustainable approach, acting to control the causes of urban floods.

In Brazil, there are still some difficulties in applying this concept, mainly because of two major factors. The first is the capability of the municipalities to design, construct, operate and maintain (including technical, legislative and institutional aspects), which are relatively low and not really up to date. This situation reflects the difficulties in making general guidelines provided by the federal laws actually work at the local scale. The second aspect is certain inertia in the case of technical personnel, who resist the breaking of traditional paradigms. However, examples have been given in this chapter, whereby SuDS concepts have been evolving in Brazil and producing interesting examples of changing the country's history of urban flood control.

References

Aroeira, R.M. (2010) Recuperação ambiental de bacias hidrográficas, Belo Horizonte, Brasil. In: A.T. Gonzaga da Matta Machado, A.H. Lisboa, C.B. Mascarenhas Alves, D. Alves Lopes, E.M. Andrade Goulart, F.A. Leite, M.V. Polignano (Org.), *Revitalização de Rios no Mundo: América, Europa e Ásia*. Belo Horizonte: Instituto Guaicuy. Cap, 221–240. (in Portuguese)

Baptista, M., Nascimento, N. and Barraud, S. (2005) *Técnicas Compensatórias em Drenagem Urbana*. ABRH, Porto Alegre, 2005, 266 pp. (in Portuguese)

Brazil (1979) *Federal Act 6,766, of* 19 December 1979. Federal Official Gazette of Brazil, Brasília, DF, 20 December 1979. Section 1, Brasília, Brazil. (in Portuguese)

Brasil (2001) Federal Act 10,257, of 10 July 2001. Federal Official Gazette of Brazil, Brasília, DF, No 133, 11 July 2001. Section 1, Brasília, Brazil. (in Portuguese)

Brasil – Ministério das Cidades (2004) *Manual de Drenagem Urbana Sustentável*, Brasília, DF. (in Portuguese)

Brasil (2007) Federal Act 11,445, of 5 January 2007. Federal Official Gazette of Brazil, Brasília, DF, No 8, 11 January 2007. Section 1, Brasília, Brazil. (in Portuguese)

Club of Rome (1968) About the Club of Rome. http://www.clubofrome.org

Genz, F. and Tucci, C.E.M. (1995) Controle do escoamento em um lote urbano. *Revista Brasileira de Engenharia, Caderno de Recursos Hídricos*, 13(1) 129–152, Rio de Janeiro, RJ. (in Portuguese)

LABHID – Laboratório de Hidrologia/COPPE/UFRJ (1996) Plano Diretor de Recursos Hídricos da Bacia dos Rios Iguaçu/Sarapuí: Ênfase no Controle de Inundações. Rio de Janeiro: SERLA. (in Portuguese)

LABHID – Laboratório de Hidrologia/COPPE/UFRJ (2009) Plano Diretor de Recursos Hídricos, Controle de Inundações e Recuperação Ambiental da Bacia do Iguaçu/Sarapuí. Análise do comportamento hidrológico e hidrodinâmico da bacia hidrográfica do rio Sarapuí, na Baixada Fluminense e estudo de intervenções estruturais em quatro de suas sub-bacias. Rio de Janeiro: SERLA. (in Portuguese)

Leopold, L.B. (1968) *Hydrology for Urban Land Planning – A Guidebook on the Hydrologic Effects of Urban Land Use*. US Geological Survey Circular 554. Washington: USA Department of Interior.

Mendiondo, E.M. (2008) Challenging issues of urban biodiversity related to ecohydrology. *Brazilian Journal of Biology*, 68 (4, Suppl.) 983–2002.

Miguez, M.G., Mascarenhas, F.C.B. and Magalhães, L.P.C. (2007) Multi-functional landscapes for urban flood control in developing countries. *International Journal of Sustainable Development and Planning*, 2 (2) 153–166.

Miguez, M.G., Mascarenhas, F.C.B., Magalhães, L.P.C. and D'Altério, C.F.V. (2009) Planning and design of urban flood control measures: assessing effects combination. *Journal of Urban Planning and Development*, 135 (3) 101–109.

Miguez, M.G., Rezende, O.M. and Veról, A.P. (2014) City growth and urban drainage alternatives: sustainability challenge. *Journal of Urban Planning and Development*, 10.1061/(ASCE) UP.1943-5444.0000219, 04014026

Nascimento, N.O., Baptista, M.B., Ramos, M.H. and Champs, J.R. (1997) Aspectos da evolução da urbanização e dos problemas de inundações em Belo Horizonte. Anais do XII Simpósio Brasileiro de Recursos Hídricos, CD-ROM, art. 335, Vitória, ES. (in Portuguese)

Pompêo, C.A. (1999) Development of a state policy for sustainable urban drainage. *Urban Water*, 1, 155–160.

Pompêo, C.A. (2000) Drenagem Urbana Sustentável. Revista Brasileira de Recursos Hídricos, Porto Alegre, RS, 5, 1, 15–24, (in Portuguese)

Righetto, A.M., Moreira, L.F.F. and Sales, T.E.A. (2009) Manejo de Águas Pluviais Urbanas. In: Righetto, A.M. (ed.), Manejo de Águas Pluviais Urbanas, Projeto PROSAB, Natal, RN: ABES. Cap. 1, 19–73. (in Portuguese)

Souza, F.C., Tucci, C.E.M. and Pompêo, C.A. (2005) Diretrizes para o Estabelecimento de Loteamentos Urbanos Sustentáveis. VI Encontro Nacional de Águas Urbanas. Belo Horizonte, Brazil. (in Portuguese)

Tucci, C.E.M., Porto, R.L.L. and Barros, M.T. (1995) Drenagem Urbana. UFRGS Ed. da Universidade/ ABRH, Porto Alegre, 430 pp. (in Portuguese)

UN-United Nations (1993) *Agenda 21: Earth Summit – The United Nations Programme of Action from Rio*. Rio de Janeiro: United Nations, Department of Public Information.

United Nations Conference on Environment, and Development (UNCDE) (1992) Earth Summit. http://www.un.org/geninfo/bp/enviro.html

UNEP – United Nations Environment Programme (1972) Stockholm. Report of the United Nations Conference on the Human Environment. http://www.unep.org/

Interim Measures Towards Sustainable Drainage in the Informal Settlements of South Africa

Kevin Winter

Slum settlements, which are commonly referred to as informal settlements in South Africa, are renowned worldwide as places where the provision of basic utility services are absent or limited at best and often dysfunctional. They are also places where housing structures are makeshift, being made of whatever materials are locally available, such as corrugated iron or wood, and are also places where the local inhabitants have no claim to the land. In 2012, the UN Habitat Programme estimated that 863 million people were living in informal settlements worldwide, and that a third of the people living in cities in the developing world were in informal settlements (UN Habitat, 2013). The figures are even more staggering in the case of sub-Saharan Africa where 62 per cent of the urban population live in slum settlements (ibid.). For example, Kibera in Nairobi, Kenya, is the largest informal settlement in the world with approximately one million people living in an area of 2.5 km^2 (Engleson, 2010). Statistics reveal the magnitude of the issue but fail to express the harsh reality of living conditions on the ground. Residents of informal settlements generally lack a host of basic services including drainage, access to safe sanitation, safe sewage disposal and safe drinking water. Most settlements are characterised by high population densities, which indirectly is a factor that is responsible for surface water runoff being a vector for the spread of diseases and the general deterioration of the physical environment. Despite the conditions, the number of people living in informal settlements continues to rise, not only as a manifestation of urbanisation, population explosion, demographic change and globalization, but as a result of poor governance, corruption, failed policies, dysfunctional land markets, deficient financial systems and a lack of political will (UN-Habitat, 2003).

There is no universally agreed definition of an informal settlement, but they are typically described as dwelling places where the occupants have no legal claim or right to the land, and where the layout of the settlement is unplanned and unstructured.

Sustainable Surface Water Management: A Handbook for SuDS, First Edition.
Edited by Susanne M. Charlesworth and Colin A. Booth.
© 2017 John Wiley & Sons, Ltd. Published 2017 by John Wiley & Sons, Ltd.

These two characteristics seriously challenge the management of drainage and other water-based services. Without secure land tenure, residents are reluctant to invest in materials and devices to improve, among other things, the drainage around their dwellings. The second challenge lies in the difficulty of installing conventional pipelines and drainage systems in densely populated, unplanned settlements largely because of the lack of space that is required to lay conventional pipelines along linear corridors between dwellings.

In informal settlements in South Africa, surface water drainage comprises ad hoc arrangements that are instigated at will by residents in response to a number of factors, including upstream flows and the pooling of water near the home; the accumulation of rancid water and associated odours; and the management of seasonal flood waters during rainfall events. Residents deal with surface water by digging shallow channels, which also double up as stormwater conduits. It is common practice to use these make-shift channels daily for discharging greywater generated from activities such as cleaning of cutlery, crockery, clothes and washing of bodies. The volume of greywater disposed per household is relatively low and varies between 75 and 150 litres per day (Carden *et al.*, 2007). The discarded water often contains elevated concentrations of salts, fats, organics and toxins that accumulate in channels and shallow ponds. Foul-smelling contaminated water attracts the disposal of solid litter within it, as well as flies and mosquitoes, and also the attention of young children who have a tendency to play in pools of water lying about in informal settlements. Thus, the disposal of greywater within the confines of limited space results in the cross contamination of greywater and stormwater. Concentrations are elevated further during the 'first flush' and at other times during low flow conditions.

Solutions for managing drainage in informal settlements remain largely elusive. The answer is unlikely to be found in technology or conventional practices alone. In the case of South Africa, the problem is rooted in a host of issues that include insecurity of land tenure, the spatial layout of informal settlements, and social behaviour of residents. However, these problems are embedded in the socio-political history of South Africa, where the legacies of inequality and entrenchment of privilege during the decades of Apartheid still remain deeply entrenched in the country. The past is proving difficult to overcome. Since the beginning of the post-Apartheid period and following democratic elections in 1994, the national government has struggled to deal with the demand for social housing, which has been exacerbated by the migration of people from rural areas to cities and towns.

This chapter explores various options and processes that were implemented in a selected settlement over a period of almost ten years. It is considered a unique case because of the dearth of experience and knowledge to date towards the realisation of an incremental upgrade of informal settlement in the country. This case study illustrates the successes and failures of a series of interventions that were designed to improve drainage and other related services. The case is situated within the informal settlement of Langrug, which is located approximately 75 km north-east of Cape Town, in a region well known for its agricultural, wine-making and tourism industry. The study explains how a combination of interventions and processes were used to address the problem of contaminated surface water and drainage, which placed a high premium on involving local residents and encouraging them to engage in forms of co-management, and represents a shift from a more conventional engineering approach in which there is usually minimal consultation and involvement of end users.

24.2   Overview of the Development of Informal Settlements in South Africa

The South African national census does record informal dwellings, that is, those made from makeshift materials, and it divides them into two separate categories: those that are found in informal settlements, and those found as backyard shacks attached to or in close proximity to a formal structure. In 2001, 1.78 million shack dwellings were recorded in informal settlements (StatsSA, 2001), but by 2011 the number had declined to 1.25 million households (StatsSA, 2012). The decline is attributed to the government's Reconstruction and Development Programme (RDP), which claims to have built 2.8 million formal free basic housing structures since 1994 of which 1.5 million houses were built in the period between 2001 and 2011 (DHS, 2012). While a small decline in the number of shack dwellings is indicative of government efforts to provide free basic housing, the scale of the demand brought about by the increase in population and urbanisation continues to overwhelm the resources of the country. Between 2001 and 2011, the Gauteng and Western Cape provinces, each with the largest urban city centres and urban economies, were net recipients of over 1,000,000 and 300,000 people, respectively. Regional migration to these provinces is predominantly from those areas with a rural-based economy, which also coincides with areas once classified as 'independent homelands' area under the Apartheid legislative system of separate development. The growth of informal settlements in the two major cities of these provinces, namely Johannesburg and Cape Town, continues to stretch the resources of these metropolitan local authorities who take the responsibility for providing water-based services, among other services. The priority is to provide access to water and basic sanitation, while the safe disposal of surface runoff is usually neglected in favour of leaving these arrangements to local residents.

24.3   Co-Management of Drainage

Literature on service delivery in developing contexts frequently emphasizes collaboration and cooperation between inhabitants, with local authority officials as prerequisites for improving service delivery and operations (WSSCC/Sandec, 2000; EAWAG, 2005; Eales, 2008). Solutions to water based-services, including drainage, require an integrated approach that is geared towards changing established practices in favour of a 'genuine commitment to partnership and empowerment' (DfID, 1998). This is emphasised in the Bellagio Principles, in which the household and neighbourhoods are identified as central to planned interventions in a settlement (WSSCC/Sandec, 2000). However, the realities of South Africa with its socio-political history, have resulted in residents of informal settlements feeling frustrated by poor service delivery and a general realisation that their aspirations for a better life in a post-Apartheid period have failed to materialise.

Evidence of cooperation and sustained partnerships in drainage management in informal settlements are in short supply in South Africa, despite the contention that interim solutions are most likely to be found in efforts that build on a genuine partnership that is able to integrate top-down initiatives offered by the authorities with a ground-up willingness to improve and maintain services. The reality is that institutional support for this form of partnership is limited because local authorities are largely unprepared to engage with civil society as critical agents in the transformation of these services (Alexander, 2010). Research findings in South Africa suggest that the urban poor expect service delivery and operations to be met by central government (Kruger, 2009). The potential to find sustainable solutions by entering into co-management arrangements between residents and local authorities remains elusive. The case study, therefore, examines some measures of success over a reasonable period of time and tries to explain how a succession of processes resulted in an improved level of cooperation between residents and authorities.

24.4   Langrug: A Case Study of an Informal Settlement

Langrug is an informal settlement situated a mere 3 km from the town centre of Franschhoek, a Dutch name meaning 'French corner'. The town lies in the apex of the Franschhoek Valley, where a small group of French Huguenots refugees began settling in the area in 1688. Many were given land by the Dutch government at the time, and they then used their expertise by turning the fertile soils into profitable viticulture and wine-making ventures. Today, a combination of the beauty of the Franschhoek valley in an agricultural setting, the quaintness of the town, its buildings and cuisines, have all contributed to attracting thousands of local visitors and overseas tourists to the town (Figure 24.1).

Figure 24.1   Location map of Franschhoek, South Africa.

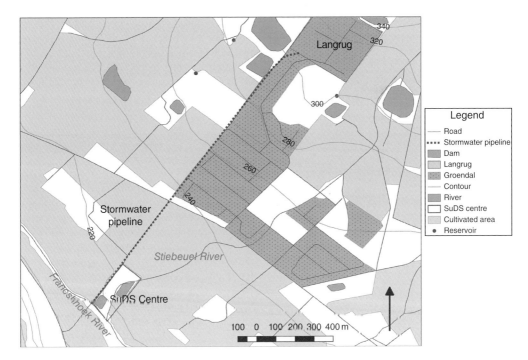

Figure 24.2 Map showing Langrug informal settlement in the north adjacent to the suburb of Groendal with the stormwater pipeline discharging into the Franschhoek River, and the proposed SuDS Centre. The town of Franschhoek is located 3 km east of Langrug.

By contrast, the history of Langrug informal settlement, is very different (Figure 24.2). The first shack dwellings were erected on government-owned land overlooking the Franschhoek Valley in 1993 by migrants from the Eastern Cape Province. By 2011, the population had grown to approximately 4100 people and the number of shacks totalled 1858 (Stellenbosch Municipality, 2011). At this stage, the utility services comprised 91 flush toilets and 57 communal tap stands. Researchers who had worked in the settlement from 2006 to 2008 regularly reported throughout their study that these water-based services were often overused or dysfunctional, and residents regularly disposed of their wastewater in the streets outside their dwellings (Carden *et al.*, 2007).

The only stormwater channel was a concrete-lined culvert adjacent to a steep gravel road that bisected the settlement from top to bottom (Figure 24.3). The stormwater culvert, therefore, conveyed a toxic mix of discarded greywater and blackwater leaking from dysfunctional communal toilets, and then stormwater, all of which finally discharged into the Franschhoek River at the bottom of the valley.

The case study examines two distinct approaches that were used in an effort to intervene and contribute to the general upgrading of drainage in Langrug. The first involved a research project undertaken by the University of Cape Town's Urban Water Management research unit in a study that was funded by the South African Water Research Commission, and the second involved an ongoing intervention by the local authority and the Western Cape provincial government, together with the facilitation and support services of a non-government organisation (NGO).

Figure 24.3   A drainage culvert alongside the road leading to the top of the settlement.

24.5   Research-Led Efforts: First Approach

One of the first initiatives to address drainage in the settlement began in 2006, in a study undertaken by academic researchers and students from the University of Cape Town. The main purpose was to explore the potential for local residents to manage greywater, but included giving attention to stormwater drainage. The study opted to use participatory action research (PAR) as a procedure for choosing a course of action that could explicitly incorporate social, political, economic and institutional factors in a participatory process (Lal *et al.*, 2001). The approach recognises that decisions should be achieved through collaboration, cooperation and consensus (ibid.).

The study began with an analysis of existing conditions and levels of service in a number of informal settlements in and around Cape Town, which included Langrug. Local inhabitants, stakeholders and interest groups, including local authority officials and councillors, were consulted to obtain their input, plans and support for the potential to improve drainage management. The specific intention of the study was to identify and strengthen local level drainage management strategies, and to support these by implementing low-cost, appropriate interventions, which included stakeholders, and most especially the local residents themselves. The researchers chose Langrug as a study site to investigate four key objectives. These were to identify existing services, stakeholders and social structures among the residents; to identify problems associated with existing management practices; to consider potential management strategies; and to explore possible intervention strategies (Armitage *et al.*, 2009).

As mentioned earlier, surface water runoff from Langrug consists of a combination of grey-, black- and stormwater that is channelled into a stormwater culvert eventually discharging into the Franschhoek River. Water samples that were collected from various points in Langrug, as well as in the Franschhoek River, showed elevated levels of bacteria (upper quartile 200,000–450,000 cts/100 ml) and nutrients (e.g. orthophosphate upper quartile 1.5–4 mg/l) (Armitage *et al.*, 2009).

An objective of the study was to propose a selection of drainage options. These options would arise following a series of site visits and workshops held with local residents from within the settlement itself. At the time of the study, field researchers were surprised at being unable to identify any social organisations or active citizen groups such as street or neighbourhood committees. In the end, it was decided that the field researchers would have to find willing, cooperative residents themselves. Meanwhile, local authority officials and politically elected councillors were prepared to meet with the researchers, but the extent of the collaboration was limited to a few interviews and discussions. Their interests were focused more on the day-to-day management of water and sanitation issues, not necessarily drainage. In the end, the researchers identified 12 households where residents showed sufficient interest and were willing to consider installing a greywater drainage device of choice. In each case, prior to any intervention, a discussion was held with a member of the household about their drainage problems and needs, and this was followed by a discussion about the potential options that could be used to introduce a rudimentary in-situ system that was affordable and one that used materials that were easily accessible such as stone, gravel and plastic crates. The crate and trench soakaways were the most popular designs selected for locations around the shack dwelling, where the soil was sufficiently permeable and where there was sufficient space between houses to install the system (Figure 24.4). A corridor of open space of at least 4 m × 1 m was required as a minimum set of dimensions for each of the soakaways. An upturned plastic milk crate with perforated sides was used as the greywater disposal point. It was covered with 'shade-cloth', a porous plastic material to capture food matter and prevent organic material from blocking the opening. The crate was located at one end of a trench, which was constructed to nominal dimensions of 3.5 m long by 0.75 m wide by 0.75 m deep. The trench was lined on the bottom and sides with a polyethylene sheet with the lower end left open

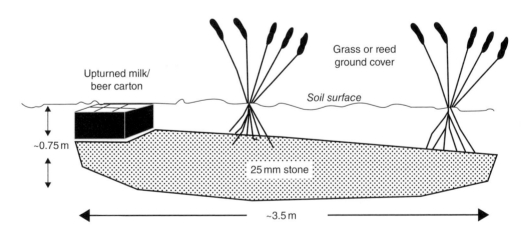

Figure 24.4 An example of a crate and trench soakaway.

to drain into the surrounding soil. The trench was then filled with small 19 mm aggregate stone and covered with infill from the hole. Reeds, ornamental flowers or rooted crops such as fruit trees or vegetables were proposed as vegetation that could be planted over the soakaway, which would aid in nutrient removal, although residents never explored the potential of a vegetated filter, simply because they were situated within a pedestrian access route. In one case, *Kikuyu* grass, an invasive grass that is ubiquitous across South Africa, grew rapidly over the soakaway, indicating a rapid uptake of nutrients that were being discharged into the system.

During each installation, an informal workshop was held with a member of the household and neighbours from dwellings alongside. However, most often it was the fieldworkers themselves who performed the demonstration and who completed the installations. Most of the soakaways failed within a couple of months and residents ceased to fix them. There was limited interest on the part of individuals to experiment further to improve the devices. The research team learnt new lessons, not just about the limitations of the technology or devices, but about social behaviour and attitudes to technologies that were perceived as experimental, interim and second class.

The research team anticipated at least four limitations of these the drainage devices. Grease and fats were likely to form a scum and block the porous shade-cloth material at the catchpit; spillage would occur when greywater was poured too rapidly into the make-shift catchpit; desirable plants and trees would be unable to grow in a high nutrient, alkaline environment from elevated contamination of greywater; and space between the shacks was too limited to build a drainage device of sufficient volume to contain more than 20 litres of water at one time. Most devices failed in less than six months, partly because residents were unwilling to experiment further or maintain the device. In one case, the local authority's mechanical road grader destroyed the device and in another the inlet became clogged with solid waste and was eventually removed. Despite these failures, it was clear that the study needed to focus attention on understanding social behaviour in this context and what might be involved in finding new ways to support the incremental upgrading of drainage in an informal settlement.

24.6 Discussion on Research-Led Approach to Drainage

Part of the failure of the PAR process was that the fieldwork team failed from the outset to find a social organisation or cohesive group of people who were prepared to be involved in drainage management. The experimental soakaways were unsuccessful largely because local participants were unwilling to engage in experimentation. In the end, the experimental devices became demonstrations that were perceived to be implemented by outsiders. The actual installation only attracted passive support from some local volunteers who sought to benefit directly from having a device near their dwelling. Thus, it was concluded that an authentic collaboration process, the kind envisaged in a PAR approach, did not materialise in this phase of the project. A PAR approach does not and cannot work in this way except in the sense that the demonstrations, if successful, might encourage residents to mobilise themselves in order to replicate the intervention and to achieve locally determined goals with the possibly of continuing collaboration with outside agents. Efforts to encourage residents to participate in building further drainage options were largely fruitless. Ultimately, the experiments failed because of limited interest, capacity and resources to install and maintain the drainage systems. Without authentic participation, it appears that any attempt to promote collective action is likely to be doomed, no matter how good the technology.

At the time of the study, Langrug informal settlement received only limited attention from local authority officials. This neglect appeared to be a product of the fact that the settlement was situated on invaded land where residents had no security of tenure, and in which the municipal authorities anticipated that residents would be removed and resettled elsewhere in due course. Future plans were uncertain and there were no discussions about incrementally upgrading the settlement. In addition, interviews held with residents indicated that they only received tacit support from elected councillors and local authority officials, which only served to raise their frustration and distrust of the local authority and elected councillors. Field researchers noted that the challenge resides in developing a partnership between residents, councillors and local authority officials. Figure 24.5 summarises the perceived understanding of social and institutional environment at the time of the study. An absence of a coherent community-based group of residents and leadership meant that the voice of the people could not be presented confidently to the elected councillor who in turn could convey these to the responsible official in the local authority. As a result there was increasing mistrust of the intentions of local government to address the plight of residents, and in turn, officials were in crisis management mode trying to address day-to-day matters in Langrug and other informal settlements in the municipality.

In theory, an effective 'bottom-up' or grassroots approach to drainage management could offer a more sustainable option, but it is conceded that such an approach is misplaced in an informal settlement context unless a series of specific conditions are met. In particular, the local authority would need to take full responsibility for service provision while the residents would require sufficient capacity to enter into a meaningful partnership with other stakeholders including the local authority. The outcome of lessons learnt during the course of this study paved the way for introducing a different approach that began five years later, in 2011.

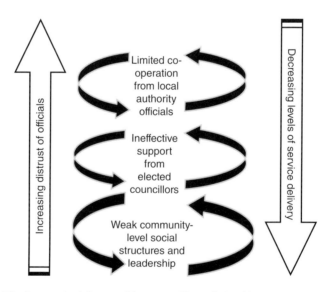

Figure 24.5   A conceptual diagram of fragmented interrelationships.

24.7   Building Partnerships: A Second Approach

The relative failure that the researchers experienced between in the earlier study highlighted the limitations of ad hoc experimental initiatives that were piecemeal, poorly resourced and lacked a formal process that engaged the local authority and politically elected councillors. The demonstration project was perceived to be more about testing ideas rather than improving drainage that would contribute to improving drainage in the settlement as a whole (Carden *et al.*, 2007). In the absence of an incremental plan, these initiatives failed soon after the researchers left the site. However, the unfolding story of what might be perceived as a hopeless situation in furthering development of Langrug, started to change.

In November 2010 a neighbouring farmer of Langrug obtained a court interdict against the local authority for allowing greywater runoff to flow into the farm's irrigation dam. The local authority responded immediately by establishing a contractual agreement with the Informal Settlements Network (ISN), an Alliance of non-governmental organisations who were tasked with the responsibility for building local capacity and leadership within Langrug local residents and to formulate a plan for the incremental upgrading of the settlement by identifying core areas of intervention and action. Whether the court interdict was instrumental in drawing the local authority into matters at Langrug, or if there had been a sudden shift in policy, was unknown, but what arose was a new intent and investment in efforts to upgrade the informal settlement. At the time, local authority of Stellenbosch was under severe pressure to deal with a housing backlog of 19,701 households required to house an estimated 20,000 families who were living in informal settlements and backyard shacks. Since the municipality only received 300 housing subsidies a year from national government, meaning that it could only afford to build this number of houses, it would take up to 130 years before the current demand could receive a formal state subsidised house. Given the magnitude of the problem and the inability to meet housing demand, the municipality decided to form an Informal Settlements Management Department with the intention of developing a strategic approach to meeting the challenges of urbanization and service delivery by establishing a people-driven, pro-poor solution to the housing crisis.

One of the first tasks of the ISN Alliance was to create a representative leadership group, and it did so by holding a series of meetings and dialogues with residents of Langrug. This was followed by an enumeration of Langrug's inhabitants and of the physical infrastructure and socio-economic profile of the settlement. At least 30 residents were drafted into the enumeration team. The information from this census provided detailed knowledge from which to formulate plans for improving services to the settlement. Upgrading sanitation facilities within the settlement was most obvious. At the time, the ratio of people to toilets was 49:1, while the number of people to a communal water point was 72:1 (SDI Alliance, 2012). The survey did not account for the state of drainage, but over many years attention to the discharge of greywater within Langrug was undeniably urgent. Greywater had long been identified as a cause of numerous negative health effects, particularly in places where children played within the settlement.

The enumeration exercise had other benefits too. Data that was collected by community members contributed to building cohesion and cooperation within the settlement and helped to focus on the pursuit of a common goal towards its upgrading (SDI Alliance, 2012). There were increasing signs of confidence and empowerment from within the community, and a shift in the way the local authority began to deal with housing and service delivery at Langrug (ibid.). These new developments agreed with the consensus internationally, which suggests informal settlements are best able to serve citizens if they are not treated as places that are waiting to be eradicated, but rather as emergent communities requiring support through various stages in an incremental upgrading process. This shift is significant in opening up opportunities for community-led participation.

The agreement between the ISN Alliance and Stellenbosch Municipality enabled the Langrug community to leverage state funds for upgrading certain projects of choice. In 2012, some solutions were implemented including opening access to streets where existing structures were blocking access; the construction of greywater channels; the provision of play parks for children; improvements to ablution facilities; and the establishment of health forums to assist with HIV/AIDS counselling (SkillsPortal, 2012). Since 2012, the Community Organisation Resource Centre (CORC), in particular, one of the NGOs in the ISN Alliance, has played an influential role in facilitating leadership development and in building capacity and skills among the local community. One of the projects involved the construction of a network of drainage pipes with points allocated along the network for the discharge of greywater. These drainage systems consist of discharge points linked to a reticulation system for the discharge of greywater into stormwater culverts (Figure 24.6).

Figure 24.6   A community-led initiative to create a greywater discharge catchpit linked to a shallow pipeline.

Local efforts to develop a network of drainage systems were followed by two further interventions, which were initiated by the Western Cape provincial government under the Department of Environmental Affairs and Development Planning (DEADP).

24.8 Provincial Government Intervention

In 2013, the Western Cape Government (WCG) began looking into the feasibility of using biomimicry principles in an informal settlement that could develop a prototype for managing wastewater, stormwater and solid waste. The concept of a 'genius of place' was born, under the management of DEADP. Langrug became the site of choice for a number of reasons but principally because social structures and community leadership were evident and it was felt that these two characteristics were essential in driving a community-based project, which would examine how nature could be used to solve human problems. In this case, the project would examine how biomimicry principles could help understand how nature could treat contaminants in surface water runoff and extract the pollutants from the water stream.

24.9 Biomimicry at Work: Greywater Swales

The WCG established a project tender and a competitive bidding process to engage a consultant in investigating the potential of using biomimicry in the treatment of greywater drainage. The successful group, in/formal South, commenced with this project in 2014 (see: http://www.informalsouth.co.za/portfolio/genius-of-place-phase-3/). The consultants chose to work closely with local residents and the leadership team at Langrug to install a series of micro wetlands that were linked to stormwater swales in order to create a 'living sewer' in the form of a bioremediation system to filter, clean and slow down the flow, and break down waste in the water.

Swales were positioned along vertical routes between the houses. Each household could dispose of its greywater at specific disposal points that were either made of buckets sunk into the ground or using the milk crate catchpits described earlier. These discharge points were all connected to shallow underground pipes, which had been installed by the community two years earlier. The swales were designed to slow the flow of the water and to reduce flooding. Trees were planted at interconnecting points along the sewer with the intention of absorbing nutrients that accumulated in the soils around the base of the tree and to filter the water at this nodal point (Figure 24.7).

The project is currently still in the development phase but the intention is to treat about 6000 litres of greywater per day from 115 households. The results of this prototype are still under investigation, but they are unlikely to be sufficient to address the daily discharge of contaminated runoff from entering the Franschhoek River; nevertheless, it is a contribution that could add value to an awareness of sustainable drainage (SuDS) options and the greening of the settlement.

24.10 Sustainable Urban Drainage Centre

The final intervention, which was also initiated by the WCG, involved the development of a SuDS Centre downstream of Langrug informal settlement. The Centre is expected to treat stormwater runoff from Langrug and the adjacent low-income settlement at

Figure 24.7   A micro-wetland surrounding the base of newly planted trees that form a node linked to a shallow stormwater pipeline.

Groendal. This project arose because of the decommissioning of the Franschhoek Wastewater Treatment Works (WWTW) and the successful diversion of sewage to a new treatment works about 12 km from Franschhoek. The existing WWTW at Franschhoek is no longer necessary, and it was decided that the disused WWTW provided a perfect opportunity to deal with contaminated stormwater from Langrug. At the same time, it could be used to showcase the value that could be extracted and recovered from stormwater. This final discussion presents a visionary example for the redevelopment of the site and a potential model for converting the disused WWTW into a stormwater treatment facility that could simultaneously support educational and research initiatives, and provide employment for local residents.

The design uses existing tanks and ponds on site, and involves minimal investment in infrastructure for the stormwater treatment. The current proposal is to divert stormwater into the site for treatment along a series of treatment trains. An existing chlorine treatment tank and a pre-release settling pond could be used as constructed wetlands to simultaneously deal with flood-level stormwater inputs during heavy rains, and to filter contaminants. The deep tank would become a vertical-flow wetland in which effluent water passes through sand or gravel, and then is filtered into an outlet (Scholz and Lee, 2005). The shallow unlined tank would become a horizontal-flow wetland that mimics the filtration in a natural wetland ecosystem. *Phragmites australis* reeds, rooted in sediments or forming floating rafts, naturally uptake nutrients, physically filter the water, and provide a surface area for beneficial microorganisms. Reed-dominated wetlands are superior to open ponds by filtering out nitrogen (Moore and Hunt, 2012). The shallow wetland also provides a natural area that can be made accessible with the construction of a boardwalk. The conversion of the WWTW follows four key principles in the management of SuDS: good housekeeping at source; managing stormwater as close to source as possible; using local or regional controls to treat water; and managing the quality at the final point of discharge before it reaches the receiving water. Figure 24.8 shows the application of selected designs and techniques that align with these four SuDS principles.

The proposed SuDS Centre includes a pipeline where stormwater is diverted along a treatment train utilising a series of existing infrastructures at the decommissioned WWTW

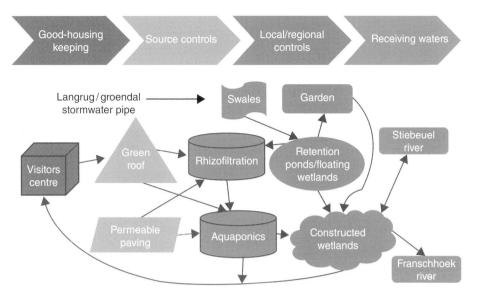

Figure 24.8   SuDS treatment principles applied to the SuDS Centre.

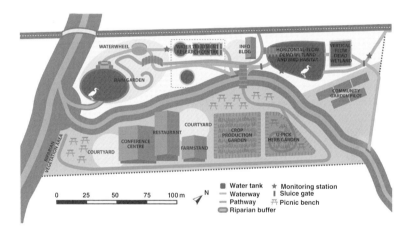

Figure 24.9   A schematic of the proposed SuDS Centre at the Franschhoek WWTW.

(Figure 24.9). The Franschhoek and Stiebeuel Rivers flow through the centre of the site. The proposed plans aim to provide an educational and research centre that will be one of the first initiatives in the country to demonstrate how resource recovery can be achieved using stormwater received from surface water runoff from an informal settlement.

24.11 Discussion

Interrelationships between various stakeholders were far more connected and cooperative since the intervention by the local authority in 2011. This perspective is summarised briefly in Figure 24.10. It suggests that the capacity and leadership of the community of Langrug

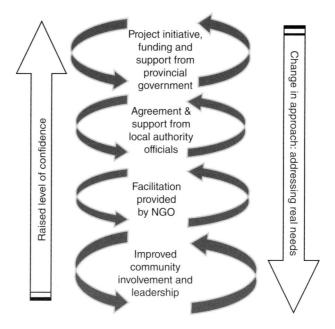

Figure 24.10   Showing connections and purposae strengthened for stakeholder engagement.

was a significant drive in enabling local residents to participate in meaningful discussion with the local authorities. The local residents' steering committee was able to leverage funding to address some pressing needs within the settlement. It was significant that the WCG project selected Langrug as the study site, because it recognised the value of working with a community social structure.

The second part in the discussion of the case study stands in contrast to the former by implying that 'top-down' interventions are necessary to ensure that those in positions of power (at the 'top') develop the capacity (and the will) to 'listen to and hear' what the needs and priorities are of those at the 'bottom'. Evidence from the most recent projects suggests that some form of collaborative participatory action is working both in terms of infrastructural development required for greywater management in high density informal settlements, and in enabling local-level administrative structures to grow in ways that they can sustain themselves by ensuring popular involvement in service delivery and maintenance by actively engaging with local authorities.

One of the salient factors emerging from the case study is the gap in understanding between local authority officials and residents, symptomatic of weak collaboration and cooperation between stakeholders, even in the later period of upgrading at Langrug. Social engagement in the implementation of solutions is critical: interventions must be socially acceptable and should attempt to establish a 'genuine commitment to partnership and empowerment' (DfID, 1998). To realise such a vision requires an integrated framework consisting of multi-sector planning (accounting for water supply, sanitation, drainage and solid waste management) and multi-actor participation (participation of all stakeholders, beginning at the household and neighbourhood scale) for successful implementation (EAWAG, 2005).

Lowndes and Wilson (2001) stress the importance of local authorities using their capacity as (local) agents of the state to develop social capital, if for no other reason than to influence democratic performance in processes of ensuring adequate service delivery. In this way, social capital can be used to facilitate a two-way relationship between civil society and government. Such a relationship is one in which government can establish social capital and simultaneously mobilise forms of service delivery that are supported by public–private-sector partnerships of a kind that Eales (2008) argues can overcome the challenges of providing viable, affordable services in informal settlement-type populations. Eales' (2008) argument is that multiple partnerships have the potential to leverage the combined strengths of government, civil society and non-government organisations/service providers – something acknowledged at the World Summit on Sustainable Development in Johannesburg (UN, 2002) in meeting the goals of sustainable development, and specifically Millennium Development Goals water and sanitation targets.

24.12 Conclusions

Local authorities in South Africa are incapacitated by a funding allocation that prioritises formal housing over the immediate needs in informal settlements. The result, in regard to drainage management, is a toxic mix of polluted water that creates a public health risk and impacts on freshwater systems, which in turn has the potential to threaten commercial export agriculture in the region of the Franschhoek Valley.

One of the objectives of this study was to explore the process of delivering low-cost, acceptable drainage devices in an informal settlement, and to ensure that in so doing there was collaboration and genuine cooperation that would include the capacity to learn collectively and to modify the management options to achieve a desirable outcome. The chosen case study presented two contrasting stories. In the earlier stages, the process and results were disappointing. This was attributed to the fact that the devices installed became dysfunctional in a matter of weeks, and more importantly, that 'on the ground' social organisations and structures were weak. It was also apparent that a PAR methodology is not going to deliver solutions without adequate commitment from local authorities, and that more effort and attention is required for developing the capacity of the end users of drainage devices themselves. In contrast, the second part of the discussion suggests that local authorities were starting to embrace the concept of being critical agents of local level change, and had the foresight to establish an agreement with an NGO and to give them the mandate of building capacity towards the development of a well-functioning, truly representative local social structure. While success of the most recent projects has not been measured as yet, and it may take some time before the results become clear, nonetheless there are already encouraging signs in the selection of the Langrug as a site for two prototype projects because the existence of a coherent social structure was recognised within the settlement. It points to emerging acceptance in South Africa that the incremental upgrade of informal settlements requires far more than technical knowledge of water-based service infrastructure, but to a new emphasis in the co-management of the incremental upgrading of informal settlements by leveraging the combined strengths of government, local residents, NGOs and service providers.

References

Alexander, P. (2010) Rebellion of the poor: South Africa's service delivery protests – a preliminary analysis. *Review of African Political Economy*, 37 (123) 25–40.

Armitage, N.P., Winter, K., Spiegel, A. and Kruger, E. (2009) Community-focused greywater management in two informal settlements in South Africa. *Water Sci Technology*. 59(12) 2341–50.

Carden, K., Armitage, N., Winter, K., Sichone, O. and Rivett, U. (2007) Understanding the use and disposal of greywater in the non-sewered areas in South Africa. WRC Report 1524/1/07, Water Research Commission (WRC), Pretoria, South Africa.

Department of Human Settlements (DHS) (2012) Annual Report 2011/2012, RSA Government, Pretoria. ISBN: 978-0-621-40896-6

Department for International Development (DfID) (1998) *Guidance manual on water supply and sanitation programmes*, Water and Environmental Health at London and Loughborough (WELL), Water Engineering and Development Centre (WEDC), Loughborough University, UK.

EAWAG (2005) Household-centred environmental sanitation – Implementing the Bellagio principles in urban environmental sanitation. Provisional guideline for decision-makers. EAWAG, Swiss Federal Institute of Aquatic Science and Technology, June 2005. ISBN 3-906484-35-1

Eales, K. (2008) Partnerships for sanitation for the urban poor: Is it time to shift paradigm? *Proceedings of IRC Symposium – Sanitation for the urban poor*, 19–21 November 2008, Delft, The Netherlands.

Engleson, E. (2010) Informal Settlements – the illegal city of Kibera. Available at: http://www.resilientcity.org/index.cfm?id = 23147

Kruger, E. (2009) Grave expectations – Participatory greywater management in two Western Cape shack settlements. MA thesis, Department of Social Anthropology, University of Cape Town, Cape Town, South Africa.

Lal, P., Lim-Applegate, H. and Scoccimarro, M. (2001) The adaptive decision-making process as a tool for integrated natural resource management: focus, attitudes, and approach. *Conservation Ecology*, 5 (2), Available at: http://www.consecol.org/vol5/iss2/art11/

Lowndes, V. and Wilson, D. (2001) Social capital and local governance: exploring the institutional design variable. *Political Studies*, 49, 629–647.

Moore, T.L.C. and Hunt, W.F. (2012) Ecosystem service provision by stormwater wetlands and ponds – a means for evaluation? *Water Research*, 46, 6811–6823.

Scholz, M. and Lee, B.-H. (2005) Constructed wetlands: a review. *International Journal of Environmental Studies*, 62, 421–447.

SDI South African Alliance (2012) Informal settlement upgrading. http://sasdialliance.org.za/projects/langrug/.

SkillsPortal (2012) HIV/AIDS: Understanding the impact it has on your company. Available at: http://www.skillsportal.co.za/content/hivaids-understanding-impact-it-has-your-company.

Stellenbosch Municipality (2011) Langrug Settlement enumeration report, Franschhoek. Stellenbosch Municipality. Available at: http://tinyurl.com/ompjub2.

United Nations (2002) *Report of the United Nations Report of the World Summit on Sustainable Development, Johannesburg, 26 Aug – 4 Sept 2002* (United Nations publication, A/CONF.199/20, Sales No. E.03.II.A.1, ISBN 92-1-104521-5), United Nations, New York.

UN-Habitat (2003) *Squatters of the World: The Face of Urban Poverty in the New Millennium*. Nairobi. Available at: http://www.unhabitat.org/publication/slumreport.pdf

UN-HABITAT (2013) *Streets as public spaces and drivers of urban prosperity*, United Nations Human Settlements Programme, Nairobi.

Water Supply and Sanitation Collaborative Council (WSSCC)/Sandec. (2000) Summary report of Bellagio expert consultation on environmental sanitation in the 21st Century. Swiss Federal Institute for Environmental Science and Technology, Switzerland. Available at: http://www.eawag.ch/forschung/sandec/gruppen/clues/approach/bellagio/index_EN

25

Low Impact Development in the USA
Bruce K. Ferguson

25.1 Introduction

American cities use a broad suite of stormwater management practices, linked by evolving collective concepts, which point in directions that are broadly parallel to those in other developed countries, encompassing water quality and quantity, environmental health and human amenity. A common concept today is 'low-impact development', which favours solving drainage and environmental problems near the source where rain falls, in numerous dispersed facilities integrated with urban design and tending to use naturally restorative environmental processes. Continuing evolution is stimulated and unified to a great degree by federal water-quality standards. However, in application, individual solutions are diverse, because they are chosen and configured to fit diverse hydroclimatic conditions, urban land-use densities and pre-existing infrastructures. As agendas for healthy, liveable cities rise parallel to those of water quality and quantity, multiple disciplines are called upon to develop creative applications that satisfy multiple types of values.

25.2 Unifying Legislation

In the USA, federal water protection laws of various kinds have existed since the 19th century. As each successive law has encouraged the removal of obvious point sources of pollution, such as industrial and wastewater discharges, the remaining diffuse sources that had not yet been addressed, have become more apparent. The diffuse sources are referred to as 'non-point-source' pollution, meaning that they discharge into waterways at multiple points, such that the effluents from many different activities merge into a single effect.

The federal law known today as the Clean Water Act originated in the 1970s; amended in 1987, it included specifically, for the first time in federal law, urban non-point-source

Sustainable Surface Water Management: A Handbook for SuDS, First Edition.
Edited by Susanne M. Charlesworth and Colin A. Booth.
© 2017 John Wiley & Sons, Ltd. Published 2017 by John Wiley & Sons, Ltd.

pollution. On that legal foundation, the federal Environmental Protection Agency (EPA) established a large set of urban runoff water-quality standards, and delegated most of the necessary administrative processes to the separate states. Municipalities discharging urban runoff into waterways are now required to maintain permits for such discharges, and to do so must meet, or show movement toward meeting, EPA's water-quality standards.

Prior to the 1987 amendments, the polluted character of urban runoff had been acknowledged, but little was being done due to unfamiliarity with the non-point-source problem, lack of authoritative standards and wariness of unknown construction and maintenance costs. The new federal law required everyone involved in urban stormwater management for the first time, at all levels, to start learning concepts and technologies that they had never been compelled to learn before. It was instrumental in making stormwater management's environmental objectives, and the accompanying costs of construction and maintenance, universally accepted.

Since long before EPA established water quality standards, municipalities had drained urban districts in order to protect property, health, and essential services (American Public Health Association, 1960). Since the 1970s they had increasingly practised or required detention: the use of reservoirs with controlled outlets to limit downstream peak flow, abide by downstream drainage capacities and prevent flooding (Poertner, 1974). Their accumulated drainage and detention systems still exist today, and are part of most stormwater management practice, alongside contemporary water quality requirements.

Manufacturing companies responded to the new combination of initiatives by inventing and offering new types of permeable pavements, storage vaults and runoff filters and separators. The new initiatives in effect stimulated the rise of a new industry, the activity of which is exhibited at the annual WEFTEC conference (www.weftec.org). The application of new materials and technologies is increasingly supported by national industry standards.

25.3   Stormwater Management Practices

Stormwater management practices that have evolved and accumulated into use today are diverse in function, configuration and appropriateness for specific applications. An early term, 'best management practices' (BMPs) referred to all of them implicitly favourably, but without indicating any particular practice's specific character or performance role. For such general use, that term is now replaced by the more neutral term 'stormwater control' (Water Environment Federation, 2012, p 4).

Tables 25.1 and 25.2 list today's common stormwater controls. The tabulated information was compiled from Water Environment Federation (2012), Debo and Reese (2002) and Cahill (2012). Detailed technical descriptions of these controls, and the guidelines for their use, are in those same sources. Where different sources used different terms to refer to fundamentally similar practices or functions, terms for the tables were chosen that were intuitively simple and general; the general term 'filter', for example, includes a variety of technically distinct water-quality transformations.

Table 25.1 shows practices that tend to have exclusively or primarily stormwater-management functions; they do not necessarily influence human accommodation in their surroundings. They carry away source-area runoff or buffer downstream areas from its quantity and quality impacts. Some of them originated in the older municipal agendas of drainage and detention, while others have developed more recently in response to modern water-quality requirements. They tend to be either set aside in single-purpose spaces or

Table 25.1 Common types of stormwater controls that have exclusively or primarily stormwater management functions, and do not necessarily influence human accommodation.

Practice	Stormwater function	Distinctive feature
Sewer	Convey and direct	Pipe or channel network
Detention	Reduce peak flow rate	Reservoir or tank, with or without permanent pool
Floodplain reserve	Reduce peak flow rate	Open lowland
Infiltration	Filter, reduce volume	Basin, trench, vault, dry well
Bioretention	Filter, reduce volume	Vegetated cell
Vegetated swale	Filter, reduce volume, reduce peak flow rate	Vegetated swale
Sand filter	Filter	Sand cell
Filter strip	Filter	Grass strip
Stormwater wetland	Filter	Wetland
Inserts	Filter or settle	Traps, vaults, separators
Extended detention	Settle	Reservoir, with or without permanent pool

Table 25.2 Common types of stormwater controls that are selected and arranged at least partly for human accommodation.

Practice	Stormwater function	Distinctive feature
Permeable pavement	Filter, detain, reduce volume	Permeable surface
Green roof	Filter, reduce volume	Vegetated soil on roof
Bioretention ('rain garden')	Filter, reduce volume	Vegetated soil at ground
District water harvesting	Reduce volume, capture supply	Vegetated soil at ground
Roof water harvesting	Reduce volume, capture supply	Downspout cistern

hidden in underground vaults; their implementation is often a narrowly technical concern. Some of them are implemented by shaping of topography and control structures; others are devices supplied by manufacturers. Their capacities are designed to meet one or more design flows.

Table 25.2 shows stormwater practices that are selected and arranged at least partly for human use and comfort, with stormwater management only an equal or secondary concern. They are useful urban structures which are modified to add stormwater function without compromising human accommodation. For example, permeable paving combines absorption and treatment of stormwater through the pavement structure, and city traffic across the ground plane. Green roofs and water harvesting reduce urban districts' runoff volumes and pollutant loads. 'Rain gardens' implement bioretention in detailed pockets of urban space without disrupting nearby circulation. Most of these practices have been developed in the last few decades, pursuing solutions in complex urban situations with multiple application objectives. They are most typically designed for relatively small, frequent flows; larger flows discharge without management.

To analyse the performance of stormwater controls and the overall drainage systems of which they are part, there are a number of hydrologic calculation methods (Cahill, 2012, p 123). The Curve Number method was developed originally by the US Department of Agriculture; it estimates runoff volume and, with its extensions, peak rate. The Rational Method estimates peak rate; versions of it have been adapted to estimate total volume. Hydrologic methods such as these are set into a variety of public- and private-domain computer models. A prominent computer model featuring continuous simulation is from

Table 25.3 Contrasting hydroclimatic conditions in six cities (data from US NOAA National Weather Service, Precipitation Frequency Data Server, hdsc.nws.noaa.gov).

City	10-year, 24-hour storm rainfall (mm)	Average annual precipitation (mm)	Average number of days per year with measurable precipitation
Atlanta, Georgia	131	1263	113
Philadelphia, Pennsylvania	122	1055	118
Los Angeles, California	100	326	36
Portland, Oregon	94	1104	164
Denver, Colorado	75	396	87
Phoenix, Arizona	54	208	30

the US Army's Hydrologic Engineering Center; as of 2015 its current version was HEC-HMS 4.1 (www.hec.usace.army.mil). Designed to simulate all hydrological processes associated with dendritic catchment systems, it utilises information such as event infiltration, unit hydrographs and hydrologic routing. It also includes evapotranspiration, snowmelt and soil moisture accounting. Further information on the modelling of urban drainage and SuDS is given in Chapter 20.

America's diverse geography presents different hydrologic conditions within which types of stormwater practices must be selected and their capacities determined. Table 25.3 contrasts conditions in six different cities. The 10-year, 24-hour rainfall is presented as an example of a design storm for which a stormwater facility might be required to supply adequate capacity. The annual precipitation and the annual number of wet days indicate the overall abundance or scarcity of water over time and thus its ability to support naturally healthy vegetation, or the value of collecting it for use. Eastern cities such as Atlanta and Philadelphia experience frequent intense storms. The resulting large runoff volumes require facilities with large capacities which take up substantial space in urban districts, and which are structurally secure amid erosive flows. In contrast, in south-western desert cities such as Phoenix, the overall scarcity of water encourages its concentration into planted areas where irrigation is needed, and into cisterns for capture and use (Phillips, 2003). In north-western Pacific cities such as Portland, design-storm rainfall is of low intensity, so facilities with adequate capacities can fit relatively easily into small pockets of urban space, while the benignly frequent precipitation naturally supports healthy vegetation. Portland's favourable hydroclimate is one of the factors that have helped it to become a centre of creative urban design with stormwater.

25.4 Low-Impact Development

Today's collective concept of 'low-impact development' (LID) refers to managing stormwater as close as possible to where the rain falls, as an alternative to concentrating it into a single corrective structure. It often employs the preservation or re-creation of natural landscape features, to minimise impervious cover and to enable natural infiltration and treatment. Some of LID is what is called 'green infrastructure' – systems that use or mimic natural processes to infiltrate, evapotranspire, or reuse stormwater (Wise, 2008).

The concept now called LID originated in the 1970s, when McHarg's (1969) 'design with nature' paradigm produced the first simple techniques for moulding development to preserve natural stormwater processes: clustering of development to avoid floodplains and riparian habitats; open drainage systems in contact with soil and vegetation; and infiltration and recharge of natural aquifers in place of disposal into surface streams (McHarg and Sutton, 1975). Subsequent contributions from a variety of sources have penetrated further into the ways urban development can be adapted to reduce runoff and pollution at the source, such as those listed in Table 25.2. The collective term 'low-impact development' originated among planners in Prince George's County, Maryland, in the 1980s or 1990s. According to one of them (Larry Coffman, pers. comm. 1999), its intent is to give regulatory credit to all of the favourable techniques that were being discovered; in his words, 'Everything you do, counts'.

Rain gardens have become public symbols of LID's agenda to merge environmentally restorative processes into the midst of cities. Rain gardens pool collected runoff in a planted area, where it infiltrates the soil, beneath which it may further infiltrate underlying soils, or be discharged gradually through drainage pipes. They mitigate runoff volumes and peak rates; their plants and soils filter suspended solids and absorb and break down dissolved pollutants (Cahill, 2012, p 146–147). The informal name 'rain garden' reflects a facility's arrangement for human accommodation as much as for stormwater management. Early rain gardens were modified portions of vegetated swales. More recently, the city of Portland, Oregon (2006 awards page at www.asla.org) discovered how to embed them in detailed pockets of space among existing city streets and sidewalks, like the ones shown in Figure 25.1.

Figure 25.1   Rain gardens in Portland, Oregon.

Figure 25.2   LID at the Morton Arboretum in Lisle, Illinois.

Figure 25.2 shows a combination of LID features at the Morton Arboretum visitors' centre in Lisle, Illinois. The permeable pavement of open-jointed blocks infiltrates rainwater rapidly, and its base course has a large reservoir capacity to hold the water from large storms. In the event of still larger storms, or pavement clogging, curb cuts admit runoff into rain gardens where water ponds on vegetated soil while infiltrating. All rainwater is treated by naturally occurring microorganisms in the pavement and soil, before discharging slowly downstream.

25.5   Stormwater and Urban Agendas

At the same time that America's stormwater management has been evolving, the field of urban design has been evolving too, with growing public expectations for comfort, convenience, health and environmental quality. Much of this thinking comes under the conception of 'new urbanism' (Katz, 1994; Congress for the New Urbanism, www.cnu.org). Today's urbanist agenda urges amenity and ecological health, together with high water quality, in all parts of the urban stormwater course.

In low-density suburban locations, or in other locations with abundant open space, it is relatively economical to manage stormwater using natural drainage courses and lowlands, as long as provision is made to do so in the land-use plan. Approximately 15% of a site area may need to be reserved in which to locate swales, reservoirs and wetlands, like those listed in Table 25.1. Within that reserved space, only earthmoving and control structures need to be added, to make these 'conventional methods' work. In such drainage systems, designers have found recreational and scenic functions, usually using naturalistic effects fitting the topography and vegetation at low cost. An example is shown in Figure 25.3, where an infiltration basin is lined with trees, grass, gravel and boulders; residences look down on it from the edges.

In contrast, in densely built-up areas space is not available for specialised land reservation, so stormwater facilities are either integrated with the city through design, or they are in conflict with it. 'Unconventional methods' must be called upon, such as those listed in Table 25.2, or those from Table 25.1 that can be sequestered in underground vaults. They

Figure 25.3   Infiltration basin in a suburban residential neighbourhood near Boise, Idaho.

are designed for hydraulic capacity and simultaneously to fit their urban settings. These involve higher construction and maintenance costs than conventional methods to achieve the same stormwater performance. However, in a location with high development pressures, they represent investment toward dense land use, with correspondingly high income, and enable 'infill' development on small properties, which would otherwise remain vacant (Ferguson, 2010). Consolidating human and stormwater objectives in a single application requires multiple disciplines including engineers, landscape architects and urban planners, working side by side and communicating positively (Water Environment Federation, 2012, pp. 1–19).

Figure 25.4 shows highly integrated stormwater and human facilities being retrofitted into a busy commercial street in St Louis, Missouri (2011 awards page at www.asla.org). The foreground pavement is pervious concrete; it absorbs both direct rainfall and runoff from the adjacent pavement near the buildings, which had to be impervious to protect leaky old basements. The pervious concrete's base course, visible in the open trench, is structural soil, which retains and treats infiltrated stormwater while supporting the street trees that will be planted here. To the left is a new parking lane, which will give the street a more 'lived in' character. The remaining street lanes have been narrowed in order to 'calm' (naturally slow down) the vehicular traffic. In the background is a disabled parking space and ramp, making the street more universally accessible. At the end of the block is a curb 'bulbout', which safely terminates the parking lane, reduces street crossing distance for pedestrians, and shapes excess space into a rain garden. Highly integrated combinations of shape and performance like this can be produced only by teams of cooperative multi-disciplinary designers.

Figure 25.4 Retrofit construction in South Grand Boulevard, St Louis, Missouri.

25.6   Choices in Challenging Urban Districts

Particularly arduous stormwater management applications are called for in urban districts with combined-sewer overflows (CSOs). Sewers that collect stormwater and wastewater together are today considered obsolete. However, hundreds of American cities have large districts still served by combined sewers constructed in the 19th and early 20th centuries. During wet weather, large volumes of runoff exceed the sewers' hydraulic capacity, and the combined flow escapes at numerous points to surface waters, producing one of America's most serious contemporary water-quality problems (Cahill, 2012, pp. 68–71). EPA requires reduction in combined-sewer overflows under the Clean Water Act, and exerts litigation to encourage it. However, CSO districts are characteristically densely built up, supporting settled residential neighbourhoods and commercial economies, so retrofitting them with new infrastructure to manage the large amounts of runoff they produce can be expensive and disruptive. The types of choices made in these districts illustrate the range of different cities' attitudes toward urban values and bureaucratic culture.

One approach to CSO reduction is to replace the old combined sewers with modern separate sewers, staying within the single-purpose types of provisions listed in Table 25.1. Most cities cannot afford the huge cost of district-wide replacement. Some have rebuilt small portions of their combined-sewer systems as parts of larger, multi-faceted programmes.

The second approach is to add detention reservoirs within the combined-sewer network, reducing the wet-weather flow to rates within the capacities of sewer pipes and treatment plants. The reservoirs are among the single-purpose technologies of Table 25.1; they tend to take the form of underground tanks or tunnels. They are typically quite large and expensive. However, they have been attractive to technical city agencies because they can be conveniently and accurately analysed for design and maintenance. For example, Washington, DC has undertaken a 20-year programme to construct a 21 km system of three tunnels, each 7 m in diameter (Allen, 2011).

Table 25.4 Examples of websites presenting ongoing research and technical guidance.

Institution	Website address
Center for Watershed Protection	www.cwp.org
Interlocking Concrete Pavement Institute	www.icpi.org
Low Impact Development Center	www.lowimpactdevelopment.org
North Carolina State University	www.bae.ncsu.edu/stormwater
US Environmental Protection Agency	http://water.epa.gov/polwaste/npdes/stormwater/
University of New Hampshire	www.unh.edu/unhsc
Water Environment Research Foundation	www.werf.org/stormwater

A third approach is to divert urban runoff away from combined sewers, and release it instead into local vegetation, soils and aquifers. This is equivalent to a retrofitting of urban districts with LID and green-infrastructure provisions like those listed in Table 25.2, with reduction of stormwater flow volume an explicit performance requirement. The cost is believed to be lower than that of sewer reconstruction or detention, and the benefits more multi-faceted.

Portland, Oregon, has emphasized this type of diversion since the 1990s (www.portlandoregon.gov/bes). In planning projects like this, the city has benefitted from an excellent planning culture since the 1970s (Abbott, 2001), including scrupulous public participation and city agencies that cooperate with each other's agendas. The least expensive of Portland's diversion practices per unit of flow reduction has been disconnection of roof downspouts from sewers and onto garden soil, where the water evapotranspires or recharges natural aquifers (Cahill, 2012, pp. 68–71). Private homeowners have welcomed this approach when its purpose is explained to them, with or without 'rain barrels' to store water for irrigation in dry periods. Another favourable practice has been diversion of runoff from street drainage inlets into excavated sumps from which it disperses into the surrounding soil. Rain gardens, vegetated roofs and permeable pavements are installed where the conditions of streets and land uses make them favourable alternatives.

More recently, Philadelphia has undertaken a 25-year programme of this type, as an alternative to the cost of four enormous tunnels (www.phillywatersheds.org). The programme foresees replacing one-third of the city's impermeable surfaces with 3800 ha of green areas and permeable pavements (Allen, 2011). The programme is administered by the city's Water Department, in a new Office of Watersheds, which merged previously separate programmes for combined-sewer overflows, stormwater management and drinking-water source protection (McIntyre, 2014). In partnership with all kinds of city and neighbourhood groups, the office undertakes hundreds of retrofit projects. Retrofit diversions include: permeable pavements in alleys, parking lots and recreational surfaces; rain gardens, infiltration beds, downspout cisterns, planter trenches, and green roofs; and replacement of playground asphalt with lawns and plantings. Many of them cool streets with tree shade, and make more useful and attractive environments for people; all their designs seek social, economic and environmental benefits fitting each particular neighbourhood. Every project has an immediate incremental effect on CSO reduction. A programmatic risk is the relative uncertainty of long-term cost and performance associated with dispersed monitoring and maintenance.

Over time, America's stormwater practices evolve and accumulate new technologies, directions and artful applications. As knowledge and diversity grow, the system as a whole grows more complex and adaptable. Challenging urban districts with combined environmental and urban agendas are testing grounds for future capabilities and directions. Table 25.4 lists examples of universities, industrial associations and other organisations that make ongoing research and updated technical guidance available.

References

Abbott, C. (2001) *Greater Portland: Urban Life and Landscape in the Pacific Northwest*, Philadelphia: University of Pennsylvania Press.

Allen, A. (2011) Green City, Gray City, *Landscape Architecture*. 101 (9) 72–80.

American Public Health Association (1960) Committee on the Hygiene of Housing, 1960, *Planning the Neighborhood, Standards for Healthful Housing*, Chicago: Public Administration Service.

Cahill, T.H. (2012) *Low Impact Development and Sustainable Stormwater Management*, Hoboken: Wiley.

Debo, T.N. and Reese, A.J. (2002) *Municipal Stormwater Management*, second edition, Boca Raton: CRC Press.

Ferguson, B.K. (2010) Porous Pavements in North America: Experience and Importance, in *Conference Proceedings, Novatech 2010*, Lyon, France.

Katz, P. (1994) *The New Urbanism*, New York: McGraw-Hill.

McHarg, I.L. (1969) *Design with Nature*, Garden City: Doubleday.

McHarg, I.L. and Sutton, J. (1975) Ecological Plumbing for the Texas Coastal Plain: The Woodlands New Town Experiment, *Landscape Architecture*. 65 (1) 78–89.

McIntyre, L. (2014) The Infiltrator, *Landscape Architecture*. 104 (1) 38–46.

Phillips, A. (2003) A Good Soaking: An Introduction to Water Harvesting in the South West, *Landscape Architecture*. 93 (8) 44–50.

Poertner, H.G. (1974) Practices in Detention of Urban Stormwater Runoff, Special Report 43, Chicago: American Public Works Association.

Water Environment Federation (2012) *Design of Urban Stormwater Controls*, WEF Manual of Practice No 23 and ASCE/EWRI Manuals and Reports on Engineering Practice No 87, New York: McGraw-Hill.

Wise, S. (2008) Green Infrastructure Rising: Best Practices in Stormwater Management, *Planning*. 74 (8) 14–19.

26

Sustainable Drainage Systems in Spain

Valerio C. Andrés-Valeri, Sara Perales-Momparler, Luis Angel Sañudo Fontaneda, Ignacio Andrés-Doménech, Daniel Castro-Fresno and Ignacio Escuder-Bueno

26.1 Introduction

Over the last 50 years, the huge economic growth observed in Spain has led to massive migrations from rural to urban areas, producing a rapid growth of urban centres. The uncontrolled urban sprawl, especially in the touristic Mediterranean regions (García *et al.*, 2014), has resulted in the waterproofing of natural soil in urban environments, increasing the problems related to urban stormwater management. Flooding is the major problem detected so far, being caused by increased runoff volumes that exceed the capacity of the sewer and drainage systems. This flooding effect then brings with it diffuse pollution, since the greater the waterproofed area, the greater the area washed by runoff and so the greater the non-point pollution effects (Castro-Fresno *et al.*, 2013).

Increasing runoff volumes along with the predominance of combined sewerage systems in Spain have increased the frequency of occurrence of combined sewer overflows (CSOs) (Castro-Fresno *et al.*, 2013). Moreover, due to the geographic position of Spain, different climatic patterns meet (Segura-Graiño, 1994), leading to different problems related to urban water management depending on the specific geographic area of the country. While in the north of Spain, high rainfall volumes that are spread throughout the year can lead to flooding problems, in some regions of the south of Spain it is normal to find periodic droughts in summer periods (Segura-Graiño, 1994). A special case is the eastern coast of the Mediterranean region, which suffers the effects of cut-off lows during autumn resulting in torrential rainfall events, producing flooding problems (Perales-Momparler *et al.*, 2014).

Stormwater-related problems were first studied in Spain in the 1990s (Dolz and Gómez, 1994; Malgrat, 1995; Temprano *et al.*, 1996; Jimenez-Gallardo, 1999). Thenceforth,

Sustainable Surface Water Management: A Handbook for SuDS, First Edition.
Edited by Susanne M. Charlesworth and Colin A. Booth.
© 2017 John Wiley & Sons, Ltd. Published 2017 by John Wiley & Sons, Ltd.

and during the last 20 years, great efforts have been made to increase knowledge of SuDS techniques and their performance. In this sense the contribution of different research centres such as the universities of Cantabria, Madrid, La Coruña and Zaragoza; and the polytechnic universities of Catalonia and Valencia is very notable.

The progressive social awareness of sustainability and the increasing frequency of water-related problems has resulted in a growing interest of the main national authorities in establishing control measures for correct water management. The approval of the Water Framework Directive in 2000 (EU 2000/60/EC) marked a turning point in the management of water bodies by the European Union authorities, integrating sustainable principles in water management. After this step forward by the European authorities, sustainable principles were integrated into other directives related to water management such as the Flood Directive in 2007 (EU 2007/60/CE). The particular and variable climatic conditions across the whole of Spain, with specific water-related problems, led to the development of specific regulations to manage the different runoff problems in the country. Spanish Royal Decree RD 1620/2007 established the minimum water quality standards for water reuse, depending on its final purpose. In 2012, Spanish Royal Decree RD 1290/2012 imposed the necessity of reducing the contribution of new urban developments to stormwater runoff volumes. Finally, in 2013, the recently approved Spanish Royal Decrees RD 233/2013 and RD 400/2013 encouraged the use of sustainable drainage techniques in the management of stormwater runoff, especially in new urban developments, which can influence the drainage behaviour of the watershed.

However, although the integration of SuDS in the urban planning process for new developments can help to mitigate stormwater-related problems (Dietz, 2007), it is necessary to increase its application in already developed urban areas. In fact, the integration of SuDS techniques in the retrofitting process in city centres can be one of the main solutions for their hydrological rehabilitation (Andrés-Valeri *et al.*, 2014a). During the past decade different retrofit applications of SuDS were developed in some Spanish regions (Table 26.1), mainly in Madrid but also in the Mediterranean region and in the north.

Table 26.1 Main SuDS application in Spain (Castro-Fresno *et al.*, 2013).

Region	SuDS application
Aragon	EXPO Zaragoza campus
Asturias	La Guia Park
	La Zoreda parking area
Basque Country	Cristina Enea Park
	Ametzagaina Park
	Philip IV Sports Arena
	Mount Ulia.
Cantabria	Las Llamas Park
Catalonia	Joan Reventós Park
	Torre Baró
	University Park
Galicia	Oleiros
Madrid	Gomeznarro Park
	Castellana extension
Valencia	Xativa
	Benaguasil

26.2   SuDS Case Studies in the Northern Regions of Spain

The first research projects carried out in Spain that were mainly focused on SuDS techniques were developed at the Civil Engineering School of the University of Cantabria by the GITECO Research Group. Knowledge gained was also transferred to the productive sector, through the development of some exploitation patents and by collaboration with some contractors in applying SuDS for stormwater management in new urban developments. As a part of the developed research projects, and with the close cooperation of different public agents and contractors, some experimental areas were built in new urban developments or as a part of retrofitting processes in run-down urban areas, mainly in the northern regions of Asturias (Figure 26.1a and b) and Cantabria (Figure 26.2a and b). The first real application of pervious pavements (PPS) for research purposes was the construction of a 22,000 m² PPS area in Gijón city (Asturias region) through a collaboration established between Gijón City Council, Atlantis corp® and GITECO. As part of the

(a)

(b)

Figure 26.1   (a) Sports Centre, Gijón car park (b) La Zoreda car park (Oviedo).

(a)

(b)

(c)

Figure 26.2   (a) General view of Las Llamas Park (Santander) and the experimental parking area integrated in the park (b) artificial wetland and retention pond.

retrofitting process in the outdoor parking area of the Sports Centre, Gijón (Figure 26.1a) in 2005, 798 permeable parking bays of grass reinforced with plastic cells were built with 15 fully monitored parking bays in which the biodegradation processes of hydrocarbons inside the permeable pavement structures were studied (Bayón *et al.*, 2005; Sañudo-Fontaneda *et al.*, 2014a).

In 2008, an experimental permeable parking area of 1100 m² was built in Santander (Cantabria region). In 2006, Santander City Council began the rehabilitation of Las Llamas wasteland, a degraded urban area near the main beaches of the city and the university campus. The objective of this regenerative action was to build an urban park with a green area of 300,000 m², which included sustainability principles in its design (Figure 26.2). Different SuDS techniques were applied for correct stormwater management of the area, mainly green infrastructure, a retention pond and an artificial wetland. With the collaboration of Santander City Council, Hanson-Formpave®, Coventry University and GITECO, an experimental permeable parking area (Figure 26.2a), with 45 fully monitored parking bays (Figure 26.3), was included in the new Las Llamas urban park. In order to study the suitability of PPS for rainwater harvesting purposes,

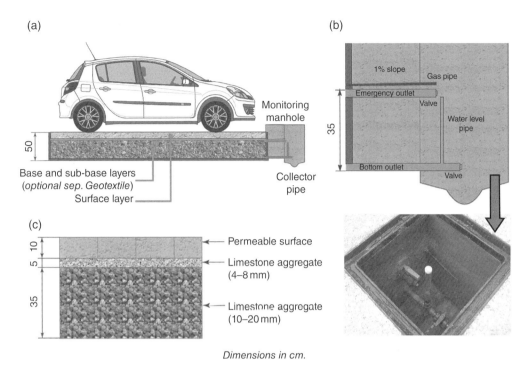

Dimensions in cm.

Figure 26.3 (a) Permeable pavement scheme (Sañudo-Fontaneda *et al.*, 2014a) (b) monitoring manhole (Sañudo-Fontaneda *et al.*, 2014a) (c) Cross-section of the permeable parking bays.

four different types of geofabrics: Inbitex®, One-Way®, Polyfelt® and Danofelt® and five different permeable surfaces were used: interlocking concrete blocks (ICB), porous asphalt, pervious concrete and reinforced grass with concrete and plastic cells.

After two years of monitoring, the water chemistry results showed a good quality of stored water for all parking bays, especially for pervious concrete surfaces, allowing the water to be reused according to Spanish regulations RD 1620/2007 (Gomez-Ullate *et al.*, 2011a; Sañudo-Fontaneda *et al.*, 2014a). However, high solid content levels were found in the outflow of the parking bays during the first months of monitoring due to initial scouring of fine particles and few rainfall events (Gomez-Ullate *et al.*, 2011a). On the other hand, the water quantity data showed that for high rain areas like Santander the parking bays remained full of water during most of the year, with the exception of summer periods due to rapid evaporation that took place because of the high temperatures (Gomez-Ullate *et al.*, 2011a, b). Each parking bay had a reservoir layer with 115.5 l · m⁻² storage capacity, thus each individual bay could manage the requirements of irrigating 10 m² of garden for nearly a complete month of drought (Gomez-Ullate *et al.*, 2011b). The preliminary results of the monitoring programme showed that it was possible to group the PPS surfaces into three with similar water storage performance, and no significant differences observed between them: ICB, porous asphalt and grassed surfaces (Gomez-Ullate *et al.*, 2011b). Geotextile properties also influenced the storage behaviour of the parking bays, modifying their infiltration rate and the evaporative processes that took place (Gomez-Ullate *et al.*, 2010). Finally, after five years of continuous use without maintenance, an important reduction in permeability values was observed (Sañudo-Fontaneda *et al.*, 2014b), confirming the clogging effect observed in laboratory-scale measurements (Rodríguez-Hernández *et al.*, 2012; Sañudo-Fontaneda *et al.*, 2013; Sañudo-Fontaneda *et al.*, 2014c).

In 2009, El Castillo de La Zoreda Hotel situated in La Zoreda Forest on the outskirts of the city of Oviedo (Asturias) was built. Sustainability principles for stormwater management were integrated in the main project, specifically the inclusion of green infrastructure and high permeability areas for increasing water infiltration into the soil. As part of this new urban development, three 20 m-long experimental linear drainage areas were built on the roadside of the parking area for the hotel (Figure 26.1b).

Each reach corresponded to a different linear drainage system, two SuDS (Figure 26.4): filter drain and swale, and a conventional concrete ditch, the most commonly used roadside drainage system in Spain. For three years, general water quality parameters were monitored in the outflow from each reach, showing an important water quality improvement when using the SuDS systems, especially the filter drain (Andrés-Valeri *et al.*, 2014b), allowing water reuse for some non-potable uses according to Spanish RD 1620/2007. The results obtained showed that the inclusion of a geotextile layer increased the retention of the solid particles in SuDS providing outflow concentrations of total suspended solids in the range of 10–15 mg/l.

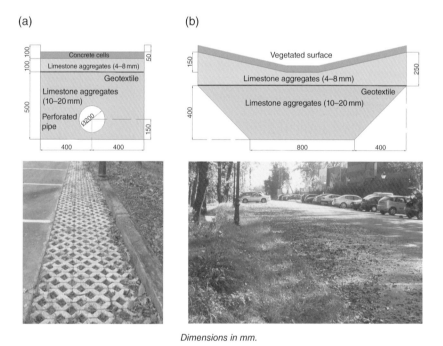

Dimensions in mm.

Figure 26.4 Cross-section and picture of the SuDS used in La Zoreda car park (Oviedo): (a) filter drain (b) swale.

26.3   Integration of SuDS into New Urban Developments

Nowadays, the Basque Country and Catalonia are probably the most advanced regions in Spain in terms of sustainable development, with many real applications of SuDS techniques in urban areas. In 2006, the Barcelona Urban Management Company (BAGUR SA) asked for GITECO's advice about integrating SuDS in the Joan Reventós Park, a new urban development of 28,000 m² on the outskirts of the City of Barcelona.

For the management of stormwater runoff, sustainability principles were followed, integrating low impact development (see Chapter 25) practices in the design, which provided marked aesthetic value. For the selection of the most appropriate SuDS techniques, the area was divided into 16 sub-catchments, studying the hydrological performance of each sub-catchment and looking for local solutions for managing the stormwater runoff. Finally, the selected SuDS techniques were integrated into a SuDS management train to manage the stormwater runoff of the area (Figure 26.5).

Filter drains and filter trenches were used as a source control system for the collection of runoff from the surrounding impervious areas and from the main pedestrian roads of

Figure 26.5   (a) Filter drains and infiltration pond (b) Retention pond at the end of filter drains (c1) Starting of construction of infiltration well (c2) Finishing of infiltration well (c3) Infiltration well integrated in the park. (Photos courtesy of Roberto Soto of BAGUR SA).

the park. These systems were connected to various swales, which conveyed the runoff through the park, transporting it to the main swale, which was constructed over a general collector for the sewage system. The green areas were designed to convey the runoff produced inside the park to the swales, which performed as natural water channels, transporting the water to the retention pond constructed in the floodplain of the park. With this design, the most polluted water from surrounding impervious areas was treated by the whole treatment train, while the less polluted runoff generated inside the park only used one or two of the systems. Finally, in order to limit the contribution from parking areas to the stormwater runoff, PPS technology was applied, allowing water infiltration into the ground.

26.4 SuDS Retrofitting Case Studies in the Mediterranean Region

This section features retrofitting case studies, designed to show the efficiency of SuDS in southern Europe. They were conducted as part of two projects, both supported by ERDF funding of the European Union:

1. Aquaval: Sustainable urban water management plans, promoting SuDS and considering climate change in the province of Valencia (Life08ENV/E/000099, www. aquavalproject.eu)
2. E²STORMED: Improvement of energy efficiency in the water cycle by the use of innovative stormwater management in smart Mediterranean cities (1C-MED12-14, www.e2stormed.eu)

The inner-city areas described below are four out of eight that were retrofitted with various SuDS typologies in two cities located in the region of Valencia (Spain): Xàtiva and Benaguasil, with 29,400 and 11,300 inhabitants, respectively. Average annual rainfall is 432 mm/year in Benaguasil and 690 mm/year in Xàtiva, in both cases with intense events usually at the end of the summer. Like many Mediterranean urban areas, they suffer from pluvial flooding due to the inability of their combined sewer networks to appropriately manage runoff, which also causes frequent untreated water overflows into receiving waters. Their city councils are proactive in looking at non-conventional ways of managing stormwater, not only with regard to technical solutions but also to management and educational aspects, among others (Perales-Momparler *et al.*, 2013; Jefferies *et al.*, 2014; Perales-Momparler *et al.*, 2015, 2016). As such, notice boards are present at each one of the sites, and water quantity and quality tests were conducted during a full hydrological year, showing encouraging results (Perales-Momparler *et al.*, 2014).

26.4.1 Infiltration Basins in Costa Ermita, Benaguasil

In Costa Ermita Park (Figure 26.6), located in a topographically high area of Benaguasil, under the Aquaval project, three interconnected vegetated basins were retrofitted to attenuate surface water runoff and sediments emanating from the hill. The main objective was to reduce the quantity of runoff that flows down the streets, causing flooding damage in garages and houses in the lower part of the town, as well as reducing sediments accumulating in the combined network.

Figure 26.6   Costa Ermita park before (left) and after (right) retrofitting three infiltration basins.

An old wall at the park entrance was removed to allow runoff to enter the park, and footpaths were elevated to divert water into the infiltration basins. These have been formed by excavating existing flat soil between trees, and providing extra attenuation volume underground. In the upper basins, one on each side of the path, runoff filters through the topsoil and is temporarily stored in a gravel layer, before infiltrating into the ground and/ or being channelled by interconnecting pipes into the third basin, which utilises a buried geocellular tank (formed of polypropylene drainage boxes with a storage volume of 18 m³). An overflow device located at this lower infiltration basin conveys exceedance flow to the municipal combined sewer. Sediments are deposited mainly at the entrance of the park, where they are easily accessible for removal when necessary.

The overall storage volume of these basins is approximately 22 m³, and it is estimated that they will remove (within a short period of time) approximately 1400 m³ of water per annum. Monitoring equipment installed at the overflow manhole consisted of a level-probe and a V-notch weir. Of the 19 rainfall events that were registered in Benaguasil during the monitoring period (from October 2012 to September 2013), only one produced overflow into the combined sewer. Considering that the catchment area is approximately 11,520 m², this highlights the efficiency of this site in removing runoff and sediments from the combined sewer system.

Water quality assessment was performed at the basin inlet; the data produced was highly variable, reflecting the fact that runoff water quality strongly depends on rainfall intensity and dry periods between storm events. For example, COD values ranged from 100 mg/l to 2000 mg/l, and the suspended solids concentrations showed similar variations.

26.4.2 Rainwater Harvesting Tank at Benaguasil Youth Centre

In this case study from the Aquaval project (Figure 26.7), the innovation comes from recovering an ancient practice: storing rainwater in tanks for non-drinking water reuse. The key drivers of the SuDS design at this location was education, and the communication of alternative drainage designs, with construction details that make it possible to see how rainwater is conveyed from the downspouts of the building roof into the underground tank.

Figure 26.7   Notice board located at Benaguasil Youth Centre retrofitted site.

To that end an aesthetically harmonious marble channel was retrofitted in the court-yard of the building and covered with high-strength glass, permitting people to see the flow of water from a section of the building roof (approximately 100 m²). The tank was made of reinforced concrete, with a storage volume of 11 m³. The cistern was covered with a concrete slab with a stainless steel grating to facilitate maintenance and water observation. It also incorporates a cleaning outlet at the bottom, and overflow pipes into an adjacent garden.

Stored water was used for gravity irrigation and cleaning of the public gardens and square located further down, saving approximately 43 m³ of drinking water per year, and providing a self-sufficient water supply for these uses nearly all year round. Water quality tests that were carried out on the stored water support its use for irrigation and cleaning of public spaces with counts of nematode eggs <1 u/10 l and Escherichia coli <10 cfu/100 ml.

26.4.3 Green Roof in Xàtiva

Gozalbes Vera public school is located in the city centre of Xàtiva and was chosen during the Aquaval project to raise awareness of SuDS among pupils from an early age. A green roof was retrofitted to assess its ability to manage runoff under Mediterranean climatic conditions. The retrofitted area consisted of a 475 m², 1060 kg/m³ density, extensive green roof, which included fragments of clay bricks to improve its drainage capacity. The soil layer was rich in organic matter (29%), total nitrogen (0.27%) and phosphorus (0.57% as P_2O_5). The 10 cm substrate was planted with a diverse variety of Sedum spp.: 10% S. album, 15% S. acre, 20% S. floriferum, 15% S. spurium, 20% S. reflexum,

10% S. sediforme and 10% S. sexangulare (Charlesworth *et al.*, 2013). Due to budget restrictions, only part of the former cobbled, conventional roof was retrofitted. This fact presented an opportunity for comparison at the same site of the storm attenuation performance of both the green retrofitted and cobbled areas. Monitoring activities comprised water quantity and quality measurements (October 2012 to September 2013) from a section of the retrofitted green roof (218 m²) as well as from the untouched conventional roof (107 m²). Flow discharges from both roofs were monitored with tipping bucket flow gauges installed at both downpipes. In addition, two sample bottles were connected to each flow gauge device to collect water samples for quality testing (Perales-Momparler *et al.*, 2014).

Even though the green roof was periodically irrigated during its start-up period to ensure proper development of vegetation, volumetric efficiencies (detained runoff over rainfall volume) of 52–100% were achieved during the monitoring period. Figure 26.8 shows the hydraulic performance of both roofs during a long duration event recorded in April 2013. Total rainfall volume was 88 mm and the maximum intensity in 10 minutes was 11 mm/h. Only 31% of the rainfall volume was detained by the conventional roof, whereas 80% efficiency was achieved for the green roof. Peak flow reduction was also significant, recording 4–5 times lower for the green roof in comparison with the cobbled roof.

Water quality tests showed a clear washing effect of the substrate of the green roof during the start-up period (in particular, September 2012 to February 2013). However, the results became gradually more similar to those obtained for a typical roof (mean COD concentrations of 41 mg/l and TN of 5 mg/l). Water from the vegetated roof was very brown in colour (because of the brick fragments) but clear (turbidity <20 NTU). All the measured pollutant concentrations in the green roof were higher (especially COD, which was much higher) than those for the conventional roof (Perales-Momparler *et al.*, 2014).

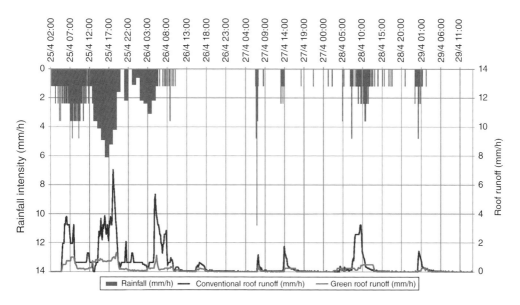

Figure 26.8   Hydraulic performance during the 25–29 April 2013 rainfall event.

Due to the substrate characteristics, the organic fraction was very high and not easily bio-degradable. During the start-up period, the green roof increased pollutant concentrations but, as reported by Rowe (2011), as the vegetation became well established, they decreased over time.

26.4.4 Green Roof in Benaguasil

One year later and under the E²STORMED European project, a 315 m² green roof was retrofitted on a public building of the Municipality of Benaguasil, using lessons learnt from the earlier experience in Xàtiva. A mineral soil, poor in nutrients and without brick debris, was used to avoid the pollutant problems previously observed. To preserve drainage capacity in the soil and to reduce runoff colour and turbidity, clay debris was substituted by volcanic gravel (40%) and silica sand (20%), the remaining 40% being made up of compost substrate. Organic matter constituted only 13.3%, total nitrogen 0.06% and phosphorus 0.04% (as P_2O_5), much lower than those used in Xàtiva. Nitrogen and phosphorus content fulfilled the requirements adopted in Büttner *et al.* (2002), and while organic matter was slightly higher than the maximum amount recommended, it was reduced by more than a half. With this new composition, washing effects during the start-up period were expected to be reduced. Figure 26.9 shows the construction stages and different layers of the green roof. After removal of the cobbles and the insulation layer, the latter was preserved and reused in the retrofitted roof. A soil layer 10 cm thick was placed over a storage and drainage plastic layer. As in Xàtiva, vegetation consisted of a mix of Sedum spp.: 20% S. album, 18% S. acre, 34% S. floriferum, 17% S. spurium, 3% S. rupestre, 3% S. sediforme and 5% S. sexangulare. Also following lessons learnt in Xàtiva, irrigation during the start-up period was done only when strictly necessary, and this was controlled by a soil humidity sensor. Thus, for events under 10 mm of total rainfall volume, volumetric efficiencies of the green roof were always higher than 90%. For the largest event recorded (125 mm), the volumetric efficiency was 57%, still significant considering the high amount of rainfall.

26.5 Conclusions

This chapter has shown that SuDS can be both included in the design of new build, and retrofitted in the diverse Mediterranean climates found across Spain, from the rainy north to areas in the south where droughts are a problem at certain times of year. The case studies presented here have shown that a variety of SuDS devices, individually and as management trains, can reduce the storm peak, even in long duration intense rainfall events, and also improve water quality, if designed correctly. Thus, lessons learnt, for example from monitoring of the green roof installed on the Gozalbes Vera public school in Xàtiva, were successfully applied in Benaguasil where the design of the green roof itself was improved, and also initial and long-term management.

Emphasis on education in the installation of these various devices and trains is ensuring that both school children and the general public are made aware of the structure and functioning of SuDS, thus going some way to making sure that every opportunity is taken to encourage the uptake of sustainable drainage to provide flood resilience, and also all the other multiple benefits that a sustainable approach to drainage can provide.

Roof before the green roof construction

Retrofitted green roof

7. Sedum
6. Mineral substrate
5. Filter layer
4. Storage and drain layer
3. Protection mat
2. Insulation layer
1. Safeguard root barrier
Waterproof membrane
Porous concrete

Floor

Figure 26.9 Retrofitting a green roof in Benaguasil: construction stages and roof layers.

References

Andrés-Valeri, V.C., Castro-Fresno, D., Sañudo-Fontaneda, L.A. and Rodríguez-Hernández, J. (2014a) Comparative analysis of the outflow water quality of two sustainable linear drainage systems. Water Science and Technology, 70 (8) 1341–1347.

Andrés-Valeri, V.C., Castro-Fresno, D., Sañudo-Fontaneda, L.A., Rodríguez-Hernández, J., Ballester-Muñoz, F. and Canteras-Jordana, J.C. (2014b) Rehabilitación Hidrológica Urbana. REHABEND 2014, Congreso Latinoamericano – Patología de la Construcción, Tecnología de la Rehabilitación y Gestión del Patrimonio. Santander (España).

Bayón, J.R., Castro, D., Moreno-Ventas, X., Coupe, S.J. and Newman, A.P. (2005) Pervious Pavement Research in Spain: Hydrocarbon Degrading Microorganisms. In: Proceedings of the 10th International Conference on Urban Drainage (ICUD), Copenhagen, Denmark.

Büttner, T., Rohrbach, J. and Schulze-Ardey, C. (coordinators) (2002) Guidelines for the planning, execution and upkeep of green-roof sites. FLL, Forschungsgesellschaft Landschaftsentwicklung Landschaftsbau e.V., Bonn, Germany.

Castro-Fresno, D., Andrés-Valeri, V.C., Sañudo-Fontaneda, L.A. and Rodríguez-Hernández, J. (2013) Sustainable drainage practices in Spain, specially focused on pervious pavements. Water (Switzerland), 5 (1) 67–93.

Charlesworth, S.M., Perales-Momparler, S., Lashford, C. and Warwick, F. (2013) The sustainable management of surface water at the building scale: preliminary results of case studies in the UK and Spain. J. Water Supply Res, 62, 8, 534–544 10.2166/aqua.2013.051

Dietz, M.E. (2007) Low impact development practices: A review of current research and recommendations for future directions. Water, Air, and Soil Pollution, 186 (1) 351–363.

Dolz, J. and Gómez, M. (1994) Problems of stormwater drainage in urban areas and about the hydraulic study of collector networks [in Spanish]. Dren. Urbano, 1, 55–66.

Directive, E.U. (2000) /60/EC of the European Parliament and the Council of 23 October 2000 Establishing a Framework for Community Action in the Field of Water Policy; Official Journal of the European Communities: Brussel, Belgium, 2000.

Directive, E.U. (2007) /60/EC of the European Parliament and of the Council of 23 October 2007 on the Assessment and Management of Flood Risks Text with EEA Relevance; Official Journal of the European Communities: Brussel, Belgium, 2007.

García, X., Llausàs, A. and Ribas, A. (2014) Landscaping patterns and sociodemographic profiles in suburban areas: Implications for water conservation along the Mediterranean coast. Urban Water Journal, 11 (1) 31–41.

Gomez-Ullate, E., Bayón, J.R., Coupe, S. and Castro-Fresno, D. (2010) Performance of pervious pavement parking bays storing rainwater in the north of Spain. Water Science and Technology, 62 (3) 615–621.

Gomez-Ullate, E., Novo, A.V., Bayón, J.R., Hernández, J.R. and Castro-Fresno, D. (2011a) Design and construction of an experimental pervious paved parking area to harvest reuseable rainwater. Water Science and Technology, 64 (9) 1942–1950.

Gomez-Ullate, E., Castillo-Lopez, E., Castro-Fresno, D. and Bayón, J.R. (2011b) Analysis and Contrast of Different Pervious Pavements for Management of Storm Water in a Parking Area in Northern Spain. Water Resources Management, 25 (6) 1525–1535.

Jimenez-Gallardo, B.R. (1999) Pollution from Urban Runoff; SEINOR Collection No 22; Spanish Civil Engineering Association: Madrid, Spain.

Malgrat, P. (1995) Overview of the stormwater runoff as a source of contamination: Possible actions [in Spanish]. In: Proceedings of Workshop Benicassim. Benicassim, Spain, 28 November – 1 December.

Perales-Momparler, S., Andrés-Doménech, I., Andreu, J. and Escuder-Bueno, I. (2015) A regenerative urban stormwater management methodology: the journey of a Mediterranean city. Journal of Cleaner Production, http://dx.doi.org/ 10.1016/j.jclepro.2015.02.039

Perales-Momparler, S., Andrés-Doménech, I., Hernadez-Crespo, C., Valles-Moran, F., Martin, M., Escuder-Bueno, I. and Andreu, J. (2016) The role of monitoring sustainable drainage systems for promoting transition towards regenerative urban built environments: a case study in the Valencian region, Spain. Journal of Cleaner Production. Article In-Press.

Perales-Momparler, S., Hernández-Crespo, C., Vallés-Morán, F., Martín, M., Andrés-Doménech, I., Andreu Álvarez, J., Jefferies, C. (2014) SuDS efficiency during the start-up period under Mediterranean climatic conditions. Clean – Soil, Air, Water, 42 (2) 178–186.

Perales-Momparler, S., Jefferies, C., Periguell-Ortega, E., Peris-García, P.P. and Munoz-Bonet, J.L. (2013) Inner-city SuDS retrofitted sites to promote sustainable stormwater management in the Mediterranean region of Valencia: Aquaval (Life + EU Programme). Novatech Conference, Lyon, France.

Rodríguez-Hernández, J., Castro-Fresno, D., Fernández-Barrera, A.H. and Vega-Zamanillo, Á. (2012) Characterization of infiltration capacity of permeable pavements with porous asphalt surface using cantabrian fixed infiltrometer. Journal of Hydrologic Engineering, 17 (5) 597–603.

Rowe, D.B. (2011) Green roofs as a means of pollution abatement. Environ. Pollut. 159, 2100–2110.

Sañudo-Fontaneda, L.A., Charlesworth, S.M., Castro-Fresno, D., Andres-Valeri, V.C.A. and Rodríguez-Hernández, J. (2014a) Water quality and quantity assessment of pervious pavements performance in experimental car park areas. Water Science and Technology, 69 (7) 1526–1533.

Sañudo-Fontaneda, L.A., Andrés-Valeri, V.C., Rodríguez-Hernández, J. and Castro-Fresno, D. (2014b) Field study of infiltration capacity reduction of porous mixture surfaces. Water (Switzerland), 6 (3) 661–669.

Sañudo-Fontaneda, L.A., Rodríguez-Hernández, J., Calzada-Pérez, M.A. and Castro-Fresno, D. (2014c) Infiltration behaviour of polymer-modified porous concrete and porous asphalt surfaces used in SuDS techniques. Clean – Soil, Air, Water, 42 (2)139–145.

Sañudo-Fontaneda, L.A., Rodríguez-Hernández, J., Vega-Zamanillo, A. and Castro-Fresno, D. (2013) Laboratory analysis of the infiltration capacity of interlocking concrete block pavements in car parks. Water Science and Technology 67 (3) 675–681.

Segura-Graiño, R. (1994) The water in Spain: Problems and solutions [in Spanish]. Bol. Asoc. Geogr. Esp. 18, 29–38.

Temprano González, J., Gabriel Cervigni, M., Suárez López, J., Tejero Monzón, J.I. (1996) Contamination in sewer systems with rainy weather: Source Control [in Spanish]. Revista Obras Publicas, 3352, 45–57.

Sustainable Drainage at the City Scale: A Case Study in Glasgow, Scotland

Neil McLean

27.1 Introduction

Glasgow is Scotland's largest city with a population of about 600,000 and a further 2,250,000 in the wider metropolitan area (National Records of Scotland, 2014). It has a strong industrial history with the claim that 30,000 ships have been built on the River Clyde. This industry brought workers into the city during the late 19th and early 20th century, with a resulting rapid increase in population. The need for housing was recognised, and the city, including its infrastructure grew rapidly. The sewer system was largely integrated as the city grew but by today's standards suffers, as many other historic cities do, with a lack of capacity to serve the requirements of modern society. The additional burden of climate change and urban creep adds to this lack of capacity. The city's largely combined sewer system was exposed to this weakness on 30 July 2002 when a slow-moving convection storm stalled over the east end of the city dropping an average month's rainfall (75 mm) in just 10 hours (Scottish Government, 2013a).

The ageing sewer system was rapidly inundated, and large areas to the east of the city centre suffered severe flooding. Post event analysis resulted in several organisations coming together to form the Glasgow Strategic Drainage Plan (GSDP), which later became the Metropolitan Glasgow Strategic Drainage Partnership (MGSDP), and comprises Scottish Government, Scottish Water (the drainage authority), Scottish Environment Protection Agency (SEPA, the environmental regulator), local government, planning and enterprise authorities and latterly railway and canal authorities (Scottish Government 2008a; MGSDP, 2009). The stated MGSDP vision is: 'To transform how the city region thinks about and manages rainfall to end uncontrolled flooding and improve water quality' (MGSDP, 2009).

To understand where the weaknesses are in the drainage system and to comprehensively understand why flooding occurs requires time and resources, and much effort has been used to establish where, why and when problems may arise across the city. Prior to the

Sustainable Surface Water Management: A Handbook for SuDS, First Edition.
Edited by Susanne M. Charlesworth and Colin A. Booth.
© 2017 John Wiley & Sons, Ltd. Published 2017 by John Wiley & Sons, Ltd.

event of 2002, planning authorities across Scotland were becoming aware of their role in flood risk management and of sustainable drainage systems (SuDS) as an alternative to conventional drainage. The unsustainable approach of simply constructing larger pipes and tanks to contain flood waters was becoming more and more unrealistic in most situations as new development and regeneration opportunities occurred. At the time, enlightened planning authorities began to install planning policy and conditions to require SuDS for new developments, resulting in SuDS being used in certain parts of the country. However, Glasgow City Council (GCC) was slow to promote the use of SuDS as evidenced by the SuDS Database (SNIFFER, 2002). This was a database established and maintained by SEPA until 2001, after which the task of keeping it updated became too onerous. However, useful data emerged: of the 32 local authorities in Scotland, GCC was ranked 16th when considering the number of SuDS locations in the council's jurisdiction. However, when the number of new SuDS locations was compared to the population in the council's area (an approximate approach to evaluate SuDS in potential development, and thereby the opportunity to install SuDS), GCC ranked a lowly 26th. Since the 2002 event, the authority has become one of the more forward thinking, as the vision of the MGSDP spread across all the departments of the different authorities making up the Partnership.

One of the major benefits of the MGSDP approach is the communication that has been enabled across the member authorities who now meet regularly. This is not simply between different authorities but, critically, within and across each authority. Planners now speak to roads engineers! Evidence of this is the Design Guide for New Residential Areas (GCC, 2013), which was written by a planner and road engineer. The process of writing the guide brought a wider understanding for planners about road development and construction, and for roads engineers about the matters involved in planning and permitting new development.

27.2 SuDS and Legislation

From a slow start, good progress has been made and continues. This progress was ensured by the introduction of new regulations: The Water Environment (Controlled Activities) (Scotland) Amendment Regulations (2013), or CAR Regulations, (Scottish Government, 2013b), which contained General Binding Rules requiring SuDS to be used in new developments. Essentially this meant that virtually any new development that had surface water draining to the water environment was required to be, 'drained by a SUD system equipped to avoid pollution of the water environment' (Scottish Government, 2013c). This requirement has been in place since the original regulations of 2006 (now amended) and curiously was driven by the European Water Framework Directive or WFD (European Commission, 2000). The initial momentum to establish SuDS in Scotland was to address flood risk, but this was overtaken by the water quality concerns of the WFD and the formation of the CAR Regulations. The driver for other parts of the UK largely remains flood risk (DCLG/Defra, 2015) and this is clearly a concern in Glasgow, but without losing focus on water quality.

As knowledge and understanding about SuDS spread, an appreciation grew of the triangle of benefits with its three distinct sides: water quality, water quantity and amenity (CIRIA, 2000, 2007, 2015). Given the driver of legislation, water quality generally received most attention once the CAR Regulations were enacted. Later, when the European Floods directive (European Commission, 2007) was pending and later transposed into Scottish Law as the Flood Risk Management (Scotland) Act (2009) (Scottish Government, 2009), the second side of the SuDS triangle, water quantity, became more of a focus, leaving amenity as the poor cousin of the SuDS triangle. Interesting analysis may be made about this, but a general assumption may be that with the carrot of incentive to install SuDS (whatever the incentive may be, but will generally result in fiscal benefit) against the stick of legislation, and the implementation of legislation for water quality and water quantity, it seems that the stick gives better results. This is not to say that there is no legislation for amenity – there is, considering biodiversity (Scottish Government, 2004) and planning guidance for open space (Scottish Government, 2008b) among others – but there has been a clear drive behind water quality and water quantity enforcement, which has not been as strong for amenity.

This has been the case as development has grown in Glasgow and while the financial downturn of 2008 resulted in a decline in development and regeneration and, therefore, the number of houses and other buildings being constructed, there was a greater focus on what was absolutely necessary to construct new developments, including legislative requirements, rather than *perceived* add-ons such as amenity and biodiversity.

As financial recovery began from around 2010, and especially as house building became more active, there has been a greater awareness of all three sides of the SuDS triangle, especially among planners, who play a critical role for the success of any development. Multi-functionality is now considered desirable, probably beyond the three sides of the SuDS triangle. There are further opportunities to engage with different sectors when considering SuDS, and this has become a target for Glasgow to aim towards.

27.3   The Importance of Multi-Functionality

As stated previously, Glasgow has a proud history. However, it also has a poor health record and, very well aware of this, government authorities including GCC want it improved. One step towards this was to achieve better green networks across the city, which resulted in the Glasgow and Clyde Valley Green Network Partnership (GCVGNP) becoming established with the claim that, 'Affecting the lives of thousands the GCV Green Network will transform people's quality of life, making a much more vibrant sustainable city for the 21st century' (GCVGNP, 2008). Comprising eight local authorities and five government agencies, the Partnership has a long-term vision of transforming the metropolitan area of Glasgow into a vibrant and thriving region that has healthy and active communities. The recognition that green space, green infrastructure and green networks can make a difference to people's lives through improved health and wellbeing is a driver, and the GCV Green Network Partnership considers the requirement for SuDS as a suitable opportunity to install green infrastructure. This also dovetailed with the multi-functional approach, which is now recognised as part of the delivery of SuDS, and the realisation that it also made good economic sense.

Sharing resources to deliver several functions within the same space takes careful design and often requires innovation to achieve more than a single (water quality) or double (water quantity) function. It clearly makes good sense to attempt to achieve several benefits from the same asset. The problem arises where those who drive the different functions – flood risk officers, planners, roads engineers, etc. – do not talk to each other, and this has been a longstanding difficulty with individuals and collective departments who are busy getting on with their 'day job' and not looking beyond their own delivery. This is a common occurrence in many agencies, authorities and municipalities across the globe.

GCC's earlier East End Local Development Strategy (GCC, 2008) sets out to, 'create a health promoting community in the East End of the city'. Quality open spaces should form an integral part of future communities, encouraging physical activity, supporting mental wellbeing and attracting economic activity (Figure 27.1). The strategy recognised the need for SuDS associated developments in the East End and, crucially, enabled opportunities for

Figure 27.1   A new swale that will form part of a central amenity area in a new development Dalmarnock in the East End (courtesy MGSDP).

this to provide multi-functional spaces, which contributed to a wider green network throughout the area (Scottish Government, 2008b).

The aforementioned Design Guide for New Residential Areas (GCC, 2013) is a success story that has enlightened roads and planning departments respectively within GCC. The need for a central coordinating facility is now realised, encouraging the sharing of knowledge across the relevant departments. Planning is an essential and central role to this, and a dedicated unit within the GCC has been established to serve this requirement. The 'Place, Strategy and Environmental Infrastructure' team has a key role in optimising the performance of green, grey and blue spaces, using an integrated environmental infrastructure to deliver multiple balanced benefits for people, businesses and the environment. This role is to support colleagues (and enlighten any that are left in their silos) as well as external organisations.

This is an evolutionary process, and with the increasing understanding and experience that is being gained as more development and regeneration takes place, and with the necessary SuDS being installed, it is becoming clear that the one piece of land that would previously have been amenity through a planning condition, could become a multi-functional area serving the community in different ways.

27.4   Design Studies

Several design studies have been produced as part of the journey towards a fuller understanding of how developments can accommodate this multi-functional approach. An initiative between SEPA, GCC, Scottish Water, GCV Green Network Partnership, Scottish Natural Heritage and Forestry Commission (Scotland) saw two of the initial four design studies delivered within the city boundary (Cowlairs Urban Village and The Burgh of Pollockshaws) with the remaining two studies (Jackton and the Gill Burn Valley and Johnstone South West) within the metropolitan region (GCVGNP, 2012a). These four studies set out to provide a water and integrated urban masterplan, the aim being to maximise land for several purposes, with each study in a different setting: inner city regeneration (Cowlairs), suburban regeneration (Burgh of Pollockshaws), town outskirts location (Johnstone South West) and green belt development (Jackton and Gill Burn Valley).

Each of the studies had a team of specialists examining the prospects of integrating green and open spaces throughout its study area. Critically, landscape architects were involved at an early stage and had a central role throughout the exercise. Hydrologists and flood risk modellers were also involved at the early stages to establish if and where flooding might arise. Given the numbers (volume of flooding, frequency and critical storm, etc.) and locations of where runoff might begin from, flow through and materialise as flooding, the next logical step was to understand where buildings could be constructed and where they should not, allowing for floodable areas to be designed as such, and also to be designed to return to normal after any flooding.

Earlier attitudes were to dedicate areas for flooding and to cordon off these areas as flood zones using fences (or perhaps barrier vegetation in better examples) for safety reasons. Discussions with UK's Health and Safety Executive to overcome this clumsy and inefficient layout strategy were straightforward: if a prescribed area was designed to flood, was managed to flood and was safe to flood with overall low risk and where no public health concerns may arise, this would be acceptable. This would require sensible design to achieve areas that would not hold deep or fast flowing water and where runoff would contain low concentrations of pollution (Figure 27.2).

Figure 27.2   A new SuDS pond under construction on the south side of Glasgow (courtesy A. Duffy).

The four studies were collectively entitled Integrated Urban Infrastructure Design Studies, and the collaboration agreed on a mission statement to provide integrated solutions, provide a shift in perception, inform best practice, promote biodiversity, improve population health, encourage economic development, reinforce partnership working and transform each site (GCVGNP, 2012a).

These earlier studies were followed by an additional two (GCVGNP, 2012a), which were driven by the GCV Green Network Partnership and received European funding under the Sigma for Water banner (Sigma for Water, 2013) within the Interreg-4c funding allocation. They follow the GCV Green Network Partnership's approach, which evolved from integrated urban design to integrated *green* infrastructure (IGI), that adds green infrastructure to the four established 'critical' infrastructure components of water, energy, transport and waste (GCVGNP, 2012b). The approach argues for green infrastructure to be given the same standing in urban design as the other infrastructure components. The studies delivered this approach with high level analysis.

One of the studies was performed for Spango Valley (GCVGNP, 2012a) on the outer edge of the metropolitan area with the second study at Nitshill (GCVGNP, 2012a) on the edge of GCC's boundary. The following section focuses briefly on this second study.

27.5 Nitshill Design Study

Nitshill lies to the south-west edge of Glasgow and is an area that includes large pockets of social deprivation and could, because of this, be argued to be an area with priority for attention. It was GCC's intention to regenerate the area over a period of time, making Nitshill ideal for consideration as a candidate site for masterplanning analysis (Figure 27.3). The degree of analysis for this study was at a higher (less detailed) level than the earlier studies using less hydrological modelling, and thus saving expense at this stage. The scrutiny required for a full flood risk analysis would of course be delivered but at the appropriate time and at a more precise level of development.

The study established baseline conditions including soil and geology, topography and hydrology but also included travel and other geographical considerations. It was found that while green infrastructure was evident, it was largely disconnected, meaning that pockets of green space existed in isolation, serving only very local residents with limited habitat and potential for biodiversity. Regeneration would require SuDS to be implemented and conveyance of runoff to manage flood risk through prescribed routes. New green corridors could be established to convey runoff and also connect these isolated sub-areas of green space. Careful planning would be needed to overcome opportunities for anti-social behaviour and the very real circumstance of community safety especially at night when many of Nitshill's residents would be careful to avoid certain areas known for trouble. Again, prudent planning and design could assist in overcoming this nuisance – multi-functionality!

What is often overlooked when planners and developers consider residential developments are the residents themselves and the environment that residents will experience in their living community. Where people live has a bearing on what they are and on their mental and physical condition. The approach that considers wellness, rather than factors causing illness or disease is called *salutogenesis*, and Antonovsky (1979) stated that unless people find the world around them comprehensible, manageable and meaningful they would experience a state of chronic stress. Thus, surroundings are critical to wellbeing. The residents of Nitshill, and of course in many, many similar locations across the planet, will have improved health with an improved living environment – another opportunity for multi-functionality.

Figure 27.3   An artist's impression of part of regenerated Nitshill with flood zones designed and green infrastructure installed (courtesy GCVG Green Network Partnership).

It may sound very grand to consider SuDS as beneficial to health, but many authorities now accept that this as a contributing opportunity towards creating better places. Ian Gilzean, the Scottish Government's Chief Architect said of the Nitshill Design Study, 'integrating green infrastructure can provide multiple benefits by helping to develop places that are designed to deal with climate change, reduce our carbon footprint and support bio-diversity as well as providing surroundings that are safe, pleasant and supportive of healthier lifestyles' (GCVGNP, 2012a).

The Nitshill design study has attempted to establish baseline conditions of critical infrastructure and anticipate opportunities to connect isolated areas of varying land uses, promote efficient use of land for future regeneration and development for the future communities of Nitshill who, ideally, will begin to take ownership of their surroundings and appreciate and value what is essentially their environment. There is a long way to go to overcome the problems that exist in the area, but doing nothing should not be an option, and creating somewhere that residents enjoy is a major step forward.

27.6   City Centre Surface Water Management

Glasgow's Buchanan Street in the city centre has the claim of being the most expensive street in the UK outside of London (BBC, 2014) and proudly boasts being the 'Style Mile'. This has great economic potential for the city region and has successfully increased tourism in the area. Concerns were that if a storm such as the one that hit the east end in 2002 should happen just a little further west, there would be even greater consequences. GCC, through their duties in the Flood and Water Management Act (Scottish Government, 2009), commissioned a surface water management study to examine what problems flooding might cause and how these could be addressed. The city centre surface water management plan (GCC, 2014) established baseline conditions for several severe and extreme rainfall events. The options to address these included water management features and of course SuDS integrated into the city centre environment. As with many urbanised centres, the degree of green infrastructure included in the location is fairly minimal. This is especially the case in Buchanan Street where high profile retailers compete to gain the attention of shoppers. Installing SuDS ponds and basins is clearly difficult, and innovative design is required. The installation of a 'green route' along the approximate route of the historic, and long since eliminated watercourse, the St Enoch Burn, was considered one such opportunity and, together with similarly configured green routes in other parts of the city centre, forms the basis of the surface water management plan.

With many complex physical and, indeed, political difficulties to overcome, this will not be an easy task and may never come about, but the will of the individuals within the council and the general attitude of the authority makes this a much stronger possibility than it would have been only a few years previously. Flood risk management by itself is unlikely to be a strong enough argument, despite the value of the surroundings, but further opportunities, especially those driven by improved economic potential including tourism, will increase the likelihood of new green infrastructure such as the green routes proposed in the surface water management plan.

The creation of pleasant blue (water) and green (vegetation) landscape features weaving among high value shop frontages with subtle functional but aesthetically pleasing crossing points adding to the appeal of the area is now a feasible and worthy consideration and may by itself result in a draw for visitors and the essential retailers.

27.7 Funding

Not surprisingly, the major reason that such grand steps take time to implement is finding the funding resource to deliver. GCC, which is still recovering from the global economic downturn and unlikely to see any growth in funding for its foreseeable future, has been pioneering alternative routes for capital. In addition, there have been ambitious attempts to promote the city as a whole and with good success. The city hosted the 2014 Commonwealth Games, a multi-disciplinary sports competition held over nearly two weeks with a similar format to the Olympic Games, comprising 71 nations of the British Commonwealth. Several thousand athletes and organisers had to be housed for the event and a new athletes' village was constructed. This included green infrastructure and a central SuDS 'canal' (MGSDP, 2014), and it has since been rearranged for the more permanent communities of Glasgow's people (Figure 27.4). Using the vehicle of the Commonwealth Games, the intention for regeneration, the desire for betterment and additional sources of funding to enable the games, a highly successful development has been constructed – and a highly successful Commonwealth Games has been held!

Additional attempts for funding have been made with the two most prominent and financially rewarding being the Rockefeller Foundation's 100 Resilient Cities award (Rockefeller, 2014), when Glasgow was selected as one of the first 40 cities in the world to merit the award and the grant that went with it, and the UK and Scottish Government's 'City Deal' (DPMO, 2014). This will see around half a billion pounds awarded to the Glasgow metropolitan area projects, including the introduction of street tree avenues in the

Figure 27.4   The new linear water feature and pond in the Athletes' Village (courtesy MGSDP).

city centre and the 'smart' management of the Forth and Clyde Canal as a flood storage reservoir throughout an available 22 miles (~35 km) of water between canal locks.

27.8 The Future

Nitshill and the other five related design studies, together with many other 'hoped for' projects, may now evolve into less abstract ideas and become reality with the funding streams that are beginning to flow.

It can be seen that there has been a growing confidence since the city was reminded of the consequences of flooding. Despite difficult financial times, with essential council priorities jostling for attention, a sincere intent to deliver efficient, integrated, cross-disciplinary and multi-functional assets into new developments and regeneration programmes has resulted, with SuDS at the core of this delivery.

The necessary funding will enable a more resilient, climate-ready and robust infrastructure and will, GCC hopes, provide strengthening and resilience of the city's economy, with healthier and prosperous communities to thrive across the region.

References

Antonovsky A. (1979) *Health, Stress and Coping.* Jossey-Bass, San Fransisco.

BBC (2014) News article based on Cashman and Wakefield Estate Agents Research. Available at: http://www.bbc.co.uk/news/uk-scotland-scotland-business-30192055.

CIRIA (2000) *Sustainable Urban Drainage Systems; Design manual for Scotland and Northern Ireland*. Sustainable Urban Drainage Scottish Working Party, Stirling and Construction Industry Research and Information Association, London.

CIRIA (2007) *The SUDS Manual*. Construction Industry Research and Information Association, London.

CIRIA (2015) *The SUDS Manual*. Construction Industry Research and Information Association, London.

DCLG/Defra (2015) *Consultation Outcome – Planning application process: Statutory consultee arrangements*. Department for Communities and Local Government and Department for Environment Food and Rural Affairs, London.

DPMO (2014) *Glasgow and Clyde Valley City Deal*. Deputy Prime Minister's Office, London.

European Commission (2000) *Directive 2000/60/EC of the European Parliament and of the Council of 23rd October 2000: Establishing a Framework for Community Action in the Field of Water Policy. Official Journal 22nd December 327/1* European Commission, Brussels.

European Commission (2007) *Directive 2007/60/EC, The EU Floods Directive*. European Commission, Brussels.

GCC (2008) *East End Local Development Strategy*. Available at: http://tinyurl.com/o6e3tpw

GCC (2013) *Design Guide for New Residential Areas*. Glasgow City Council. Available at: http://www.glasgow.gov.uk/designguide

GCC (2014) *Glasgow city centre surface water management plan*. Glasgow City Council. Available at: TBC unpublished.

GCVGNP (2008) *The Glasgow and Clyde Valley Green Network Partnership; Our Vision*. Available at: http://www.gcvgreennetwork.gov.uk/about-us/our-vision

GCVGNP (2012a) *Integrated Urban Infrastructure/Integrated Green Infrastructure Design* Studies Glasgow and Clyde Valley Green Network Partnership. Available at: http://tinyurl.com/p6 cmyr8

GCVGNP (2012b) *Integrated Green Infrastructure*. Glasgow and Clyde Valley Green Network Partnership. Available at: http://www.gcvgreennetwork.gov.uk/igi/introduction

MGSDP (2009) *Metropolitan Glasgow Strategic Drainage Partnership, Our Vision Objectives and Guiding Principles*. Available at: http://www.mgsdp.org/index.aspx?articleid = 2009

MGSDP (2014) *Briefing Note 14 Winter 2014/15* The Metropolitan Glasgow Strategic Drainage Partnership. Available at: http://www.mgsdp.org/CHttpHandler.ashx?id = 28353&p = 0

National Records of Scotland (2014) *Council area profiles*. Available at: http://tinyurl.com/ooubzhz

Rockefeller (2014) *100 Resilient Cities*. The Rockefeller Foundation. Available at: http://www.100resilientcities.org/#/-_Yz43NjgyOCdpPTEocz5j/

Scottish Government (2004) *Nature Conservation (Scotland) Act (2004)*. Scottish Government, Edinburgh.

Scottish Government (2008a) *The Future of Flood Risk Management in Scotland: A consultation document*. Scottish Government, Edinburgh.

Scottish Government (2008b) *Planning Advice Note 65 – Planning and Open Space*. The Scottish Government, Edinburgh.

Scottish Government (2009) *Flood Risk Management (Scotland) Act (2009)*. Scottish Government, Edinburgh.

Scottish Government (2013a) *Flood Risk Management (Scotland) Act 2009, Surface Water Management Planning Guidance*. Scottish Government, Edinburgh.

Scottish Government (2013b) *The Water Environment (Controlled Activities) (Scotland) Amendment Regulations (2013)*. Scottish Government, Edinburgh.

Scottish Government (2013c) *The Water Environment (Controlled Activities) (Scotland) Amendment Regulations (2013), Activity 10, General Binding Rule (d), (ii)*. Scottish Government, Edinburgh.

Sigma for Water (2013) Website. Available at: http://www.sigmaforwater.org/

SNIFFER (2002) *SUDS in Scotland – The Scottish SUDS Database Project Reference SR(02)09*. The Scotland and Northern Ireland Forum for Environmental Research, Edinburgh.

Water Sensitive Design in Auckland, New Zealand

Robyn Simcock

Water sensitive design (WSD) has been implemented across New Zealand (NZ), with a concentration in Auckland – the fastest growing city in Australasia. Many sites use individual stormwater treatment devices, such as rain gardens or swales (Figure 28.1). Treatment trains, combined with retention of sensitive areas, minimisation of impervious surfaces and avoidance of high-contaminant building materials, are increasingly common, particularly in larger, masterplanned or civic developments. These are showing that WSD can deliver on its core promise of enhancing resilience of aquatic and terrestrial ecosystems to the acute and chronic impacts of urban stormwater runoff. WSD greenfield and brownfield developments are also delivering superior public open spaces. Such spaces have been called a 'hybrid park typology', where stormwater treatment is integrated with high aesthetics, recreational values and ecosystem restoration, which respects the values of Maori, NZ's indigenous people.

Maori values stress the importance of water, regarded as a living being with unique *mauri* (life force or spiritual health), and part of people, expressed in the saying 'I am the river and the river is me'. The passing of urban runoff through soil can restore the mauri of stormwater. Also fundamental to restoration of mauri is reconnecting waterways (e.g. by daylighting piped streams or removing blockages to fish passage), and return of species, especially tuna (short and long-finned eels), kokupu (galaxid fish) and native riparian or wetland plants. Such actions allow expression of *kaitiakitanga* (environmental stewardship). Conversely, discharging stormwater directly into water with higher mauri (cleaner water) without passing it through the earth is inconsistent with Maori values.

In NZ, there is often great potential to implement WSD in ways that not only restore mauri of water, but also promote *mātauranga Maori* (traditional knowledge). It is increasingly common for WSD projects to draw on mātauranga Maori through consultation with *tangata whenua* – Maori with traditional connections to an area. This approach also

Sustainable Surface Water Management: A Handbook for SuDS, First Edition.
Edited by Susanne M. Charlesworth and Colin A. Booth.
© 2017 John Wiley & Sons, Ltd. Published 2017 by John Wiley & Sons, Ltd.

Figure 28.1   Densely planted rain gardens and swales are the most common devices used in Auckland projects with WSD. Native plants are usually used in public spaces to reinforce sense of place.

informs landscape design that reinforces local 'sense of place' through using native plants, materials, patterns and sculpture.

In this chapter, the drivers of WSD in Auckland are identified: the biophysical features that include a near-perfect climate, the impacts of stormwater on sensitive receiving environments, and a new regulatory approach to stormwater management through the Proposed Auckland Unitary Plan (PAUP) (Auckland Council 2013). WSD is defined in the PAUP as

> An approach to freshwater management… applied to land use planning and development at complementary scales including region, catchment, development and site. Water sensitive design seeks to protect and enhance natural freshwater systems, sustainably manage water resources, and mimic natural processes to achieve enhanced outcomes for ecosystems and our communities.

Auckland Council technical reports provide guidance and specification for WSD (Table 28.1). These reports are subject to independent and external review processes, and are freely available. Key reports are combined in General Design Guidance 1 (GD01), covering the design, construction, operation and maintenance of stormwater treatment devices (Table 28.1), GD02 covers rainfall-runoff calculations, and GD04 covers water sensitive design: *He tauira aronga wai*. The design section of GD01 (Volume 2) is currently being updated to reflect the new approach of the Proposed Auckland Unitary Plan. An updated living roof chapter has been published (Fassman-Beck and Simcock 2013), and other chapters will be released in 2016/17.

This chapter concludes with a case study of an international-award-winning brownfield development constructed in 2011. WSD was embedded in the site to deliver a highly aesthetic environment creating a uniquely 'Auckland' sense of place. Boulevards of indigenous plants were used to anchor the development where all buildings must meet high sustainability ratings.

Table 28.1   Technical guidance and specification for water sensitive design in Auckland, New Zealand.

TR number	Technical report title
TR2009/083	Landscape and ecology values within stormwater management (Vol 5 GD01 – Lewis *et al.*, 2009)
TR2010/052	Construction of stormwater management devices in the Auckland Region (Vol 3 of GD01)
TR2010/053	Operation and Maintenance of stormwater devices in the Auckland Region (Vol 4 of GD01)
TR2013/024	Hydrologic basis of stormwater device design (Fassman-Beck *et al.*, 2013a)
TR2013/018	Inlet, outlet and energy dissipation design for stormwater treatment devices (Buchanan *et al.*, 2013)
TR2013/045	Living roofs review and design recommendations for stormwater management (Fassman Beck and Simcock, 2013)
TR2013/020	Caring for urban streams (Kanz, 2013)

28.2 WSD in Auckland: Drivers of Design

Auckland, Tamaki Makaurau (Tamaki of many lovers) and 'city of sails', lies on a narrow isthmus between two harbours. Stormwater runoff is recognised as a predominant contributor to water quality and stream and coastal ecosystem health. In older areas of the city, combined sewer overflows occur. In the central city, 122 active overflow points discharge 1.2 million m³ of wastewater on an average annual basis into streams and harbours (Watercare, 2012). Sediment, metals and other contaminants degrade some estuaries and fisheries, while gross pollutants, such as plastics, impact the harbours' diverse sea birds and marine mammals. The Manukau Harbour is an internationally important feeding and breeding ground for migratory birds, including tens of thousands of bar-tailed godwits and NZ endemic species. The Waitemata Harbour includes the Hauraki Gulf Marine park which supports a local population of Bryde's whale among 22 recorded dolphin and whale species. The PAUP has increased the focus on reducing the generation of stormwater and managing it at or near source using WSD (Auckland Council 2013).

28.2.1 Ideal Biophysical Conditions for WSD

Auckland's geography and climate are ideally suited to WSD using plant-based, at-source devices. Auckland's estimated 16,500 km networks of streams (Storey and Wadhwa, 2009) are typically small, short and low-gradient, fed by small catchments (Kanz, 2013). The majority of individual rainfall events are small storms: 80% of individual events are less than 22 mm on average across the region, while 90% of events are less than 31 mm (Shamseldin, 2010). A long growing season also favours bioretention using living roofs or trees. Year-round plant growth and water uptake occurs when native tree species are used, as all are evergreen in Auckland. Evapotranspiration is responsible for up to 40% v/v of annual rainfall of extensive living roofs in Auckland (Voyde *et al.*, 2010). Runoff from roofs with just 70 mm media depth is effectively negligible for rain events less than 10 mm depth (Fassman-Beck *et al.*, 2011).

Large areas of the Auckland region are underlain by soils with high clay and silt content and slowly permeable subsoils with poor connection to groundwater. Most bioretention devices are therefore under-drained. Subsoil infiltration devices are restricted to the small areas of recent volcanic deposits, sand-dunes and peat swamps. Many of the slowly permeable soils are highly vulnerable to degradation through compaction and loss of soil structure because they have low bearing strength when moist, and weak structure. These soils are also difficult

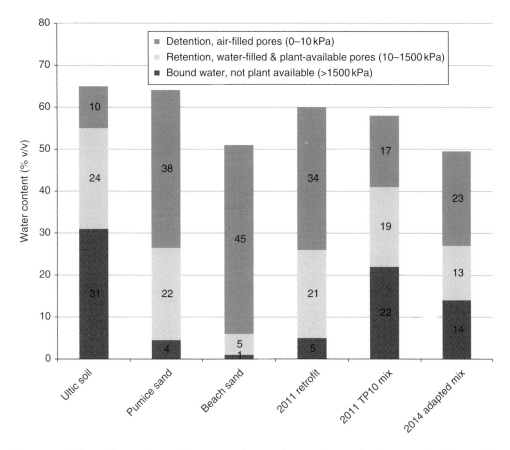

Figure 28.2   Volume of water detained (uppermost, deep grey bars) and retained and transpired by plants (middle, white bars) for, from left to right, natural (ultic) soils, pumice sand, non-vesicular beach sand, and three manufactured bioretention blends using pumice sands.

to rehabilitate over large areas by ripping or subsoiling due to a narrow moisture content at which they fracture into a fine tilth (Simcock, 2009). Such soils are unsuitable for use in rain gardens, even when mixed with sand in a 1:1 ratio, as they collapse. This leads to inadequate permeability and low aeration.

Auckland biofiltration media for rain gardens, tree pits and swales are increasingly engineered from blends of specific sands, composts and soil. Auckland has ready access to large quantities of pumice sands. These vesicular, lightweight materials hold large volumes of plant-available water. Using pumice sand enhances the retention performance of very sandy bioretention mixes (Figure 28.2, Fassman-Beck *et al.*, 2013).

28.2.2 Sensitive Receiving Environments Impacted by Stormwater

Urban development has caused large-scale loss and modification of Auckland's complex and extensive stream network (Kanz, 2013). Most streams receive stormwater from extensive hard surfaces. These streams often have naturally soft-bottomed beds and are

Table 28.2 Technical reports underpinning water sensitive design in Auckland, New Zealand.

TR number	Technical report title
TR2009/112	Environmental condition and values of Manukau Harbour (Kelly, 2008)
TR2010/021	Effects of stormwater on aquatic ecology in the Auckland region (Kelly, 2010)
TR2013/033	Ecological responses to urban stormwater hydrology (Storey et al., 2013)
TR2013/015	Auckland's urban estuaries – management opportunities (Marshall et al., 2013)
TR2013/002	Pharmaceutical residues in the Auckland estuarine environment (Stewart et al., 2013)
TR2013/044	Temperature as a contaminant in streams in the Auckland region (Young et al., 2013)
TR2013/017	Classification of stormwater-borne solids: a literature review (Semandeni-Davies, 2013)
TR2013/035	Technical basis of contaminant and volume mitigation requirements (Auckland Council 2013)
TR2013/040	Stormwater disposal via soakage in the Auckland region (Strayton and Lillis, 2013)
TR2013/043	Stormwater management provisions cost and benefit assessment (Kettle and Kumar, 2013)

vulnerable to erosion from peak flows of urban stormwater (Auckland Regional Council, 2010). The PAUP identifies sediment, heavy metals (copper and zinc) and temperature as key contaminants to be managed to reduce the impact of urban stormwater on streams. Suspended and deposited sediments degrade the waterways though physical smothering, reducing water clarity and light, and reducing food quality and the feeding efficiency of aquatic fish and invertebrates. Auckland's streams support 21 native fish species (Stevenson and Baker, 2009), 14 of which are diadromous, undergoing migrations between fresh- and saltwater as part of their life cycle, so they can be affected at many points by stormwater. Copper and zinc are toxic at low aquatic concentrations and increase vulnerability of native fish to other stressors, such as disease and other contaminants. Strong evidence indicates that the health of benthic communities is affected by the cumulative impact of chemical contaminants at levels lower than predicted by available guidelines (Auckland Council, 2010).

Surveys of Auckland's estuary sediments show high metal concentrations in the upper reaches of estuarine creeks receiving runoff from older, highly urbanised catchments (Mills et al., 2012). Chemical contaminants are also increasing most rapidly in these areas (Auckland Council, 2010). Zinc is the metal that most often reaches concentrations where adverse effects on benthic ecology are expected to occur. Although road runoff contributes zinc and copper (Depree, 2008), galvanised steel roofs are probably the major source (Timperley et al., 2005; ARC, 2004). Zinc/aluminium alloy coated steel was introduced into NZ in 1994, and is replacing galvanised zinc. Zinc loading from steel roofing is therefore predicted to reduce over the next 25–50 years as these roofs are replaced. The reduction should be accelerated by new PAUP controls on runoff.

Small urban streams are also vulnerable to high-temperature discharges. Effects are exacerbated by reduced base flows and depleted riparian vegetation (Young et al., 2013). Stormwater is considered a minor direct contributor to nutrient loads (Kelly, 2010). Auckland monitoring on emerging contaminants (e.g. fire retardants, preservatives such as tricolosan and pharmaceuticals) indicates environmental concentrations are similar to those reported worldwide (Stewart et al., 2014; Stewart, 2013; TR2013/0002 – see Table 28.2 for TR numbers).

28.2.3 Socio/Political Drivers of WSD

Auckland is a region under pressure. It is home to more than a third of NZ's population. In 2012, its population of 1.5 million was forecast to nearly double over the next 30 years, requiring up to 400,000 more homes (Auckland Council, 2013). This growth is the main regional driver of environmental change. Swimming, boating and marine sports are important Auckland recreational activities. Auckland has half of NZs' marina berths and an estimated 132,000 boats (in 2011) with 15–19% of households owning one or more boats, canoes or windsurfers (Beca, 2012).

The Auckland Plan (Auckland Council 2013) sets a vision of becoming the world's most liveable city. Auckland has been consistently placed in the top ten in international city comparisons. Auckland's environment underpins its liveability. Protection of the natural environment was identified as the number one priority for residents of Auckland's North Shore, one of the five cities that make up Auckland (Malcolm and Lewis, 2008). Protection of stream health and the receiving harbour from stormwater discharges, including combined sewer overflows, drove large-scale stream riparian enhancement in North Shore through the 2000s. Comprehensive bioretention guidelines were released (Malcolm and Lewis, 2008). These were updated bioretention chapters in the Auckland Region guidelines known as Technical Publication 10 (TP10, Auckland Regional Council 2003). TP10 is still used across NZ as a guide to the design of individual stormwater management devices.

In March 2015, Water Sensitive Design for Stormwater guidelines were released by Auckland Council (as GD04). This 192-page resource is part of the Auckland Design Manual. It summarises Council's guidance on the role and application of water sensitive design within the planning, design and development of Auckland. GD04 emphasises taking into account the multiple objectives influencing project outcomes, so stormwater is managed to achieve the greatest benefit, for both community and land developer.

28.2.4 Stormwater Mitigation Requirements: Volume and Quality

The PAUP has specific objectives requiring stormwater management devices to provide detention (peak flow attenuation) and retention (reduced volume), together with maximum contaminant concentrations in discharges, depending on the sensitivity of the receiving waters (Tables 28.3 and 28.4). Catchments with sensitive streams are identified based largely on slope, catchment imperviousness and macroinvertebrate community index. This catchment classification determines the specific flow and volume mitigation required (Table 28.3).

This approach is a change from managing stormwater through peak flow control and extended detention. The immediate impact of the new requirements was a reduction in use of traditional ponds due to their ineffectiveness at reducing total stormwater volume, the potential for catchments with multiple ponds to cause an extended period of peak flows (hence erosion of streams), and the potential for elevated temperature of discharges from unshaded open water. Auckland Council is developing design guidance for more natural wetlands that have reduced operational costs (by ensuring efficient sediment removal), reduced temperature of discharges and enhanced mauri (by increasing vegetation cover and diversity and using 'cooling' outlet designs).

Prior to 2013, stormwater treatment devices were generally designed to achieve the performance standard of removing 75% of total suspended solids. This has been changed for two reasons. First, using 'percentage removal' does not guarantee effluent quality from a device because this is dependent on the influent quality. Second, an exclusive focus on

Table 28.3 Proposed design requirements for stormwater detention and retention in the Auckland region (Auckland Council 2013).

Stormwater mitigation	Flow and volume mitigation requirement for the impervious area for which mitigation is required
Level 1 – more sensitive	Retention (temporary storage) of a 10 mm, 24 hr rainfall
	Detention (volume reduction) of the 95th percentile, 24 hr rainfall
Level 2 – more sensitive	Retention (temporary storage) of an 8 mm, 24 hr rainfall
	Detention (volume reduction) of the 90th percentile, 24 hr rainfall

Table 28.4 Proposed design effluent quality requirements for stormwater discharged to rivers and streams (Auckland Council, 2013).

Contaminant	Design effluent quality requirement
Sediment	TSS < 20 mg/l
Metals	Total Cu < 10 μg/l, Total Zn < 30 μg/l
Temperature	Temperature < 25 °C*

*Temperature is not regarded as a contaminant for receiving environments such as lakes, estuaries and harbours.

sediment is now not considered to adequately address all contaminants of concern. New standards have been derived, based on the performance that can be reliably expected from properly sized, installed and maintained devices.

28.2.5 Stormwater Mitigation Requirements: Source Control

The PAUP prioritises treatment close to source, and source control. High contaminant generating activities have been identified, and from September 2013, requires either the materials restricted, or treatment of stormwater passing over these surfaces. High contaminant-yielding roofing, spouting, cladding and architectural features include copper, galvanised metal and materials with exposed surface of metallic zinc or any alloy containing >10% zinc or copper. 'High use' roads are also considered high-contaminant yielding surfaces. These are defined as roads carrying more than 10,000 vehicles per day, and car parks carrying more than 50 vehicles per day (Auckland Council, 2013).

28.2.6 Auckland Design Guidance

By mid 2015 the new, PAUP-driven regulatory approach was being applied to 97 'special housing areas'. These areas have fast-tracked consent processes to build residential housing for the influx of new Auckland residents and expanding population in the expectation that quickly increasing house numbers, including affordable homes, will supress spiralling house prices. WSD is being applied in these areas. Some of the larger, masterplanned subdivisions in particular are using WSD in accordance with GD04 and the Auckland Design Manual. They are creating places and spaces that are recognisably Auckland as a city in the south Pacific.

The Auckland Design Manual supports delivery of the world's most liveable city, by showing how to create uniquely Auckland places. Such places are strongly influenced by Maori culture and identity. The manual includes the *Te Aranga Maori* design principles, which outline a process to engage with *mana whenua* (Auckland Council and Nga Aho, 2014). The seven principles include *kaitiakitanga*, being the management and conservation of the environment as part of a reciprocal relationship (i.e., environmental stewardship).

Water sensitive design is also embedded in national highway design guidance published (in 2014) by the NZ Transport Agency (Bell *et al.*, 2014), as one of ten design principles required to be applied to all large transport projects. Common WSD devices used to improve the contribution of state highways to the environmental and social wellbeing of NZ are swales, bioswales and wetlands. NZ Transport Agency's design specification for stormwater management was updated (in 2010) to incorporate WSD devices (NZTA, 2010). The specification recommends wetlands whenever peak control or stream erosion protection is a priority, if space and catchment size are suitable. The 256-page specification emphasises designing for efficient, safe maintenance over the whole life of a road, and using stormwater mitigation practices in series to maximise benefits.

The least common WSD devices in Auckland are living roofs or green roofs (Figure 28.3). This is partly due to their significant additional structural costs in a very immature market, where most residential roofs are lightweight steel. The cost of maintenance necessitated by working at heights is also a factor. In some projects, aspects of WSD continue to be 'value engineered out', based on the precedence given to capital cost reduction over public, non-monetisable benefits. In the case of swales, this is reflected in an abundance of grassed swales that have frequent, ongoing mowing costs and high risk of damage from vehicle compaction and scalping. In contrast, swales in public areas that are dangerous to mow, such as fast-flowing, high-capacity roads, are increasingly established in frangible, perennial native plants that do not need mowing. In such cases, perennial swales are more cost- and risk-effective in the medium term.

Figure 28.3   Green roofs are relatively uncommon in New Zealand. The New Zealand green roofs shown increase in depth and cost from left to right: Auckland commercial building with lightweight sedum and succulent roof; Auckland café turf roof with sculpture and lift; Wellington commercial building roof garden.

28.3   Case Study: Wynyard Quarter

Wynyard Quarter is a precinct of retail, hotel, office and intensive housing being developed within and alongside established light marine industries enabled by the removal of bus and bulk petro-chemical storage areas – locally known as 'the tank farm'. The transformation of this reclaimed brownfield area to desirable work and play spaces started with Jellico Street and should be complete by 2030. The land is owned by Auckland Council through its development agency, Water Front Auckland. The agency was tasked with leading and delivering projects in the 45 ha of Auckland's waterfront that are consistent with the city's vision. It built on the 2007 urban design framework for Wynyard Quarter, developed by Sea + City Projects Ltd. The first new public places were completed for the 2011 Rugby World Cup (Figure 28.4).

The Wynyard development aimed to be 'NZ's premier example of environmentally responsible development', and to showcase world-class strategic and design responses to local and global environmental issues. An overarching objective was to create a blue–green, public waterfront recognised for the natural environmental quality of public spaces. Managing stormwater was a critical component. The site is almost flat, close to sea level and some parts are contaminated, so water is unable to be drained into the ground. Stormwater discharges to the harbour. Public spaces encourage direct contact with the harbour through broad steps leading into the water. The sea is also used for public events such as international multi-sport triathlons.

Figure 28.4   Densely planted rain gardens are used at Wynyard Quarter's Karanga Plaza, Auckland, to create the vision of 'a pedestrian-focused boulevard that uses Auckland's indigenous flora to create a memorable enveloping landscape'.

The intent of the rain garden design was to implement the latest, best practice, water sensitive urban design. Specific success measures were a 42% reduction in stormwater discharge volume and exceeding the (then) statutory requirement of 75% removal of total suspended solids. The stormwater mitigation approach adopted was to capture and reuse roof-water within buildings (e.g. for flushing toilets) and to treat runoff at the surface in rain gardens. The rain gardens are very large, almost-continuous and indistinguishable from landscaping. Continuous simulation modelling of the 10-year, 6-minute rainfall indicated that 78% TSS removal could be achieved; the oversized rain gardens helped compensate for areas that would not get treated.

A feature of Wynyard Quarter is high in-built resilience. Resilience was delivered in the rain gardens by including: a large volume of media per tree; highly diverse tree and groundcover species, with a complex planting pattern; and a three-year maintenance requirement during which the growth of each tree was measured annually. Along Jellico Street and Tītoki Plaza, rain gardens extend under the impervious area of car parks and some pedestrian pavements to achieve about 10 m³ potential root volumes per tree. The overlying impervious areas are supported by interlocking, rigid plastic 'strata cells', avoiding compaction of the rooting media and allowing plant-friendly soils to be used. The additional root volume under impervious areas allowed a high density of trees needed to deliver the intended vision.

Vegetation selection and placement also underpin the vision. Unusually, six native evergreen tree species were used together, in clusters of different species, as they might be found in a native forest. Many of the species had rarely been used in rain gardens, nor planted so close to the harbour. The approach reduced the risk of any one tree species not performing; individual trees could be removed without affecting the overall landscape. In Auckland, most rain garden and street tree plantings use evenly spaced, evenly aged and single-species of trees, risking failure of the overall design. At Wynyard Quarter's Karanga Plaza, some older trees were salvaged from streets being renovated and were craned into rain gardens. This was particularly effective at creating a variation of canopy and giving a more 'natural' outcome. The largest tree (left in Figure 28.4) supported insect and lichen diversity absent from the nursery-raised trees.

The rain garden groundcover plants are particularly diverse, dense and textured. They range from 1.4 m tall native lilies to 50 mm carpets. About half these plants have produced seedlings that have themselves successfully established, as have four tree species. The careful analysis of probable human movements was used to place rain garden crossings, and seating is used in places to protect the rain gardens from foot traffic. The dense plantings also discourage ingress. However, the absence of boards at the end of parking bays has meant cars continue to damage vulnerable corners; these areas need regular replanting and re-mulching to reduce weed establishment; adjacent streets have incorporated low bollards that prevent such vehicle damage.

28.4 Conclusions and Parting Thought

WSD is not yet mainstream in NZ, despite having been implemented in some regions, such as Auckland, for over a decade. WSD is generally regarded as capital-intensive and requiring higher maintenance costs than conventional piped infrastructure. Outstanding, international-award-winning WSD developments, such as Wynyard Quarter, tend to reinforce this assumption. The environmental co-benefits of WSD in NZ are both poorly quantified and undervalued because they do not accrue to the developer. Co-benefits are not 'counted' by the narrow performance indicators that tend to drive investment in stormwater and local road infrastructure. Minimal local data is available on public health benefits associated with WSD, whether through encouraging cardiovascular-benefitting exercise, reducing stress or decreasing exposure to ultraviolet radiation. Australasia has the highest rate of melanoma in the world, but unlike Melbourne, Adelaide or Sydney, Auckland's mature tree canopy is low (estimated at about 8%), and anecdotally decreasing under twin pressures of intensification and relaxation of tree protection statutes.

A regime change is probably needed to achieve the majority political buy-in that will implement stronger regulatory approaches requiring WSD and broad uptake across the urban development sector. This shift has occurred in the state highway network, facilitated by NZ Transport Agency legislation that requires state highways to contribute to the environmental and social wellbeing of NZ (Bell *et al.*, 2013, 2014). NZ towns also need this transformational shift to achieve the outcomes wanted by many communities. Such outcomes are expressed in the Auckland Plan (Auckland Council 2012) as 'the world's most liveable city', and in Christchurch's Central City Plan (2011) as 'a city in a garden'. Positive changes are proposed in the Auckland Unitary Plan for new developments. Auckland Council is also supporting WSD by providing detailed guidance documents based on a rigorous research programme. These research outcomes are publically available as technical reports (free of charge) on the Council website. Across NZ, the absence of an impervious surface charge, or similar incentive to reduce impervious areas or encourage stormwater treatment, is probably a key barrier to retrofitting established areas that have conventionally managed stormwater discharges.

References

Auckland Council (2010) *State of Auckland Region report* 2009. http://www.aucklandcouncil.govt.nz/EN/planspoliciesprojects/reports/technicalpublications/Pages/stateaucklandregionreport2010.aspx

Auckland Council (2013) *Auckland Unitary Plan Stormwater Management Provisions: Technical Basis for Contamination and Volume Management Requirements*. Auckland Council Technical Report 2013/035.

Auckland Council (2014) GD04 http://content.aucklanddesignmanual.co.nz

Auckland Council and Nga Aho (2014) *Te Aranga Principles*. www.aucklanddesignmanual.co.nz/design-thinking/maori-design/te_aranga_principles

Auckland Regional Council (2003) *Technical Publication 10 (TP10): Stormwater Treatment Devices Design Guideline Manual*.

Auckland Regional Council (2004) *A study of roof runoff quality in Auckland, New Zealand: Implications for stormwater management*. Technical Publication 213.

Beca (2012) Auckland Recreational Boating Study http://www.aucklandcouncil.govt.nz/EN/planspoliciesprojects/plansstrategies/unitaryplan/Documents/Section32report/Appendices/Appendix%203.332.pdf.

Bell, J., Desrosiers, L. and Lister, G. (2013) Bridging the Gap New Zealand Transport Agency Urban Design Guidelines. NZ Transport Agency. New Zealand Government. www.nzta.govt.nz/assets/resources/bridging-the-gap/docs/bridging-the-gap.pdf.

Bell, J., Bourne, S., Collins, C. and Lister, G. (2014) Landscape Guidelines Final Draft Sept 2014, Final Draft September 2014. NZ Transport Agency. New Zealand Government. www.nzta.govt.nz/assets/resources/nzta-landscape-guidelines/docs/nzta-landscape-guidelines-20140911.pdf.

Buchanan, K., Clarke, C. and Voyde, E. (2013) *Hydraulic energy Management: inlet and outlet design for treatment devices*. Prepared by Morphum Environmental Limited for Auckland Council. Auckland Council Technical Report 2013/018.

Christchurch City Council (2011) Draft Central City Plan. August 2011. Adopted on 11 August 2011. ISBN978-0-9876571-1-4. Downloadable from http://canterbury.royalcommission.govt.nz/documents-by-key/20111006.33

DePree C. (2008) *Contamination characterisation and toxicity testing of road sweepings and catchpit sediments: towards a more sustainable reuse option. Land Transport NZ Research Report 345.*

Fassman-Beck E. and Simcock R. (2013) *Living roof review and design recommendations for stormwater management*. Auckland UniServices Technical Report to Auckland Council: Auckland Council Technical Report 2013/045.

Fassman-Beck, E., Voyde, E. and Liao, M. (2013a) *Defining Hydrologic Mitigation Targets for Stormwater Design in Auckland*. Auckland UniServices, Auckland Council Technical Report 2013/024.

Fassman-Beck, E.A., Simcock, R. and Wang, S. (2013) Media Specification for Stormwater Bioretention Devices. Auckland UniServices Technical Report to Auckland Council. Auckland Council Technical Report 2013/011. Auckland, New Zealand.

Fassman, E. and Simcock, R. (2012) Moisture measurements as performance criteria for extensive living roof substrates. *Journal of Environmental Engineering*, 138(8) 841–851.

Kanz, W. (2013) Caring for urban streams. Auckland Council Technical Report 2013/020.

Kelly, S. (2010) *Effects of Stormwater on Aquatic Ecology in the Auckland Region*. Prepared by Coast and catchment. Auckland Regional Council Technical Report 2010/021.

Kettle, D. and Kumar, P. (2013) *Auckland Unitary Plan stormwater management provisions: cost and benefit assessment*. Auckland Council Technical Report 2013/043.

Lewis, M. (2008) Stream Daylighting Identifying Opportunities for Central Auckland: Concept Design. Technical Report TR2008/027, Boffa Miskell for Auckland Regional Council.

Lewis, M., Simcock, R., Davidson, G. and Bull, L. (2009) *Landscape and Ecology Values within Stormwater Management*. Auckland: Auckland Regional Council, Technical Report 2009/083.

Malcolm, M. and Lewis, M. (2008) *North Shore City Bioretention Guidelines*. First edition. C. Stumbles Editor. https://www.northshorecity.govt.nz/Services/Environment/Stormwater/Documents/bioretention-guidelines.pdf.

Marshall, G., Kelly, S., Easton, H., Scarles, N. and Seyb, R. (2013) *Auckland's Urban Estuaries: Management Opportunities*. Auckland Council Technical Report 2013/015.

Mills, G., Williamson, B., Cameron, M. and Vaughan, M. (2012) *Marine sediment contaminants: status and trends assessment 1998 to 2010*. Prepared by Diffuse Sources Ltd. for Auckland Council. Auckland Council Technical Report 2102/041.

New Zealand Transport Agency (NZTA) (2010) *Stormwater Treatment Standard for State highway Infrastructure*. https://www.nzta.govt.nz/assets/resources/stormwater-management/docs/201005-nzta-stormwater-standard.pdf.

Semadeni-Davies, A. (2013) *Classification of stormwater-borne solids: a literature review*. Prepared by NIWA for Auckland Council. Auckland Council Technical Report 2013/017.

Shamseldin, A. (2010) *Review of TP10 Water Quality Volume Estimation*. Auckland: Auckland UniServices Ltd, Auckland Regional Council Technical Report 2010/066.

Simcock R (2009) *Hydrological effect of compaction associated with earthworks: soil infiltration, permeability and water storage*. Prepared by Landcare Research Manaaki Whenua for Auckland Regional Council: Auckland Regional Council Technical Report 2009/073.

Simcock, R. and Dando, J. (2013) *Mulch specification for stormwater bioretention devices*. Prepared by Landcare Research for Auckland Council. Auckland Council Technical Report 2013/056.

Stevenson C. and Baker C. (2009) *Fish passage in the Auckland Region – a synthesis of current research*. Prepared by NIWA for Auckland Regional Council. Auckland Regional Council Technical Report 2009/084.

Stewart, M., Olsen, G., Hickey, C.W., Ferreira, B., Jelic, A., Petrovic, M. and Barcelo, D. (2014) A survey of emerging contaminants in the estuarine receiving environments around Auckland, New Zealand. *Science of the Total Environment*, 468, 202–210.

Stewart, M., Aherns, M. and Olsen, G. (2009) *Field Analysis of Chemicals of Emerging Environmental Concern in Auckland's Aquatic Sediments*. Prepared by NIWA for Auckland Regional Council. Auckland Regional Council Technical Report, 2009/021.

Stewart, M. (2013) Pharmaceutical residues in the Auckland estuarine environment. Prepared by NIWA for Auckland Council. Auckland Council Technical Report 2013/002.

Storey, R., Brierley, G., Clapcott, J., Collier, K., Kilroy, C., Franklin, P., Moorhouse, C. and Wells, R. (2013) *Ecological responses to urban stormwater hydrology*. Prepared by NIWA for Auckland Council. Auckland Council Technical Report 2013/033.

Storey, R. and Wadhwa, A. (2009) An assessment of the lengths of permanent, intermittent and ephemeral streams in the Auckland Region. Prepared by NIWA for Auckland Regional Council. Auckland Regional Council Technical Report 2009/028.

Strayton, G. and Lillis, M. (2013) *Stormwater disposal via soakage in the Auckland region.* Prepared by Pattle Delamore Partners Ltd for Auckland Council. Auckland Council Technical Report 2013/040.

Te Aranga (2008) Te Aranga Maori Cultural Landscape Strategy. www.tearanga.maori.nz.

Timperley M., Williamson B. and Horne B. (2005) *Sources and loads of metals in urban stormwater.* Auckland, New Zealand, Auckland Regional Council.

Voyde, E.A., Fassman, E.A. and Simcock, R. (2010) Hydrology of an extensive living roof under sub-tropical climate conditions in Auckland, New Zealand. *Journal of Hydrology*, 394, 384–395.

Watercare Services Limited (2012) Central Interceptor main project works resource consent applications and assessment of effects on the environment. Part A.AEE Report. August 2012.

Young, D., Afoa, E., Wagenhoff, A. and Utech, C. (2013) *Temperature as a contaminant in stream in the Auckland region, stormwater issues and management options.* Prepared by Morphum Environmental for Auckland Council. Auckland Council Technical Report 2013/044.

Section 7   Summary of the Book

Challenges for the Future: Are Sustainable Drainage Systems Really Sustainable?

Susanne M. Charlesworth and Colin A. Booth

29.1 Introduction

Sustainable drainage badges itself as being 'sustainable' and, as many of the chapters in this book have detailed the multiple benefits of this approach, it clearly supports that description. The newly published SuDS square (Woods Ballard, 2015), mentioned in Chapter 1 and developed from the original triangle (CIRIA, 2001), gives amenity and biodiversity their own separate categories, emphasising their importance individually. However, there is far more to sustainable drainage than these four traditional areas covering reduction in water quantity (Chapter 5), improvement in water quality (Chapter 6), increased opportunities for biodiversity (Chapter 7) and for amenity and recreation (Chapter 8). Sustainable drainage is truly multi-, cross- and trans-disciplinary, covering such subjects as chemistry and microbiology (Chapters 9 and 10), policy and governance (Chapter 3), archaeology (Chapter 2), materials (Chapter 11) and issues around management (Chapter 4).

SuDS is not all about draining *urban* areas, which was reflected in the contraction of the term sustainable *urban* drainage systems (SUDS) to sustainable drainage systems (SuDS) (CIRIA, 2000; Fletcher *et al.*, 2014; Woods Ballard, 2015). They can also be used in rural areas, and there has been some attempt to introduce the term 'rural SuDS' (RSuDS) into the lexicon (Avery, 2012), although now 'natural flood management/flood resilience measures' would appear to be gaining ground instead (Chapter 12), with case study catchments in the UK making headline news by '*working with nature, not against it*', for example Pickering, Yorkshire, UK (Iacob *et al.*, 2012; Nisbet *et al.*, 2015). Strengthened by the publication of SEPA's 2016 handbook, the challenge here is in engaging with landowners and farmers to ensure that the benefits to settlements downstream do not disbenefit those on whose land these devices are proposed to be installed.

Modelling, monitoring and evaluation (Chapter 20) of the different devices, and also management trains (Chapter 1) have provided data and information on the functioning

Sustainable Surface Water Management: A Handbook for SuDS, First Edition.
Edited by Susanne M. Charlesworth and Colin A. Booth.
© 2017 John Wiley & Sons, Ltd. Published 2017 by John Wiley & Sons, Ltd.

and efficiency of these interventions. However, there are very few studies of the monitoring of in-use management trains to validate models, and the challenge here is in being able to identify suitable sites to instrument and gain suitable data from.

As the chapters in this book have shown, it is possible to use the approach regardless of climate; techniques similar to SuDS have been used for millennia (Chapter 2), and in Chapter 3 it is stated that the author is of the opinion that 'SuDS have undergone a transition from new, to an accepted technology in many countries'. The fact that they are gaining ground and acceptance is illustrated in Chapters 22–28, in which countries, with many climatic types, are willing to design and install devices and trains appropriate to their environmental conditions. Chapter 24 is unique because it describes efforts to drain informal settlements in South Africa, using sustainable infiltration and slow conveyance, utilising available materials and redesigned and reimagined devices, without resorting to traditional approaches or technological fixes. In common with attempting to use natural flood management in the UK (Chapter 12), engagement with communities is particularly vital in these informal settlements as, unlike farmers and landowners in the UK, they have minimal, if any, support from local government and are essentially on their own regarding design, installation and maintenance.

SuDS are without doubt flexible. As well as offering multiple benefits, which accord with one of the definitions of sustainability, it is possible to construct them to include other sustainable techniques such as harvesting renewable energy (Chapter 13). They have a substantial part to play in adapting and mitigating changes due to global climate change – one example of this is the potential to sequester and store carbon in SuDS devices (Chapter 14). Furthermore, rainwater harvesting (Chapter 15) has multiple benefits on its own: it can reduce potable water use and thereby reduce the energy associated with the production of drinking water but, when designed carefully, it can also attenuate storm peaks by storing excess water and then releasing it slowly. Vegetated SuDS devices have several ecosystem services benefits (Chapter 15), from improving aesthetics and providing pleasant living spaces to reducing flooding, reducing energy use (in the case of green roofs, walls and strategically placed trees) and addressing the urban heat island effect in towns and cities.

One challenge in promoting SuDS is in acknowledging their multiple benefits and flexibility because, to many, they are still considered as being designed and constructed purely for flood attenuation, with other advantages simply part and parcel of their existence. To move forward, their value (Chapter 16), and that of the ecosystem services they provide, has to be expressed in monetary terms to be able to engage with governments, or local authorities, and sometimes this is inappropriate. It is difficult to assign a pecuniary value to recreational opportunities, amenity and aesthetics because they cannot be objectified commercially, things to profit from, apart from improved quality of life and environmental health. Quite obviously, some of this can be quantified, but the real value of the view of a retention pond with fringing vegetation treating and reducing the volume of stormwater from a development is both invaluable and un-valuable.

Since SuDS can quite obviously be of enormous benefit, why then is it still not the go-to method for drainage in both urban and rural areas locally and/or globally?

29.2   Barriers and Drivers

The issue of what promotes the use of SuDS – and what hinders it – has been addressed in many publications over the years and, fortunately, the list of barriers has been shrinking. However, some of the reasoning for not using SuDS generally, is due to perceptions that they take up space, particularly in new developments, where space to build more houses and make more money are paramount. However, there is some evidence to show that houses built near a SuDS pond can command a premium in terms of their sale price (Bastien *et al.*, 2009), to a certain extent offsetting their construction and land-take cost.

Chapter 19 has shown how SuDS and green infrastructure can be designed into a utilitarian construct, such as a motorway service area, reducing its hydrological footprint almost to zero and ensuring minimal impact on the receiving watercourse. However, the need for maintenance is also regularly raised as a reason not to implement SuDS, but it seems to be forgotten that traditional drainage of pipes and gullies also require regular maintenance, and by using pervious paving, for example, there would be no need for gully pots or their cleaning, or the disposal of their contents to landfill.

Retrofitting of SuDS devices to existing buildings is perceived as being difficult and expensive but, as Chapters 18, 26 and 27 show, this can be successfully undertaken in the very different climatic conditions of Australia, Scotland and Spain. In the UK, with 70% of housing required by 2014 already in existence, the challenge of these homes being retrofitted (e.g. green roofs, pervious paving and rain gardens) needs to be grasped, as people inexorably pave over their front gardens to create parking for their cars, rendering urban areas more impervious, more flood-prone, and less water-aware.

It is arguable that without positive public perceptions of SuDS they are not sustainable (Chapter 21), and the challenge in the developing world is in contesting the view that open water is unsafe in cities. It is perfectly possible to design interventions with fringing vegetation and judicious use of fencing, which can discourage access to ponds and wetlands (Figure 29.1). However, careful maintenance of these structures is required, particularly

Figure 29.1   Fencing and fringing vegetation preventing access to open water around a SuDS pond, Scotland.

during dieback of vegetation during the winter months. In the developing world (Chapters 22–24), in some cases governance and policy structures are often either absent or too weak. Here, therefore, public perceptions are secondary to the reduction of risk of flooding and the health implications of poor drainage.

29.3   What is the Future for SuDS?

There is cause for hope for SuDS, which Chocat *et al.* (2007) term 'hydro-optimism', but mainly due to reaction to emergencies (such as water shortages and rising populations across the world), rather than an acknowledgement that SuDS is efficient, flexible, multiple-benefit and cost-effective in the long term (Chapter 17). In the short term, supporting legislation (Chapter 3) is needed to encourage the use of SuDS, although legislators will have in mind the unfortunate impacts of recent floods worldwide, which may advance the cause. In the medium term, sites to demonstrate functions, construction and benefits are required as an educational tool, and also to provide evidence of their reliability and performance. Global climate change is the long-term driver, even at the small scale; for example, in the UK, CCRA (2012) quotes a potential increase (27%) in sewer flooding due to the impacts of climate change even if nothing is done.

Taking a wider view, the concepts of low impact development and water sensitive design (WSD) were introduced in Chapters 25 and 28. SuDS would be part of both of these, designed into a water sensitive city (Wong, 2007), in which the presence of water can ensure the proper functioning and health of the city and its inhabitants. Implementing SuDS makes use of 'wasted waste' (Charlesworth, 2010), transforming it into what Chocat *et al.* (2007) call 'opportunity water', or what Semadeni-Davies *et al.* (2008) call a 'liquid asset'.

The future for SuDS could include addressing new and emerging pollutants (NEPs), such as pharmaceuticals, hormones, herbicides and cosmetics. Chapters 6 and 9–11 have shown that there is much known about some of the biodegradation processes and storage of pollutants, such as hydrocarbons and particulate-associated metals. However, there is less known about how, or if, these devices can also deal with NEPs. Due to the implementation of these devices in urban environments, the fate of trapped pollutants when, for example, a pervious pavement needs refurbishing, or replacing at its end-of-life is also not known and further research is needed here.

SuDS are still used in the traditionalist sense of primarily a flood risk management solution with the smallest development footprint. As a result, below-ground attenuation tanks coupled with oversized pipes are used as routine, and the natural hydrology of development sites is poorly/rarely considered during planning. It is possible to address problems associated with this conventional thinking, in which there is no perceived need to engage with a new approach when it is felt that the old one still functions; the £3.2 billion spent on the 2007 floods in the UK proved that the latter was not the case. This hinges around re-educating developers, construction companies, planners, designers and the public, in a process that Chocat *et al.* (2007) call 'de-learning'. Non-structural SuDS (education and information) are key, and here the 'SuDS for Schools' initiative in the UK is valuable in teaching schoolchildren the importance of sustainably managing excess surface water (Duggin and Reed, 2006) – see also the Wildfowl and Wetlands Trust at: http://sudsforschools.wwt.org.uk/. Production of guidance for SuDS implementation in the UK can only help in raising the awareness of SuDS (Woods Ballard, 2015).

29.4   Conclusions

Flooding is a natural process that would happen whether humans had colonised the planet or not. The resulting problems are the many ways in which society has manipulated the environment to its detriment and the arrogance of thinking that it could control the impacts. There is not one single answer that will solve issues with drainage; however, as has been shown throughout this book, SuDS can be successfully applied worldwide in a variety of different climates and situations, and it can also be combined with conventional, pipe-based systems. However, these systems need to be better understood, requiring acceptance by both stakeholders and the general public. Until that happens, it is unlikely they will be used more widely.

So, to address the question posed in the title of this chapter, as this book shows, while the scientific evidence exists that SuDS have sustainable credentials, the challenges lie in the political, institutional and societal arenas that are currently retarding progress and acceptance.

References

Avery, L.M. (2012) *Rural Sustainable Drainage Systems* (RSuDS). Environment Agency, Bristol. 147 pp

Bastien, N.R.P., Arthur, S. and McLoughlin, M.J. (2009) Public perception of SuDS ponds – Valuing Amenity. 12th International Conference on Urban Drainage, Porto Algre/Brazil, 11–16 September 2011. http://tinyurl.com/htlk95g

Climate Change Risk Assessment (CCRA) UK (2012) *Water*. http://tinyurl.com/jfwre55

Chocat, B., Ashley, R., Marsalek, J., Matos, M.R., Rauch, W., Schilling, W. and Urbonas, B. (2007) Towards the sustainable management of urban storm water. *Indoor and Built Environment*, **16**, 3, 273–285.

Charlesworth, S. (2010) A review of the adaptation and mitigation of Global Climate Change using Sustainable Drainage in cities. *Journal of Water and Climate Change*, **1**, 3, 165–180.

CIRIA (2000) *Sustainable Urban Drainage Systems – Design Manual for Scotland and Northern Ireland*. Dundee, Scotland: CIRIA Report No. C521.

CIRIA (2001) *Sustainable Urban Drainage Systems: Best Practice Manual*. CIRIA Report C523, London.

Duggin, J. and Reed, J. (2006) *Sustainable Water Management in Schools*. CIRIA W12. http://tinyurl.com/h48huub

Fletcher, T.D., Shuster, W., Hunt, W.F., Ashley, R., Butler, D., *et al.* (2014) SUDS, LID, BMPs, WSUD and more – The evolution and application of terminology surrounding urban drainage. *Urban Water Journal*, **7**, 12. http://tinyurl.com/j3jyv2a

Iacob, O., Rowan, J., Brown, I. and Ellis, C. (2012) Natural flood management as a climate change adaptation option assessed using an ecosystem services approach. BHS Eleventh National Symposium, *Hydrology for a Changing World*, Dundee. British Hydrological Society. http://tinyurl.com/ks43stm

Nisbet, T.R., Marrington, S., Thomas, H., Broadmeadow, S.B. and Valatin, G. (2015) *Defra FCERM Multi-objective Flood Management Demonstration project*. Project RMP5455: Slowing the flow at Pickering. Phase II. http://tinyurl.com/h63zlnt

Semademi-Davies, A., Hernebring, C., Svensson, G. and Gustafsson, L, 2008. The Impacts of Climate Change and Urbanisation on Drainage in Helsingborg, Sweden: Suburban Stormwater. *Journal of Hydrology* **350** (1–2), 114–125.

SEPA (2016) *Natural Flood Management Handbook*. http://tinyurl.com/jtam3mc

Wong, T.H.F. (2007) Water sensitive urban design – the journey thus far. *Environment Design Guide*, **11**, 1–10.

Woods Ballard, B., Wilson, S., Udale-Clarke, H., Illman, S., Ashley, R. and Kellagher, R. (2015) *The SuDS Manual*. CIRIA. London.

Index